Modern physical Science

George R. Tracy
Harry E. Tropp
Alfred E. Friedl

Holt, Rinehart and Winston, Publishers
New York • Toronto • Mexico City • London • Sydney • Tokyo

The Authors

George R. Tracy
Formerly of the Science Department, American Academy for Girls, Istanbul

Harry E. Tropp
Former Science Supervisor, Hillsborough County Public Schools Tampa, Florida

Alfred E. Friedl
Professor of Education, Kent State University, Kent, Ohio

Picture credits appear on page 555.
Cover photo: © Dan McCoy from Rainbow
Cover design by Caliber Design Planning, Inc.
About the cover: Silicon photovoltaic cells, as shown in the cover photo, are the most common type of solar battery. A photovoltaic cell consists of an N-P junction between two semiconductors (see pages 535-536). Conduction in the N-type semiconductor is caused by electrons; in the P-type semiconductor it is caused by positive "holes." When sunlight is absorbed near the N-P junction, new electrons and holes are released. This released charge (the electrons and holes) moves into an electric field that exists in the N-P junction. The electric energy thus produced is available to do work.

ISBN: 0-03-059954-7

67-071-9876

contents

preface

Modern Physical Science was written primarily for students taking a unified course in the basic principles of physical science, namely chemistry and physics. However, in an age when we are surrounded by many scientific developments, all students should have an understanding of science as a great human achievement. Therefore, this textbook can be used as a part of a general education program for all students. The course is a product of many years of science teaching by its authors.

Too often, a great deal of stress is placed on the highly technical and mathematical end-products of scientific knowledge. *Modern Physical Science*, on the other hand, presents a well-balanced science course stressing the processes and activities of science as well as the basic concepts. All of these concepts are developed fully using a nonmathematical approach.

While many scientific laws are presented throughout the course, it is important to remember that science is not a book of rules. Science is a way of learning about our world. Laws and theories can be changed as a result of new discoveries.

The text is arranged into five units, each covering a major topic in physical science. The units are further divided into 29 chapters. Each chapter is written in such a way as to develop and apply basic concepts and applications. Applications include emphasis on environmental education, energy education, consumer education, and career and occupational education.

All chapters begin with a brief introduction. Each chapter contains a list of several objectives to be attained after studying the chapter. These objectives serve as a guide to the major topics discussed in the chapter. Difficult words are pronounced phonetically in parentheses. New words or terms stand out in boldface type. These words or terms are discussed or defined the first time they are used. Italic type is often used for emphasis. Sample problems provide an opportunity to experience the quantitative aspects of science. Activities that can be performed at home or at school with a minimum of equipment and material are also included. These activities are designed to develop process skills and highlight basic understanding of concepts.

Each chapter is enriched with chapter-end material to include either a biographical sketch of a famous scientist, a science process, or a career outline relating to the study of science. A short summary of each chapter is also included. A matching review exercise, questions, and problems at the end of a chapter serve as a self-test of the material contained in the chapter.

Group A questions at the end of each chapter test mostly recall information obtained from reading the chapter. Group B questions are generally more difficult, calling for interpretation and application of the material presented. Problems are included where appropriate. A list of material for further reading has been found useful to many students.

The glossary at the back of the book may be helpful in recalling new words. An appendix gives hints on how to use the textbook and its various elements. A complete index is also available for easy reference.

acknowledgements

The authors would like to thank Dr. Jerry S. Faughn, East Kentucky State University, Richmond, Kentucky, Dr. Jack Fernandez, University of South Florida, Tampa, Florida, and Rev. Mark M. Payne, O.S.B., St. Benedict's Preparatory School, Newark, New Jersey, who acted as consultants for this revision of *Modern Physical Science*.

unit 1
matter and energy

1

Put some baking soda in the thumb of a thin disposable plastic glove. Then add vinegar to the fingers. Twist the glove closed and mix the contents. Describe what you see. Can you explain what happened?

objectives:

☐ To explore career opportunities in science.

☐ To develop an appreciation for the study of science.

☐ To make measurements using the metric units for length, volume, force, and mass.

☐ To describe the difference between the mass of an object and its weight.

☐ To describe the various forms of energy and to explain how they relate to each other.

science in today's world

Science begins with curiosity. Human beings are naturally curious. They want to know what makes things "tick." That is, they want to find out what things are made of, and why they act the way they do. Scientists believe that everything in the universe behaves in an orderly way. They think that people can learn to understand this order. This desire to know about things around us is the key to our knowledge of the universe. **Science** *is the orderly search for answers to our questions about the world we live in.*

Scientists use careful methods in their work. They use common sense to study and solve problems. Look at Fig. 1-1. It shows an ordinary piece of steel wool being held in a flame. What do you think causes the steel wool to sparkle when it is heated? Why does the steel wool not melt, turning from a solid into a liquid? Are any new substances formed when the steel wool burns? The study of science will help you to find answers to questions like these.

Scientists learn through reading, thinking, and testing, and by sharing ideas with one another. To learn more about nature, scientists gather facts from many sources. These facts are then studied and are used to help solve problems. All these ways of finding out about nature make up the process of science. Remember, science starts with curiosity!

Science is a method for learning. People have always looked for knowledge about *how* and *why* things behave the way they do. They tried to find patterns that would help them predict how things will act. Science has been defined as the process by which new knowledge of the universe is gained. Science is not a "book of answers" but rather a method of learning!

Scientific discoveries have led to such useful results as polio vaccine, weather satellites, TV, jet airplanes, computers, robots, and nuclear energy. (See Fig. 1-2.) How-

1-1 Describe what you see as a piece of steel wool burns when it is heated in a flame.

1-2 Robots at this automobile plant measure openings for doors and windshields. They work about 10 times as fast as human workers. What are some other advantages of these machines? What are some disadvantages?

ever, scientists cannot predict all of the results of their discoveries. Thus, the unwise use of scientific knowledge may also cause pollution of the environment, energy shortages, and nuclear warfare.

Science is divided into various groups. Science can be divided into two major groups. One group consists of the *biological sciences,* which deal with the study of all living things. The other group is made up of the *physical sciences,* which deal with matter, force, and energy, about which you will learn in this chapter. The two main branches of physical science are *chemistry* and *physics.* Chemistry deals with the study of matter and its changes. Physics is the study of energy and the changes from one form of energy to another. You will study more about chemistry and physics later in this course.

Other branches of the physical sciences include geology (rocks and minerals), meteorology (weather), astronomy (stars and planets), and oceanography (oceans). All of these branches are related and often depend upon one another. For example, a geologist must know about the chemistry of rocks; a meteorologist must know about the physics of air movements.

Science helps us in many ways. When you drop a fork, will you have visitors? Do you carry a rabbit's foot for good luck? Are you afraid that if you break a mirror you will have seven years of bad luck? These may seem like silly questions, but many people still believe in such superstitions. Your study of science can help you to tell fact from superstition. Before the age of science, most people were frightened by comets, eclipses, and thunder and lightning. They thought that these were supernatural events. By explaining the nature of such events, science has helped to do away with these fears.

The water you use in your home has probably been treated and purified. You use electric lights and enjoy the comforts of heating or air conditioning. Machines help you perform a job faster or easier. Your clothing is durable and easy to care for because science has provided better fabrics. Science has also made possible more nourishing foods, better building materials, and more effective drugs and medicines.

Some of you may use microfilm libraries and receive computer-based instruction that permits you to "talk" with a computer. You may have heard of such terms as *computer programming*, *decision-making theory*, and *system analysis*. You live in an age of electronic "brains," "smart" missiles, replacement of human body parts, and super-drugs. The replacement of defective genes in cells by healthy genes and the use of "oil-eating" bacteria to help clean up oil spills are useful developments of modern science. See Fig. 1-3. Every one of these developments, and many others, affects your daily life. All of these developments are the result of scientific discoveries.

The study of science is important not only for its practical applications. Many people study science not because it is useful to do so, but because they find great beauty in it. For example, ancient people found more beauty in a rectangle whose sides had a 3-to-5 ratio than in a simple square! What causes a rainbow, a beautiful sunset, or the formation of chemical compounds? The search for answers to such questions is a basic function of the human mind.

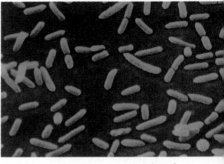

1-3 This photo shows a variety of "oil-eating" bacteria magnified 28,000 times. What are some problems that may arise from the use of these bacteria?

Science can help you make wise decisions as a citizen. Your city or town may have to decide whether to build new roads, bridges, parks, an airport, a water supply system, or a sewage disposal plant. Your study of science can help you deal with many of the problems that arise when these projects are considered by your community. Science also helps you to answer questions about chemicals in your drinking water, air and water pollution, conservation of natural resources, new sources of food or energy, and flood and pest control. Only by knowing and understanding the principles of science can you help decide these problems wisely. The future of our nation and the world depends on good decisions by well-informed citizens.

It is possible that in your lifetime you may see the partial control of weather, the complete control of diseases, and the production of simple life forms in a test tube. The decisions that you will have to make regarding these developments are great and far-reaching.

Science is useful in many careers. A knowledge of science is a basic necessity for many jobs in such fields as medicine, engineering, architecture, and agriculture. Science is also an important area of study for many sales, trade, and service occupations. Throughout this book, some of these jobs will be described. This information will help you to make up your mind about the type of work that best fits your interests and abilities. Of course, your school guidance counselor will also be able to help.

Here is one example of how a number of jobs relate to a product in common use today. Many people in this country help to design, build, sell, and repair cars, trucks, and buses. Table 1-1 lists a few questions about a school bus that a knowledge of science can help to answer.

Table 1-1:

Jobs, Science, and the School Bus	
Kind of Job	Questions That Science Helps to Answer
Designer	What is the best type of engine for the bus?
Materials tester	What is the strongest and lightest material that can be used?
Salesperson	Is the product the best bus for the money spent?
Steel worker	What is the best way to make and shape the steel used?
Mechanic	What is needed to keep the bus running safely and smoothly?
Driver	What skills are needed to drive the bus safely?

measurements and concepts of science

Scientists use exact measurements that can be reproduced. Weights and measures were among the earliest tools invented. The people of earlier times measured the length of objects by a "rule of thumb." What was later called an inch was in those times the width of a man's thumb! The old footrule measure started out as the length of a man's foot.

In the Bible, Noah used a unit of length called a *cubit* when he built the ark. The cubit of Noah's time was the distance from the elbow to the tip of the middle finger. See Fig. 1-4. In some ways, this was a handy unit of measurement. It was always with you and was easy to use. The cubit was an important unit of length to the early Babylonians and Egyptians.

King Henry I of England, in about A.D. 1120, defined a yard as the distance from the tip of his nose to the end of his thumb when his arm was held level. In the fourteenth century, King Edward II defined the inch as the length of three dry barleycorns, laid end to end. Can you suggest a reason why such standards are not suitable?

In 1793, the *metric system* of measurement was developed in France. This was an entirely new set of standards, which Congress made legal in the United States in 1866. This system is now called the *International System of Units* (SI).

The metric system is used by scientists all over the world. It is an exact, easy-to-use system, based on decimals, as is our money system. In most countries, it is also the system of weights and measures used by the people as well.

Length is measured in meters. The basic unit of length in the metric system is the *meter*. A meter is about the length of an ordinary broom handle. Scientists have defined the meter exactly, and this definition has been agreed upon by the countries of the world. In 1960, the meter was defined as equal to 1,650,765.73 wavelengths of the orange-red light emitted by krypton-86 gas.

The meter is divided into 100 equal parts, called *centimeters*. A centimeter is about the width of the tip of your little finger. The prefix *centi-* means 1/100. A centimeter is 1/100 part of a meter, just as a cent is 1/100 part of a dollar. A *millimeter* is 1/1000 part of a meter. The prefix *milli-* means 1/1000. A millimeter is about the thickness of a 10-cent coin. The fourth unit of length in the metric system is the *kilometer*. The prefix *kilo-* means 1000, and so there are 1000 meters in one kilometer. Study Table 1-2 on the following page carefully.

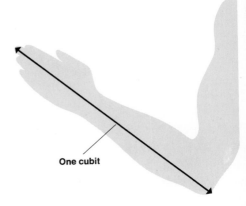

One cubit

1-4 The cubit is the earliest known measure of length. Can you think of some disadvantages of using this measure of length today?

Table 1-2:

Metric Units of Length	
10 millimeters (mm)	= 1 centimeter (cm)
100 centimeters	= 1 meter (m)
1000 meters	= 1 kilometer (km)

activity

Using a meter stick. Measure the length and width of your classroom in meters. Make three separate measurements and take an average of your three readings. Now change your average readings to millimeters and then to kilometers. Compare your results with those of your classmates. Can you explain any differences? Why do you think you were asked to take the average of three separate readings?

The liter is the basic unit of volume. All material objects in the world occupy a certain amount of space. A car takes up more space than a book, and a school building takes up more space than a car. *The measure of the space occupied by an object is called its* **volume.** The volume of a box-shaped object is found by multiplying its length, width, and height: volume = length × width × height.

sample problem

A rectangular wooden box is 2000 mm long, 0.50 m wide, and 72 cm high. Find the volume of the box.

solution:

Step 1. First write the formula for the volume:
$$\text{volume} = \text{length} \times \text{width} \times \text{height}$$
Step 2. Substitute, after converting all measurements to meters. Then multiply the numbers together and the units together.
$$\text{volume} = 2.0 \text{ m} \times 0.50 \text{ m} \times 0.72 \text{ m}$$
$$= 0.72 \text{ m}^3 \text{ (cubic meters)}$$

(Note: m × m × m = m^3)

The *liter* (*lee*-ter) is the metric unit for volume. The volumes of liquids and gases are often measured in units of liters. The same prefixes are used for parts of a liter as are used for parts of a meter. A cube measuring 10 cm on each side has a volume of 1000 cubic centimeters (10 cm × 10 cm × 10 cm = 1000 cm³). A volume of 1000 cm³ is defined as one liter (L). A box with a volume of 1000 cm³ would hold 1 L of liquid. See Fig. 1-5. Note that one *milliliter* (mL) is the same as one cubic centimeter (cm³).

To find the volume of a small irregular object such as a piece of metal, you can use the method of *water displacement*. Fill a graduated cylinder about half-full with water and note the level of the water. See Fig. 1-6. Then place the piece of metal in the water. Since the metal displaces, or takes the place of, some of the water, the water level will rise. Now take a second reading of the water level. The difference between the two readings is equal to the volume of the metal in milliliters or cubic centimeters. How accurate do you think this method is? Try it two more times on the same object to see if there are any differences. Would an average of these readings give you a more accurate value for the volume than one reading? Why?

The basic unit for measuring time is the second. All of you are familiar with the *second* (sec) as a measurement of time. For thousands of years, scientists have tried to find an accurate standard to measure time. Early clocks like the sundial and hourglass were off by as much as 30 minutes (min) in a single day. Even a good modern watch will gain or lose a second or two each month. Scientists need an even more accurate standard for measuring time.

In 1894, astronomy gave scientists a standard for keeping time. One second (1 sec) was defined as the time it took the earth to complete 1/86,400 of its daily turn on its axis. In other words, an average 24-hour (hr) day was divided into 86,400 seconds (24 hr × 60 min/hr × 60 sec/min = 86,400 sec). However, since the earth wobbles a little on its axis, its exact rotation time varies very slightly from one day to the next. Furthermore, the speed at which the earth turns is slowing down very slightly. The average length of a day was not quite the same in 1980 as it was in 1970.

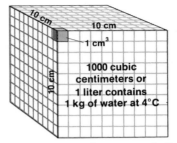

1-5 The large cube represents the volume of 1 L, or 1000 cm³. What part of a liter is represented by the small cube in the corner?

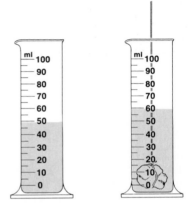

1-6 From this diagram, find the volume of the piece of metal in the graduated cylinder.

Today, scientists have found a way of using the cesium atom for producing the most exact definition of the second. This atomic device is so accurate that it is like running a clock more than 250,000 years without losing or gaining a second!

Weight is a measure of force. A book falls to the floor because it is pulled to the earth. *Any push or pull on an object is called a* **force.** The unit of force in the metric system is the *newton* (N). *A measure of the pull between the earth and an object is called its* **weight.**

The attraction of the earth for a given object is called the force of **gravity.** The pull of gravity is less on top of a mountain than it is in a valley. However, the difference is so small that it would be hard to measure. The higher an object is above the earth, the less the force of gravitational attraction and the less the object weighs.

The weight of an object depends on the following two things: (1) the distance of the object above the surface of the earth (or other body such as the moon or planets) and (2) the amount of matter in the object. Sugar, wood, water, and air are examples of matter. All of these things take up space. Therefore, **matter** *is anything that takes up space and, depending on its location on the earth, has a definite weight.*

The weight of a small object can be found by hanging it from a *spring scale.* A spring scale is a device used to compare forces. Since the measurement of weight is the same as the measurement of force, a spring scale is a handy device. A reading on the scale shows the force of gravitational attraction between the earth and the object. The greater the stretch of the spring, the greater the force, and therefore the more the object weighs.

Mass is a measure of the amount of matter. Look at a rubber sponge, a rock, or a book. You can see that each object takes up space and each has a given weight. Suppose you were to take one of the objects to the top of a high mountain, or to the moon. You now know that the weight of the object would change. However, the amount of matter contained in the object would remain the same. *A measure of the amount of matter in an object is called its* **mass.**

The mass of an object has another important property. It is harder to start or stop an object than it is to keep it in motion. Have you ever tried to push a stalled car, or to stop the same car once you got it moving? *This resistance to any change of motion or change in direction is called* **inertia.** A baseball and a cannonball of the same size do not have the same inertia. It takes more force to throw the cannonball with the same speed as the baseball. You can see that objects having greater inertia (cannonball) also have more mass than objects with less inertia (baseball). Inertia is a property of matter that opposes any change in the state of motion of matter.

Mass is measured in grams. The *gram* (g) is a unit of mass in the metric system. A paper clip has a mass of slightly less than 1 g. A United States 5-cent coin has a mass of about 5 g. The prefixes for the parts of a gram are the same as for the parts of a meter. Thus, a *centigram* (cg) is 1/100 part of a gram, and a *milligram* (mg) is 1/1000 part of a gram. The *kilogram* (kg) is the basic metric unit for mass. It contains 1000 times the mass of a gram. One kilogram equals 1000 grams.

The kilogram standard of mass was not discovered. Instead, it was defined and agreed upon by the nations of the world. The standard kilogram mass is kept in a vault in France. An exact copy of the standard is stored in the National Bureau of Standards in Washington, D.C.

Masses can be compared by using a spring scale or a platform balance. See Fig. 1-7. Since the pull of gravity is greater on a large mass than it is on a small one, the spring in the spring scale stretches farther for a large mass than for a small one. By reading the amount the spring stretches on a scale, an unknown mass can be found.

1-7 A spring scale (left) and a platform balance (right) are often used in the science laboratory. Describe how these devices are used to compare masses.

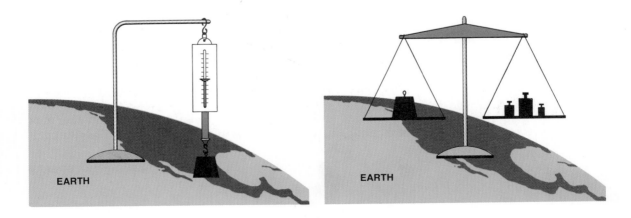

EARTH

EARTH

1-8 The student uses energy (does work) when she climbs the stairs against the force of gravity (F_g) through a vertical distance (d).

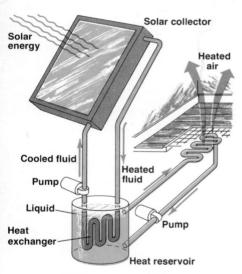

1-9 This diagram shows a home solar heating system. The sun's heat energy is collected by panels on the roof. This energy then heats a liquid, which circulates through the house. In other systems, heat energy can be stored in drums of water or bricks and used when needed.

At sea level, the weight of a 1-kg mass is about 9.8 N. A 2-kg mass weighs about 19.6 N (2 kg × 9.8 N/kg = 19.6 N), and a 60-kg student weighs about 588 N (60 kg × 9.8 N/kg = 588 N). You can see that, for a given location, the weight of an object is proportional to its mass. Since the pull of gravity on the moon is about 1/6 that of the earth, a 60-kg student would weigh only 98 N (1/6 × 588) on the moon. For objects near the earth's surface, you can use the following formula relating weight and mass:

weight in newtons = 9.8 × mass in kilograms

Using a platform balance, you can also find the mass of a small object by placing the object on the left pan and known masses on the right pan. A balance operates like a seesaw. When the pans balance, the mass of the object is equal to the total of the known masses. Would the pans still be in balance if the balance were taken to the moon? Explain.

Energy. The idea of *energy* is one that unites all the sciences. Energy does not take up space, so it is not matter. **Energy** *is defined as the ability to do work.* For example, you do work when you lift an object from the ground onto a truck against the pull of the earth's gravity. The amount of work done depends on the weight of the object and the vertical distance moved. The greater the weight or the longer the distance, the greater the work. See Fig. 1-8. This example is a form of mechanical energy. Other forms of energy include heat, light, sound, electricity, magnetism, and nuclear energy. Almost any form of energy can be changed from one form to another. You will study more about these forms of energy later in the course. You will also see that the wise use of energy is necessary to ensure better living conditions.

The sun is our chief source of energy. It provides us with heat and light. The energy that the earth receives from the sun can be changed into other forms of energy. For example, *solar cells* can change sunlight directly into electricity. These solar cells have already been used to provide the energy to power many earth satellites. Solar energy can also be used to provide heat and hot water for homes and other buildings. Large panels on the roof of a building collect the sun's heat energy. This energy

unit 1 matter and energy

is then stored for later use. See Fig. 1-9. Solar energy may someday be an important source of badly needed energy.

Albert Einstein

Albert Einstein was born in Germany in 1879. He was one of the greatest scientists of all time. Many people think that Einstein had the most creative mind in history.

When Einstein was 5 years old, his father showed him a pocket compass. The little boy was curious about the behavior of the compass needle. The needle kept pointing in the same direction no matter which way the compass was turned. Einstein later said he felt that "something deeply hidden had to be behind all things."

By the age of 12, Einstein decided to devote himself to solving the riddle of the "huge world." Three years later, with poor grades in geography, language, and history, he left school with no diploma. Later, however, he completed his schooling in Switzerland.

Einstein is best known for his special theory of relativity, which he first advanced when he was only 24 years old. One part of this theory says that when an object is moving, its mass increases. This new mass, which increases rapidly as the object approaches the speed of light, is called its relativistic mass. Most speeds, however, are well below that of light. Therefore, for most purposes, the relativistic mass is not used. This is one reason why scientists define mass more accurately as that property of an object by which the object resists any change in its motion or state of rest.

In 1921, Einstein was awarded the Nobel Prize in physics. He won this prize for his work suggesting that light is made up of a stream of tiny particles called *photons*. Shortly before the Second World War, Einstein left his homeland to live in the United States.

The brilliant scientist lived a quiet personal life in Princeton, New Jersey. He was very fond of classical music and played the violin. Einstein had a deep feeling for people and a very religious nature. The universe to Einstein was one of absolute order. Einstein died in 1955.

summary

1. Science deals with the study of all material things in nature; it is largely a process of learning by which new knowledge is found.
2. Physical science deals with matter and energy.
3. Volume is the amount of space occupied by an object.
4. Matter is anything that occupies space and has a certain weight.
5. Weight is the force of attraction between the earth and an object.
6. Mass is a measure of the amount of matter in an object. Mass is also a measure of an object's inertia.
7. Energy is the ability to do work.

review

Match each item in the left column with the best response in the right column. *Do not write in this book.*

1. meter
2. mass
3. light
4. liter
5. weight
6. platform balance
7. milli-

a. a force of attraction
b. a measure of time
c. a device for comparing masses
d. the metric unit of length
e. 1/1000
f. a form of energy
g. the metric unit of volume
h. 1000
i. a measure of inertia

questions

Group A
1. How are science and curiosity related?
2. In your own words, what is science? What are some reasons why people study science?
3. What is meant by the physical sciences? Name five branches of the physical sciences.
4. What are the basic units of length, volume, and mass in the metric system?
5. Describe how you could find the volume of a piece of coal.
6. How would you determine the mass of a small object?
7. What happens to a person's mass as the distance from the earth increases? What happens to a person's weight?
8. What is energy? Name five forms of energy.

Group B

9. What are some advantages and disadvantages of the cubit as a unit of measure?
10. List some reasons for favoring the use of the metric system in the United States for everyday purposes.
11. From the information contained in Fig. 1-5, find the mass of 1 mL of water.
12. From the information contained in Fig. 1-6, describe how you would determine the volume of a small piece of hard candy. What problem arises when you use water in this method?
13. Suggest a way of finding the volume of an irregularly shaped piece of sponge.
14. Write a paragraph describing how the study of science will help you in choosing a career or job.

problems

1. The distance from A to B is 3.5 km. Express this distance in (a) meters, (b) centimeters, and (c) millimeters.
2. How many liters are there in 4000 mL?
3. A steel rod is 25 cm in length. Find its length in (a) meters, (b) millimeters, and (c) kilometers.
4. A rectangular box is 3.0 m long, 2.3 m wide, and 150 cm high. Find its volume in cubic meters (m^3).
5. An object has a mass of 3 kg. What is its mass in grams? What is its weight at sea level?

further reading

Blackwelder, S. K., *Science for All Seasons.* Englewood Cliffs, N.J.: Prentice-Hall, 1980.

Freidman, Herbert, *The Amazing Universe.* Washington, D.C.: National Geographic Society, 1975.

Gamow, George, *One, Two, Three . . . Infinity.* New York: Bantam Books, 1971.

Kraft, B., *Careers in the Energy Industry.* New York: Franklin Watts, 1977.

Smith, W. S., *Science Career Exploration for Women.* Washington, D.C.: N.S.T.A., 1980.

Taylor, L. S., *Chemistry Careers.* New York: Franklin Watts, 1978.

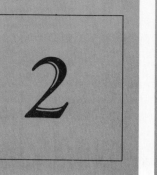

2

properties, changes, and composition of matter

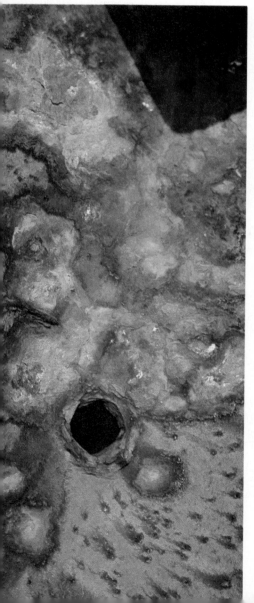

In the last chapter, you studied some of the contributions science has made in today's world. You learned that science is a method of finding answers to our questions about the world we live in. You also studied the metric units used for measuring volume, force, weight, mass, and energy.

In this chapter, you will learn about some of the many properties of matter. As you will see, matter has both physical and chemical properties. You will study the kinds of changes, both physical and chemical, that can take place in matter. You will also study the classification and composition of matter, and the laws that describe the behavior of matter.

objectives:

☐ To explain why density is a useful property of matter.

☐ To distinguish between physical and chemical properties, and between physical and chemical changes.

☐ To distinguish between atoms and molecules, and between elements, mixtures, and compounds.

☐ To recognize common chemical symbols and chemical formulas, and to use chemical equations to describe chemical reactions.

☐ To explain the law of constant proportions and the law of conservation of matter.

ways to identify matter

Physical and chemical properties. In day-to-day living, you learn to recognize many different materials. Some materials, such as a piece of iron, are heavier than others, such as a piece of cork of the same size. In order to tell one material from another, scientists look for certain traits that help identify a given material. These traits are called *properties*.

Some *physical properties* of a material include color, density, hardness, solubility (how much of it will dissolve in a certain amount of liquid), freezing temperature (the temperature at which a material freezes), and boiling temperature (the temperature at which a material boils). These properties can be easily seen or measured.

The terms *heavy* and *light* are commonly used to refer to an object's mass. In science, however, these terms refer to **density.** *The density of a substance is its mass per unit volume: density = mass/volume.* It is a measure of how close the particles of matter are packed together. The density of solids and liquids is most often expressed in grams per cubic centimeter (g/cm^3). The density of gases is expressed in grams per liter (g/L). It is useful to know that the density of water is 1 g/cm^3. See Fig. 2-1.

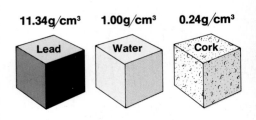

2-1 These blocks have equal volumes but different masses. Which material has the greatest density? Which is the least dense?

sample problem

A piece of brass has a mass of 32.0 g and a volume of 3.8 cm^3. Find the density of brass.

solution:

Step 1.   Write the formula for density from its definition.
$$density = mass/volume$$

Step 2.   Substitute the values given in the problem into the formula.

$$density = 32.0 \text{ g}/3.8 \text{ cm}^3$$

$$= 8.4 \text{ g/cm}^3$$

Substances that are less dense than water (cork, gasoline, wood, ice) will float on water; substances that are more dense (aluminum, zinc, iron) will sink.

2-2 Pepper is still pepper even when the pieces are ground very fine. Is this an example of a chemical or a physical change?

2-3 New substances are formed when a tree burns. Is this an example of a chemical or a physical change?

Properties other than physical properties are also used to describe matter. For example, gold resists being tarnished by water and common acids. Magnesium burns with a bright white flame as it joins with oxygen of the air to form magnesium oxide. These traits are called **chemical properties.** *Chemical properties describe the way in which one substance is changed into a different substance with an entirely different set of properties.*

changes in matter

Physical and chemical changes. Changes in matter are always taking place with the passing of time. Ice changes into water. Paraffin (wax) melts when it is heated. Gasoline evaporates quickly unless kept in a tightly closed container. A piece of chalk can be ground into a fine powder. All of these changes are alike in one way. They do not result in new substances. Such changes are called *physical changes.* Water is the same chemically whether it is in the form of ice or steam. Chalk particles remain chalk no matter what the size or shape of the pieces. Gasoline remains gasoline even after it evaporates. See Fig. 2-2.

Changes in matter that produce new substances are called *chemical changes.* The decay of dead plants, the souring of milk, the rusting of iron, and the digestion of food are examples of chemical changes. In each of these processes, new substances are formed. Their chemical makeup and properties are different from those of the original substances. Magnesium burns to produce a white powder called magnesium oxide. Magnesium oxide does not have the color or metal luster of magnesium. Iron rust is chemically different from iron. Sweet milk is not the same chemically as sour milk. See Fig. 2-3.

Whenever a physical or a chemical change takes place, it is *always* accompanied by an energy change, usually in the form of heat given off or absorbed. The energy released or absorbed can also be in the form of light, sound, or electricity. A good example of a common physical change in which heat is absorbed is the melting of ice, or the boiling or evaporation of water.

Electrolysis is a chemical change. The laboratory setup

unit 1 matter and energy

pictured in Fig. 2-4 can be used to show that water is made up of two different gases. Water is not a simple substance. It can be broken down, or *decomposed,* into simpler substances.

First, the jar is filled with water. Then a small amount of acid is added to the water to make the liquid conduct an electric current. A current separates the water into two colorless gases, oxygen and hydrogen, that collect in separate tubes. *The chemical change that takes place when an electric current passes through a liquid is called* **electrolysis** (eh-lek-*trol*-uh-sis). Electrolysis is important in both science and industry.

You can see from the diagram in Fig. 2-4 that the volume of hydrogen produced is twice the volume of the oxygen produced. Pure hydrogen burns with an almost colorless flame. The fact that hydrogen gas burns is an important chemical property of hydrogen. When hydrogen burns, it combines chemically with oxygen in the air to form water. This process is the reverse of electrolysis.

The oxygen collected in the test tube does not burn. However, if you put a glowing splint of wood into the oxygen, the wood will burst into flame. This reaction shows an important chemical property of oxygen: oxygen does not burn, but it supports the burning of other substances. You will study more about hydrogen and oxygen in a later chapter.

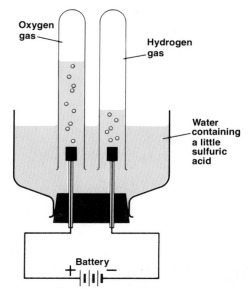

2-4 A direct current separates water (compound) into oxygen (element) and hydrogen (element).

the composition of matter

There are three classes of matter. You know that there are many different kinds of substances. Therefore, when chemists study substances, they must first classify them into similar groups. Chemists have found that all forms of matter can be classified on the basis of their properties into three major classes. These three classes are *elements, mixtures,* and *compounds.* See Fig. 2-5.

Elements are simple substances. You have just seen that water is made up of two gases, hydrogen and oxygen, that have different chemical properties. These gases were found to be simple forms of matter. Hydrogen and oxygen cannot be broken down any further into simpler substances. *A substance that cannot be broken down into*

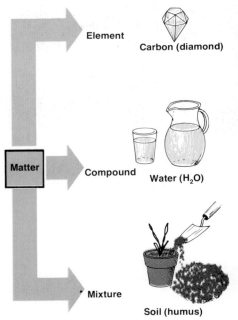

2-5 This diagram shows the three general classes of matter.

simpler substances by chemical means is called an **element.** Elements are nature's building blocks. All materials in the universe are made of either one element, or two or more elements chemically combined.

There are over 100 different elements. Most of these elements exist on the earth in varying amounts. Some are common, others are very rare. Scientists have been able to make a few of these elements under special conditions in the laboratory. You will learn about some of these elements later in the course.

Table 2-1 is a list of some of the more common elements found in the earth's crust. Remember that these elements are not spread evenly throughout the earth's crust. Thus, these percentages (by mass) may not hold for a particular part of the crust.

Table 2-1:

Distribution of Elements in the Earth's Crust			
Oxygen	49.5%	Sodium	2.6%
Silicon	25.8%	Potassium	2.4%
Aluminum	7.5%	Magnesium	1.9%
Iron	4.7%	Hydrogen	0.9%
Calcium	3.4%	Titanium	0.6%
All other elements 0.7%			

Notice that many familiar elements, such as copper, lead, zinc, silver, and gold, are not found in this list. The fact that an element is not common may make it valuable. Usable amounts of these important metals are found in only a few places near the surface of the earth. We are quickly using up many of these metal resources.

Mixtures are physical combinations of materials. From the 26 letters in the alphabet you can put together an endless number of words. In the same way, from about 100 elements an almost endless number of both physical and chemical combinations can be made. Millions of these combinations exist in nature or have been made by chemists.

unit 1   matter and energy

When powdered iron and powdered sulfur are mixed, a grayish material results. Both iron and sulfur are elements. These two elements can be brought very close together by mixing, but they will not combine with each other to form a new substance. A strong magnet can separate the powdered iron from the sulfur. In other words, each element in this mixture keeps its own properties. *A* **mixture** *contains substances that have not been joined chemically.* Air, for example, is a mixture of many gases, mainly nitrogen and oxygen. Can you suggest a way to separate a mixture of powdered iron and sugar without using a magnet?

In a mixture, the substances that make it up can each be present in any amount. You can mix a large amount of iron and a small amount of sulfur, or a small amount of iron and a large amount of sulfur.

2-6 (See below.) The elements zinc (Zn) and sulfur (S) do not react when mixed at room temperature, as shown in A. When they are heated, a chemical reaction takes place, as shown in B, giving off energy. A compound (ZnS) is formed. What forms of energy are released?

activity

Making and separating a mixture. Grind together about 20 g of table salt and about 15 g of powdered charcoal with a mortar and pestle until you obtain a gray powder. Now try to find a way to separate this mixture into the two original substances.

(A)

Compounds are chemical combinations of elements. If a sample of the iron-sulfur mixture is placed in a test tube and heated with a hot flame, the iron will unite chemically with the sulfur. The new substance formed is iron sulfide, which is a *compound.* The physical and chemical properties of iron sulfide are different from those of the two original elements. The iron can no longer be separated from the iron sulfide by a magnet. *A* **compound** *is composed of two or more elements that are joined chemically.* The iron sulfide formed in the above reaction is an example of a chemical compound. Although mixtures can be separated by physical means, a compound can be broken down only by chemical means. Table 2-2 lists some examples of elements, mixtures, and compounds. See Fig. 2-6.

(B)

Table 2-2.

Three Classes of Matter		
Elements (symbols)	Mixtures	Compounds (formulas)
Tin (Sn)	Milk	Water (H_2O)
Iron (Fe)	Butter	Sugar ($C_{12}H_{22}O_{11}$)
Aluminum (Al)	Toothpaste	Table salt (NaCl)
Silver (Ag)	Baking powder	Baking soda ($NaHCO_3$)
Oxygen (O)	Paint	Ethyl alcohol (C_2H_5OH)
Copper (Cu)	Air	Carbon dioxide (CO_2)

A molecule is the smallest part of a substance. Picture a lump of sugar. Imagine dividing it into two pieces, then these two pieces into four, and the four into eight, and so on. Finally, in your mind, you would reach a point at which the particle is so small that, if you divide it again, the two halves would no longer be sugar. This smallest particle, *the simplest unit*, would be one *molecule* of sugar. Any further division of this molecule would mean that its parts would no longer have the properties of sugar. *The smallest particle of a substance that has all the properties of the substance is called a* **molecule** *of that substance.* Molecules are so small that even with the aid of an electron microscope only a few of the largest ones have ever been seen.

The following example will give you some idea of the size of molecules. There are more molecules in one tiny grain of sugar than there are grains of sugar in a 2-kg bag. Actually, each tiny grain of sugar is made up of many billions of sugar molecules.

Perhaps another example will help you realize how small molecules really are. Water is a compound that is made up of molecules. Comparing the size of a molecule of water to the size of a drop of water is like comparing the size of a beach ball to the size of the earth.

An atom is the smallest part of an element. You have seen that water is an example of a common compound. It is composed of the elements hydrogen and oxygen.

unit 1 matter and energy

Each molecule of water must contain hydrogen and oxygen that have been joined chemically. Thus, within each molecule of this compound there must be smaller particles. These single particles of hydrogen and oxygen are called **atoms.** *An atom is the smallest unit of an element that can combine chemically with other elements.* All matter is made up of atoms in various combinations.

You will see in the next section that one atom of oxygen combines with two atoms of hydrogen to form one molecule of water. Each molecule of a compound is made of two or more atoms. Molecules of some gaseous elements, including oxygen and hydrogen, are made up of two identical atoms. Atoms of elements join in many forms. It is estimated that there are over a million compounds known today. New compounds are being made almost daily in research laboratories in every part of the world.

chemical symbols, formulas, and equations

A chemical symbol identifies an element. Each element has a **symbol.** *A symbol is a shorthand way of writing the name of an element.*

Chemists label the elements by using symbols consisting of one or two letters. For example, **O** is the symbol for oxygen, **H** is for hydrogen, **C** is for carbon, **Al** is for aluminum, **Mg** is for magnesium, **S** is for sulfur, **Cl** is for chlorine, and **Ca** is for calcium. (See Fig. 2-7.)

Many elements have been known since ancient times. The symbols for these elements were taken from the shortened form of their Latin names. Some examples of these symbols are **Au** (*aurum*) for gold, **Ag** (*argentum*) for silver, **Pb** (*plumbum*) for lead, **Cu** (*cuprum*) for copper, and **Fe** (*ferrum*) for iron. A more complete chart of the common elements and their symbols appears in Table 3-2, p. 36. (See Fig. 3-6, pp. 38–39.)

Chemical formulas identify compounds. With so many known compounds, it would be very difficult to keep track of them all if each were referred to only by name. **Chemical formulas** *are used to identify compounds by showing the kind and number of atoms that are present.*

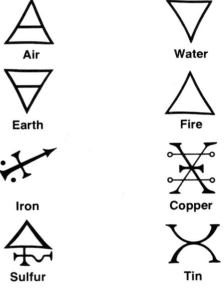

Air   **Water**

Earth   **Fire**

Iron   **Copper**

Sulfur   **Tin**

2-7 Ancient chemists believed that all matter was made up of the basic substances air, water, earth, and fire. They used triangles as symbols for these substances, as shown above. Other early symbols for several elements are also shown. How are today's chemical symbols an improvement over these earlier ones?

You probably know that H_2O is the chemical formula for the compound water. This formula can also be written as **HOH.** The simple formula tells you that water is a compound composed of the elements hydrogen and oxygen. It also tells you that one molecule of water consists of two atoms of hydrogen (H_2) and one atom of oxygen (O). Two molecules of water are written as 2 H_2O.

Notice that chemical symbols are used in writing these formulas. When a number written slightly below the line follows a symbol, it tells how many atoms of that element are found in the molecule. For instance, one molecule of table sugar, $C_{12}H_{22}O_{11}$, has 12 atoms of carbon **(C),** 22 atoms of hydrogen **(H),** and 11 atoms of oxygen **(O).** In reading this formula, say "C twelve, H twenty-two, O eleven." There are 45 atoms in one molecule of this kind of sugar. Table 2-3 is a list of several examples of chemical compounds and their formulas.

Table 2-3:

Some Chemical Compounds and Their Formulas	
Name of Compound	Formula
Magnesium oxide	MgO
Sodium chloride	$NaCl$
Sodium carbonate	Na_2CO_3
Table sugar (sucrose)	$C_{12}H_{22}O_{11}$
Sulfuric acid	H_2SO_4
Iron sulfide	FeS
Calcium hydroxide	$Ca(OH)_2$

Chemical equations show chemical changes. One way to show what takes place in a chemical change is to describe the change in words. This description is called a *word equation.* A better way of describing a chemical change is to use a *chemical equation.* Such equations state briefly what elements or compounds enter into a chemical change and what elements or compounds are produced by chemical change. For example, heating a mixture of iron and sulfur produces iron sulfide. The word equation for this reaction is written as

iron + sulfur → iron sulfide

neither sodium iodide nor lead nitrate. This solid separates from the liquid and is called a **precipitate.** The formation of an insoluble precipitate is evidence that a chemical reaction has taken place. The new compound formed is lead iodide (PbI_2). If the flask is put back on the balance, you will find that its mass remains the same. No mass has been gained or lost in the reaction.

This experiment is an example of the **law of conservation of matter:** *matter cannot be made or destroyed by ordinary chemical means*. Scientific laws are based on many careful tests. It is important to remember, however, that science is a way of knowing about the material world; it is not a book of rules. Even a so-called scientific law may be changed to fit newly discovered facts. Science is a continuing process of discovery. In a later chapter, you will learn that matter can be changed into energy as a product of nuclear reactions. Nuclear reactions are not ordinary chemical changes.

activity

Conservation of matter. Carefully balance a flashbulb on one pan of a balance. Without disturbing the mass in the other pan, remove the flashbulb and use it in a camera. When the bulb cools, carefully put it back on the balance pan again. Observe the effect, if any, on the original balance setting. (1) Has a chemical change taken place in the flashbulb? How do you know? (2) Did you observe any change in the mass of the used flashbulb? (3) Can you explain what happened?

careers in
CHEMISTRY

Careers in the area of chemistry can be divided into three general job groups: chemical technician, chemist, and chemical engineer. Sometimes it is hard to decide to which group a given job belongs.

Chemical lab technicians do tests to find out if materials have or retain certain properties. They may test certain materials to find out their chemical content.

Technicians also prepare chemical solutions used in making various products, such as drugs, textiles, soaps, foods, beauty aids, and many others. They also run tests to see if products meet purity or strength standards. Further, technicians work with many lab instruments.

Chemists are most often engaged in research and development of new products for the consumer. They set up tests and gather data. Chemists may make natural substances or combine existing elements to make new compounds. Chemists work in a wide range of industries. They sell consumer products, market new products, and consult with other chemists and business people. Many chemists teach in colleges and universities, or work for federal, state, and local government agencies.

Chemical engineers combine their skills in chemistry in the design of equipment for new plants and chemical processes. They are involved in the making of paints, plastics, and hundreds of other items that are in daily use. The chemical engineer helps to select new plant sites and to plan the layout of machines and equipment.

Training for lab technicians should include high school courses in physics, chemistry, and mathematics, and possibly additional specialized courses. A four-year degree program in a college or university is necessary for chemists and chemical engineers.

summary

1. Density equals the mass of a substance divided by its volume.
2. A physical change does not produce a new substance.
3. A chemical change produces a new substance.
4. Every physical or chemical change in matter involves an energy change.
5. The smallest particle of an element is an atom.
6. A compound is formed by a chemical union of two or more elements. The smallest particle of a compound is called a molecule.
7. Chemical equations express chemical reactions that actually take place.
8. The law of constant proportions states that every compound always contains the same proportion by mass of the elements that make up that particular compound.
9. The law of conservation of matter states that matter cannot be created or destroyed by ordinary chemical reactions.

review

Match each item in the left column with the best response in the right column. *Do not write in this book.*

1. density
2. chemical change
3. formulas
4. physical change
5. symbols

a. oxygen and hydrogen
b. MgO, H_2SO_4, H_2O
c. sharpening a pencil
d. Fe, Pb, Zn
e. mass per unit volume
f. burning coal
g. conservation of energy

questions

Group A
1. What is meant by properties of matter? Give five examples of physical properties.
2. What is density? How is it usually expressed?
3. What is the difference between a physical change and a chemical change? Give five examples of each.
4. Name three classes of matter.
5. Define an element. an atom.
6. Describe a mixture. Name five mixtures and their uses.
7. Define a chemical compound. What is the smallest particle of a compound called?
8. List five compounds and their uses.
9. Make a list of ten chemical elements and their symbols.
10. What information does a chemical formula give? a chemical equation?
11. What is meant by a scientific law or principle?

Group B
12. What is wrong with the following statement: "Lead is heavier than aluminum"? How can it be corrected?
13. Describe how you would find the density of a small irregular solid.
14. If you were given a flask of an unknown gas, how would you test for oxygen? for hydrogen?
15. What gases are produced by electrolysis of water?
16. What compound is formed when hydrogen burns in air? Write its formula.
17. What does the formula for carbonic acid, H_2CO_3, tell about its composition? How many atoms are contained in a molecule of carbonic acid?
18. The reaction between sodium chloride (NaCl) and silver nitrate ($AgNO_3$) forms

a white precipitate of silver chloride (AgCl). Write the complete chemical equation for this reaction.

19. Give an example of the law of constant proportions. Does it apply to mixtures or compounds? Explain.

problems

1. The density of an unknown sample was found to be 11.2 g/cm^3. Could you find the volume of the sample from this information? What do you need to know to find its mass?
2. What is the mass in grams of water that fills a tank 100 cm long, 50 cm wide, and 30 cm high? in kilograms?
3. The density of a piece of brass is 8.4 g/cm^3. If its mass is 500 g, find its volume.
4. A graduated cylinder is filled with water to a level of 40.0 mL. When a piece of copper is lowered into the cylinder, the water level rises to 63.4 mL. Find the volume of the copper sample. If the density of copper is 8.9 g/cm^3, what is its mass?
5. The results of an experiment showed that 24 g of magnesium (Mg) combined with 16 g of oxygen (O) to form magnesium oxide (MgO). Find the percent of magnesium and oxygen in the sample. Will these percentages always be the same for magnesium oxide? Explain.

further reading

Asimov, Isaac, *The Road to Infinity.* Garden City, N.Y.: Doubleday, 1979.

Asimov, Isaac, *View from a Height.* New York: Avon Books, 1975.

Dobbs, Frank W., *Age of the Molecule: Chemistry in the World and Society.* New York: Harper & Row, 1976.

Jaffe, B., *Crucibles: The Story of Chemistry from Ancient Alchemy to Nuclear Fission.* New York: Dover, 1976.

Keen, Martin, *Let's Experiment.* New York: Grosset & Dunlap, 1976.

Waxter, Julia B., *Science Cookbook.* Belmont, Cal.: Pitman Learning, Inc., 1980.

structure of atoms and molecules

3

This chapter will increase your understanding of matter by describing the makeup of atoms and molecules. You will see how chemists create models in order to explain their observations. You will be able to "picture" how atoms combine to form molecules, and you will see how chemists have arranged the elements in an orderly way.

objectives:

☐ To describe the structure of elements through the use of atomic models.

☐ To define atomic number, atomic mass, and valence number of an element.

☐ To explain how elements combine with other elements to form chemical compounds.

☐ To explain the use of the periodic table of the elements.

the world within the atom

Atomic theory helps you to understand atoms. As early as 400 B.C., learned Greeks believed that matter could not be destroyed. Some of them also believed that matter was made up of particles that they called "atoms," a word meaning "cannot be divided." See Fig. 3-1. It was not until the early nineteeth century that the English chemist John Dalton (1766–1844) proposed the first

3-1 Democritus (deh-*mock*-rih-tus) (460–370 B.C.), a Greek thinker, was the first to suggest that all matter could be broken down into tiny particles called atoms. What were the four basic "elements" of his time?

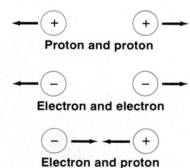

Proton and proton

Electron and electron

Electron and proton

3-2 From the diagram above, what general statements can you make about like charges and unlike charges?

useful theory of the nature and properties of matter. Although Dalton's theory included some of the old ideas about atoms, he was the first to describe the unchanging makeup of every compound.

The modern atomic theory was developed from the one proposed by Dalton. Today, scientists have a good deal of evidence for believing the following:

1. Matter is made up of very small particles called atoms.
2. Atoms of the same element are chemically alike; atoms of different elements are chemically different.
3. Although single atoms of a given element may not all have quite the same mass, they do, in any natural distribution, have a definite *average* mass that is a property of that element.
4. Atoms of different elements have a different average mass.
5. Atoms do not break down in ordinary chemical changes.

Single atoms of any element are much too small to be seen even with the best optical microscopes. Yet, indirect observations of how atoms behave have led to a firm idea of today's atomic theory. Such observations may be compared to "knowing" that a bullet was fired from an unseen gun by hearing a loud noise, seeing smoke, and then seeing a hole in a nearby target all about the same time.

Earlier it was stated that one atom of oxygen joins with two atoms of hydrogen to form one molecule of water. A drop of water contains large numbers of these molecules grouped together. Each molecule of water in the drop is formed in the same way. In other words, every oxygen atom acts like every other oxygen atom. Likewise, every hydrogen atom acts like every other hydrogen atom. It is no surprise to find that a molecule of water acts like every other molecule of water. All water molecules are always alike.

A model of an atom helps you picture its structure. Atoms are too small to be seen by any known methods. What scientists know about atoms has been learned by observing the way a large number of samples of matter behave. Scientists find it useful to make diagrams or

unit 1 matter and energy

actual physical **models** to show what an atom "looks" like. *A model is a way of explaining how or why things behave the way they do.* Models are always subject to change. Future findings may improve present models. It is unlikely that anyone will ever see an atom in the same sense that you see diagrams and drawings of atoms.

activity

Describing things you cannot see. You learn about the world you live in by using your senses. You depend on your sense of sight for much of the information you receive. However, you can also make reasonable judgments about data gathered through your other senses.

Place any object in a small box. Seal the box so no one in your class will know what you placed in the box. Write your name on the outside of the box, and pass it to someone else in the class. Then take someone else's box in exchange. You now have a sealed box with an unkown object in it.

Try to decide what is inside the box without opening it. Move the box gently, tilting it from side to side. Does the object make a noise? Can you guess its shape? Is it a sphere or a cylinder? flat or irregular? Is it heavy or light? rough or smooth? Make a list of your observations and try to identify the object from these clues. Give your list to the person whose name is on the box to find out if your findings are correct.

Basic particles of an atom. For the past 80 years, scientists have been gathering a great deal of evidence about the makeup of atoms. This evidence has led to the modern atomic model. Scientists now know that atoms of all elements contain three different kinds of basic particles: *electrons, protons,* and *neutrons.*

Electrons and protons are electrically charged particles. The electron has a negative charge (−); the proton has a positive charge (+). The neutron has no net charge (0). You have seen that a falling object is attracted to the earth because of the force of gravity. Similarly, as a result of electrical forces, *unlike charges attract each other.* However, *like charges repel each other.* These electrical forces act within the atom. See Fig. 3-2.

The mass of an electron is very small. The mass of a proton is about the same as that of a neutron. Protons and neutrons have small masses also, but each has about 2000 times the mass of an electron.

3-3 A section of a 6.4-km-long device used to move protons at very high speeds. Could uncharged particles be moved by a device of this type? Explain.

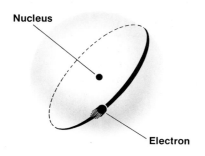

Nucleus

Electron

3-4 A diagram of an ordinary hydrogen atom (H), which is made up of one proton and one electron. What is the charge on the nucleus? What is the charge on the electron?

3-5 (top) A diagram of a helium atom (He) and (bottom) a lithium atom (Li). Explain why each atom is electrically neutral.

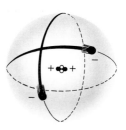

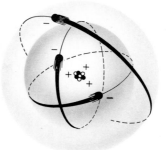

Although scientists have now identified over 200 other atomic particles, you will study only the three basic particles at this time. Figure 3-3 shows one of the huge machines used to study atomic particles. This *linear accelerator* applies electric pulses that pull charged particles through a long tube with ever-increasing speeds. These high-energy "atomic bullets" hit target atoms, causing the atoms to break apart into various types of particles.

Different elements are made up of different numbers of protons. The simplest atom is the most common form of hydrogen **(H),** which has one proton and one electron. Its single electron may be pictured as traveling in a path around the proton. See Fig. 3-4.

Except for the simplest kind of hydrogen, atoms of all other elements also contain neutrons. The protons and neutrons are grouped closely together in a central core, called the *nucleus*. Electrons move rapidly around the nucleus at some distance away from it.

The next simplest element after hydrogen is helium **(He).** This atom is made up of a nucleus containing two protons and two neutrons. Helium has two electrons that move in a region around the nucleus.

Lithium **(Li)** is the third atom in this sequence. The most common form of this element has three protons and four neutrons in its nucleus. Three electrons are in rapid motion around the nucleus. See Fig. 3-5.

Atoms of other elements have four, five, six, and more electrons in motion around the nucleus. The number of protons and neutrons in the nucleus also increases from one element to the next.

The ninety-second and most complex of the natural elements, uranium, has a total of 92 electrons in rapid motion around the nucleus. Uranium also has 92 protons and about 146 neutrons in its nucleus. Like the three atoms just discussed, the uranium atom is electrically neutral. Because of its large number of protons and neutrons, it has a very high density.

Thirteen additional elements after uranium have been made by scientists to date. The newest element contains 105 electrons in motion around a positively charged nucleus of 105 protons. All atoms of a given element have the same number of protons as electrons, and so are neutral. Atoms of the same element may have a varying number of neutrons in the nucleus.

unit 1 matter and energy

Table 3-1:

Partial List of Elements: Number and Position of Electrons								
			K-Shell	L-Shell		M-Shell		
Name of Element	Symbol	Atomic Number (protons)		1st sub-shell	2nd sub-shell	1st sub-shell	2nd sub-shell	3rd sub-shell
Hydrogen	H	1	1					
Helium	He	2	2					
Lithium	Li	3	2	1				
Beryllium	Be	4	2	2				
Boron	B	5	2	2	1			
Carbon	C	6	2	2	2			
Nitrogen	N	7	2	2	3			
Oxygen	O	8	2	2	4			
Fluorine	F	9	2	2	5			
Neon	Ne	10	2	2	6			
Sodium	Na	11	2	2	6	1		
Magnesium	Mg	12	2	2	6	2		
Aluminum	Al	13	2	2	6	2	1	
Silicon	Si	14	2	2	6	2	2	
Phosphorus	P	15	2	2	6	2	3	
Sulfur	S	16	2	2	6	2	4	
Chlorine	Cl	17	2	2	6	2	5	
Argon	Ar	18	2	2	6	2	6	

The atomic number identifies an element. *The number of protons contained in the nucleus of an atom is called the* **atomic number** *of an element.* The atomic number is also equal to *the total number of electrons around the nucleus of a neutral atom.* Table 3-1 is a partial list of the elements and a breakdown of their protons and electrons. Notice that the elements are arranged in a definite order, according to their atomic numbers.

Each element has a different atomic number. The nucleus of common hydrogen, for example, contains one proton. The atomic number of hydrogen is 1. The nucleus of the helium atom has two protons, so the atomic number of helium is 2. From lithium, with an atomic number of 3, the number of protons keeps increasing by one proton. Finally, the last natural atom, uranium, with an atomic number of 92, is reached.

Atomic mass number. The actual mass of a single atomic particle is very small. These small values are not easy to use in calculations. As a result, chemists are mostly concerned with the *relative* masses of the various kinds of atoms. That is, they want to know how atomic masses compare with each other.

The chemists of the world agreed to set the mass of the most common kind of carbon atom (known as carbon-12) equal to *exactly* 12 *atomic mass units*. Using this standard, ordinary hydrogen has an atomic mass of about 1 and oxygen, about 16. This means that the oxygen atom has a mass of about 16 times that of hydrogen. The relative atomic mass for each element is shown in the periodic table in Fig. 3-6. Also study the list of common elements shown in Table 3-2.

Table 3-2:

List of Common Elements							
Name of Element	Symbol	Atomic Number	Mass Number	Name of Element	Symbol	Atomic Number	Mass Number
Aluminum	Al	13	27	Lithium	Li	3	7
Barium	Ba	56	137	Magnesium	Mg	12	24
Bismuth	Bi	83	209	Manganese	Mn	25	55
Boron	B	5	11	Mercury	Hg	80	201
Bromine	Br	35	80	Neon	Ne	10	20
Calcium	Ca	20	40	Nickel	Ni	28	59
Carbon	C	6	12	Nitrogen	N	7	14
Chlorine	Cl	17	35	Oxygen	O	8	16
Chromium	Cr	24	52	Phosphorus	P	15	31
Cobalt	Co	27	59	Platinum	Pt	78	195
Copper	Cu	29	64	Potassium	K	19	39
Fluorine	F	9	19	Silicon	Si	14	28
Gold	Au	79	197	Silver	Ag	47	108
Helium	He	2	4	Sodium	Na	11	23
Hydrogen	H	1	1	Sulfur	S	16	32
Iodine	I	53	127	Tin	Sn	50	119
Iron	Fe	26	56	Tungsten	W	74	184
Lead	Pb	82	207	Zinc	Zn	30	65

unit 1   matter and energy

For many uses, chemists find it helpful to express mass as a *whole* number, called the **mass number.** *The mass number is the whole number closest to the atomic mass shown in the periodic table. The mass number is equal to the total number of protons and neutrons found in the nucleus of an atom of a given element.*

The common hydrogen atom consists of one proton and one electron. The mass of either a proton or a neutron is about 2000 times that of an electron. Therefore, scientists disregard the mass of the electron in calculating atomic mass. Thus, the atomic mass (in atomic mass units) of a proton, a neutron, or a hydrogen atom is very close to 1, as shown in Table 3-3. These masses can be easily rounded off to give a mass number of 1.

Table 3-3:

	Average Mass of Atoms and Subatomic Particles			
Name	Actual Mass (gram)	Atomic Mass Unit	Mass Number	Charge
Electron	9.109×10^{-28}	0.0005486	0	Negative $(-)$
Proton	1.673×10^{-24}	1.007278	1	Positive $(+)$
Neutron	1.675×10^{-24}	1.008665	1	Neutral (0)
Hydrogen	1.7×10^{-24}	1.00797	1	Neutral (0)
Carbon	2.0×10^{-23}	12.01115	12	Neutral (0)
Oxygen	2.65×10^{-23}	15.9994	16	Neutral (0)

The number of neutrons in the nucleus of an atom can be found. Since the nucleus of any atom, except the common form of hydrogen, is made up of protons and neutrons, the atomic mass number of an atom is equal to the sum of the protons and neutrons. The atomic number, however, is equal to the number of protons. To find the number of neutrons in an atom, subtract the atomic number (protons) from the mass number (protons + neutrons).

neutrons = mass number − atomic number
　　　　　(protons + neutrons)　　(protons)

Period

1.00797	1
hydrogen	
H	
1	

— Electrons in energy levels
— Atomic mass
— Name of element
— Symbol
— Atomic number

1

METALS

Note: Values in parentheses indicate the mass number of the most stable isotope or, for those marked with an asterisk, the best-known isotope.

TRANSITION ELEMENTS

	I		II																	
2	6.939 lithium **Li** 3	2 1	9.0122 beryllium **Be** 4	2 2																
3	22.9898 sodium **Na** 11	2 8 1	24.312 magnesium **Mg** 12	2 8 2																
4	39.102 potassium **K** 19	2 8 8 1	40.08 calcium **Ca** 20	2 8 8 2	44.956 scandium **Sc** 21	2 8 9 2	47.90 titanium **Ti** 22	2 8 10 2	50.942 vanadium **V** 23	2 8 11 2	51.996 chromium **Cr** 24	2 8 13 1	54.9380 manganese **Mn** 25	2 8 13 2	55.847 iron **Fe** 26	2 8 14 2	58.9332 cobalt **Co** 27	2 8 15 2
5	85.47 rubidium **Rb** 37	2 8 18 8 1	87.62 strontium **Sr** 38	2 8 18 8 2	88.905 yttrium **Y** 39	2 8 18 9 2	91.22 zirconium **Zr** 40	2 8 18 10 2	92.906 niobium **Nb** 41	2 8 18 12 1	95.94 molybdenum **Mo** 42	2 8 18 13 1	(99*) technetium **Tc** 43	2 8 18 13 2	101.07 ruthenium **Ru** 44	2 8 18 15 1	102.905 rhodium **Rh** 45	2 8 18 16 1
6	132.905 cesium **Cs** 55	2 8 18 18 8 1	137.34 barium **Ba** 56	2 8 18 18 8 2	Lantha-nide series / 174.97 lutetium **Lu** 71	2 8 18 32 9 2	178.49 hafnium **Hf** 72	2 8 18 32 10 2	180.948 tantalum **Ta** 73	2 8 18 32 11 2	183.85 tungsten **W** 74	2 8 18 32 12 2	186.2 rhenium **Re** 75	2 8 18 32 13 2	190.2 osmium **Os** 76	2 8 18 32 14 2	192.2 iridium **Ir** 77	2 8 18 32 15 2
7	(223) francium **Fr** 87	2 8 18 32 18 8 1	(226) radium **Ra** 88	2 8 18 32 18 8 2	Actinide series / (257) lawrencium **Lw** 103	2 8 18 32 32 9 2	(261) kurchatovium **Ku** 104	2 8 18 32 32 10 2	(260) hahnium **Ha** 105	2 8 18 32 32 11 2	(263) 106	2 8 18 32 32 12 2	(261) 107	2 8 18 32 32 13 2				

RARE EARTH

Lanthanide series

138.91 lanthanum **La** 57	2 8 18 18 9 2	140.12 cerium **Ce** 58	2 8 18 20 8 2	140.907 praseodymium **Pr** 59	2 8 18 21 8 2	144.24 neodymium **Nd** 60	2 8 18 22 8 2	(147*) promethium **Pm** 61	2 8 18 23 8 2	150.35 samarium **Sm** 62	2 8 18 24 8 2	151.96 europium **Eu** 63	2 8 18 25 8 2

Actinide series

(227) actinium **Ac** 89	2 8 18 32 18 9 2	232.038 thorium **Th** 90	2 8 18 32 18 10 2	(231) protactinium **Pa** 91	2 8 18 32 20 9 2	238.03 uranium **U** 92	2 8 18 32 21 9 2	(237) neptunium **Np** 93	2 8 18 32 23 9 2	(242) plutonium **Pu** 94	2 8 18 32 23 8 2	(243) americium **Am** 95	2 8 18 32 24 9 2

38

THE ELEMENTS

NONMETALS

	III	IV	V	VI	VII	VIII
						4.0026 (2) — **He** — helium — 2
	10.811 (2,3) **B** boron 5	12.01115 (2,4) **C** carbon 6	14.0067 (2,5) **N** nitrogen 7	15.9994 (2,6) **O** oxygen 8	18.9984 (2,7) **F** fluorine 9	20.183 (2,8) **Ne** neon 10
	26.9815 (2,8,3) **Al** aluminum 13	28.086 (2,8,4) **Si** silicon 14	30.9738 (2,8,5) **P** phosphorus 15	32.064 (2,8,6) **S** sulfur 16	35.453 (2,8,7) **Cl** chlorine 17	39.948 (2,8,8) **Ar** argon 18

58.71 (2,8,16,2) **Ni** nickel 28	63.54 (2,8,18,1) **Cu** copper 29	65.37 (2,8,18,2) **Zn** zinc 30	69.72 (2,8,18,3) **Ga** gallium 31	72.59 (2,8,18,4) **Ge** germanium 32	74.9216 (2,8,18,5) **As** arsenic 33	78.96 (2,8,18,6) **Se** selenium 34	79.909 (2,8,18,7) **Br** bromine 35	83.80 (2,8,18,8) **Kr** krypton 36
106.4 (2,8,18,0) **Pd** palladium 46	107.870 (2,8,18,1) **Ag** silver 47	112.40 (2,8,18,2) **Cd** cadmium 48	114.82 (2,8,18,3) **In** indium 49	118.69 (2,8,18,4) **Sn** tin 50	121.75 (2,8,18,5) **Sb** antimony 51	127.60 (2,8,18,6) **Te** tellurium 52	126.9044 (2,8,18,7) **I** iodine 53	131.30 (2,8,18,8) **Xe** xenon 54
195.09 (2,8,18,32,16,2) **Pt** platinum 78	196.967 (2,8,18,32,18,1) **Au** gold 79	200.59 (2,8,18,32,18,2) **Hg** mercury 80	204.37 (2,8,18,32,18,3) **Tl** thallium 81	207.19 (2,8,18,32,18,4) **Pb** lead 82	208.980 (2,8,18,32,18,5) **Bi** bismuth 83	(210*) (2,8,18,32,18,6) **Po** polonium 84	(210) (2,8,18,32,18,7) **At** astatine 85	(222) (2,8,18,32,18,8) **Rn** radon 86

ELEMENTS

157.25 (2,8,18,25,9,2) **Gd** gadolinium 64	158.924 (2,8,18,27,8,2) **Tb** terbium 65	162.50 (2,8,18,28,8,2) **Dy** dysprosium 66	164.930 (2,8,18,29,8,2) **Ho** holmium 67	167.26 (2,8,18,30,8,2) **Er** erbium 68	168.934 (2,8,18,31,8,2) **Tm** thulium 69	173.04 (2,8,18,32,8,2) **Yb** ytterbium 70
(247) (2,8,18,32,25,9,2) **Cm** curium 96	(249*) (2,8,18,32,27,8,2) **Bk** berkelium 97	(251*) (2,8,18,32,28,8,2) **Cf** californium 98	(254) (2,8,18,32,29,8,2) **Es** einsteinium 99	(253) (2,8,18,32,30,8,2) **Fm** fermium 100	(256) (2,8,18,32,31,8,2) **Md** mendelevium 101	(254) (2,8,18,32,32,8,2) **No** nobelium 102

sample problem

How many protons are in a sodium atom? How many neutrons? How many electrons?

solution:

Step 1. From Table 3-2, you see that sodium has an atomic number of 11, which means that there are 11 protons in its nucleus.

Step 2. The mass number (protons + neutrons) of sodium is 23. Therefore, the number of neutrons in its nucleus is $23 - 11 = 12$ neutrons.

Step 3. The total number of electrons around the nucleus is the same as the atomic number, that is, 11 electrons.

The periodic table contains useful data. To study the properties of the elements and to predict the behavior of compounds formed from various elements, chemists have grouped the elements in the order of their increasing atomic numbers in a special way. In this arrangement, nuclei of the atoms of one element differ from the nuclei of atoms of the next element by the addition of one proton, and usually by one or more neutrons. This grouping forms the basis of the *periodic table* of the elements, shown in Fig. 3-6.

All the elements except hydrogen, which stands alone, are placed in seven numbered horizontal rows, called *periods*. (See Fig. 3-7.) The vertical columns numbered from I to VIII are called *groups*. Elements in each group have chemical and physical properties that are somewhat alike.

The six elements in Group I are very active chemically. That is, they combine readily with certain other elements. For instance, Group I elements react with water as follows: lithium reacts slowly; sodium, fast; potassium, very fast; and rubidium, violently. Group VIII contains the family of noble (rare) gases. These elements are very stable; that is, they do not react readily with other elements. Chemists have been able to prepare only a few compounds containing these elements.

On the left side of the periodic table, the most active elements are at the bottom. On the right side, the most active elements are at the top.

        unit 1   matter and energy

Notice the place of **metals** and **nonmetals** in the table. You are already familiar with some of the metals, such as iron, zinc, lead, copper, tin, and gold. Well-known nonmetals include such elements as nitrogen, oxygen, and chlorine.

Metals are elements that usually have a shiny surface. They are commonly solid at room temperature and have a high density. Metals can be easily hammered or drawn into different shapes. Most metals are good conductors of heat and electric current.

Nonmetals are mostly very poor conductors of heat and electric current. They are of various colors. For example, sulfur is yellow, chlorine is greenish-yellow, and carbon is black. Nonmetals have a low density.

The electron cloud provides a good atomic model. Because an electron moves around the nucleus of an atom with such great speed, its position at any one time is uncertain. Thus, it is hard to pinpoint an electron's exact position. Instead, all of the electrons are pictured as occupying a cloudlike region around the nucleus. Think of a spinning airplane propeller or electric fan. When the blades are moving very fast, their outline is no longer clear. Similarly, the fast-moving electrons also seem to form a hazy oval area around the nucleus.

You can imagine how complex this motion becomes as you add more electrons in many different levels and think of them moving within a sphere instead of a flat area. The diagram in Fig. 3-8 shows how a hydrogen atom might "look" on a flat surface. The spherical region in which electrons move is called an *electron cloud.*

The electron cloud model of more complex atoms shows the regions occupied by all the electrons of an atom. For instance, all 92 electrons of the uranium atom do not move at the same distance, or level, from the nucleus. Instead, the electrons are in seven different levels, each at a different distance from the nucleus. These levels, in fact, are called *electron shells,* or *energy levels,* and are the subject of study in more advanced chemistry courses.

Electrons have a fixed amount of energy. As stated, electrons appear to be found in distinct shells in all atoms. These shells are at various distances from the atomic nucleus. The electrons in these shells have a

Sodium
Na

Magnesium
Mg

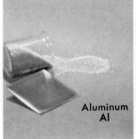
Aluminum
Al

Silicon
Si

3-7 The first four elements of period 3 of the periodic table. What are the remaining elements of this period?

Outer region (electron cloud)

Nucleus (proton)

3-8 An electron cloud model of an ordinary hydrogen atom. What causes the cloud around the nucleus?

3-9 By climbing up each rung of the ladder, the painter increases the potential energy level by a definite amount. What energy level would you assign to the ground as a reference point?

3-10 The distribution of electrons in shells, or energy levels, around the atomic nucleus. How many electrons are needed to complete the second subshell of the M shell?

Shell or energy level	Maximum total in completed shell	Maximum totals in completed sub-shell
P		
O		
N	32	
M	18	2, 6, 10
L	8	2, 6
K	2	2
Nucleus (+)		

fixed amount of energy. The greater an electron's distance from the nucleus, the greater energy the electron has. See Fig. 3-9.

According to the present atomic model, each of these shells can hold only a certain maximum number of electrons and no more. The inner shell, called the **K** shell, can hold no more than two electrons. The second shell, **L,** consisting of two subshells, can hold no more than a total of eight electrons. The third shell, **M,** consisting of three subshells, can hold no more than a total of 18 electrons. The **N** shell can hold as many as 32 electrons. Figure 3-10 and Table 3-1 show how electrons are distributed for the first 18 elements in the periodic table.

Look again at Table 3-1. Notice that oxygen has eight protons and, therefore, a total of eight electrons. These electrons are distributed as follows: two in the **K** shell (the maximum number) and six in the **L** shell. Of these six electrons, two are in the first subshell and four are in the second.

Each element is different because no two *neutral* elements have atoms with the same number of protons (or electrons). *Chemists believe that the way atoms combine to form molecules is determined by the number of electrons in the outer shells of the atoms.*

how compounds are formed

Electrons move between atoms. You are now ready to learn how two different atoms unite to form a compound. Consider a particle of sodium chloride (NaCl), which is common table salt. You have seen from Table 3-2 that sodium has an atomic number of 11 and a mass number of 23. First, imagine what a sodium atom would look like if it could be seen: there would be 11 protons and 12 neutrons in its nucleus. Of the 11 electrons around its nucleus, two would be in the **K** shell, eight in the **L** shell, and one electron left in the outer **M** shell. See Fig. 3-11.

Now, picture the chlorine atom. In Table 3-2, note that chlorine has an atomic number of 17 and a mass number of 35. This means that there are 17 protons and 18 neutrons tightly packed in the nucleus. A total of 17 electrons are in constant motion around the nucleus. Of these 17 electrons, two are in the **K** shell, eight are in the **L** shell, and the remaining seven are in the outer **M** shell. Of the outer **M** shell electrons, two fill the first subshell. The remaining five electrons are in the second subshell. The second subshell thus needs one more electron to be completely filled. You can check this structure by looking at Fig. 3-10. Also study the diagram in Fig. 3-12.

The outer shell of the sodium atom contains only one electron, whereas the outer subshell of the chlorine atom lacks only one electron to complete it. *There seems to be a strong tendency for atoms to form complete shells and subshells.* Chemists have learned that a sodium atom will give (lend) its one outer electron readily. The loss of an electron means the sodium atom has lost a negative charge. This leaves the sodium atom with a charge of +1. On the other hand, a chlorine atom will readily accept (borrow) the electron to complete its outer subshell. The addition of one electron gives the chlorine atom a charge of −1. These two oppositely charged particles attract each other to form sodium chloride (NaCl).

The property of atoms to lend or borrow electrons tells, in many cases, how one element will combine with another in a chemical reaction. *An element's ability to combine with other elements is shown by its* **valence number.**

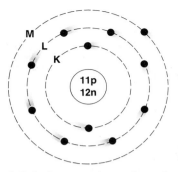

3-11 A diagram of a sodium atom (Na) where **p** represents the number of protons and **n** the number of neutrons in the nucleus.

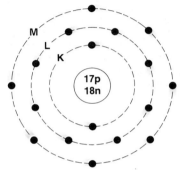

3-12 A diagram of a chlorine atom (Cl).

Table 3-4:

Valence Numbers of Some Common Elements					
Name	Symbol	Valence Number	Name	Symbol	Valence Number
Aluminum	Al	+3	Potassium	K	+1
Barium	Ba	+2	Silver	Ag	+1
Calcium	Ca	+2	Sodium	Na	+1
Copper(II), (cupric)	Cu	+2	Zinc	Zn	+2
Copper(I), (cuprous)	Cu	+1	Bromine	Br	−1
Iron(III), (ferric)	Fe	+3	Chlorine	Cl	−1
Iron(II), (ferrous)	Fe	+2	Fluorine	F	−1
Hydrogen	H	+1	Iodine	I	−1
Lead(II)	Pb	+2	Oxygen	O	−2
Magnesium	Mg	+2	Sulfur	S	−2

The valence number tells the number of electrons that an atom of an element will give up, accept, or share during a chemical reaction.

The sodium atom is said to have a valence number of +1 because it takes on a positive charge of one unit when it *gives up* its outer electron to another element. The valence number of chlorine is said to be −1 because it takes on a negative charge of one unit when it *accepts* an electron.

You can use Table 3-4 to find the valence numbers of some of the more common elements.

Notice that some of the elements listed in Table 3-4 have more than one valence number. Iron has a valence number of +2 in some compounds and +3 in others. When the valence number of iron is +2, its compounds are called ferrous, or iron(II). When its valence number is +3, its compounds are called ferric, or iron(III). Thus, there are two chlorides of iron: ferrous chloride, or iron(II) chloride; and ferric chloride, or iron(III) chloride. Copper is another element with two valence numbers: +1 and +2. Can you name the two chlorides of copper?

It is useful to know that metals have positive valence numbers. Metals often tend to lose electrons in chemical reactions. Nonmetals usually have negative valence numbers. They tend to gain electrons during chemical changes. Can you tell from Table 3-4 which elements are nonmetals?

Look also at the periodic table in Fig. 3-6. Notice that the elements in Group I have a valence number of +1. What is the usual valence number of Group II elements?

Most elements combine to form ionic bonds. The transfer or sharing of electrons between atoms forms **chemical bonds.** *Chemical bonds are electric forces that tend to hold two or more atoms together.* These electric forces hold the atoms together because unlike charges attract each other. The making of some bonds involves not only the outer-shell electrons but also those of the next inner shell as well.

As you have seen, many compounds are formed by the *transfer* of electrons from one element to another. *The bond formed when two atoms transfer electrons is called an* **ionic bond.** Sodium chloride is a typical example of a compound formed by ionic bonding. The sodium atom readily gives up its outer electron. The chlorine atom easily accepts an electron. The outer shell, or subshell, of both elements is thus completed. The electrons farthest from the nucleus of an atom are the ones that are involved in chemical changes. Therefore, when an electron from the sodium atom is transferred to the chlorine atom, a chemical change takes place. In this case, a white solid compound called sodium chloride (table salt) is formed. This compound may be written as $Na^{+1}Cl^{-1}$. However, formulas are commonly written without the valence numbers. For example, sodium chloride is written as NaCl. Study Fig. 3-13.

If you know the valence numbers of the elements, you can predict the formulas of the compounds they form. The valence number of sodium is +1 and of chlorine, −1. Since all compounds are electrically neutral, atoms of these two elements combine in a 1-to-1 ratio to form the compound NaCl.

Can you predict the formula for calcium chloride? In Table 3-4, the valence number of calcium is listed as +2.

3-13 (top) Sodium chloride is formed by the transfer of a single electron in the outer shell of the sodium atom to a vacancy in the outer shell of the chlorine atom (bottom). The loss of an electron from the sodium atom leaves it positively charged. The gain of an electron by the chlorine atom makes it negatively charged. What kind of bond is formed?

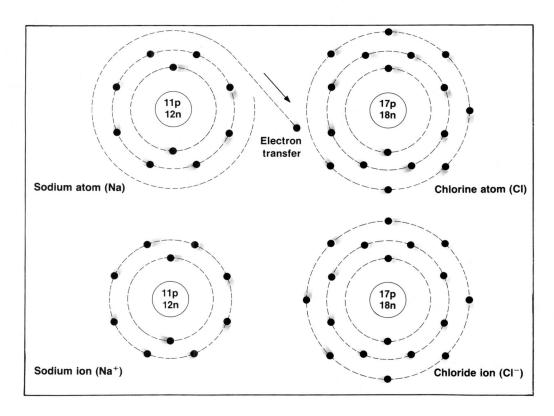

Sodium atom (Na) Electron transfer Chlorine atom (Cl)

Sodium ion (Na⁺) Chloride ion (Cl⁻)

The total valence of the positive part of a compound must equal the total valence of the negative part. Since chlorine has a valence number of -1, two atoms of chlorine are required to react with one atom of calcium. Therefore, the formula of calcium chloride is written as $CaCl_2$. In the same way, aluminum chloride is written $AlCl_3$, since aluminum has a valence number of $+3$. How would you write the chemical formula for lead(II) oxide?

Notice that the nucleus of an atom is not changed in chemical reactions. Also notice that no atoms or parts of atoms are created or destroyed by a chemical change. When most compounds are formed, a chemical change is brought about by electrons from the outer shell of one element filling the outer shell of another element. The highest energy level is occupied by the valence electrons. *Energy is always exchanged whenever chemical bonds are made or broken.*

Some elements combine to form covalent bonds. The use of valence numbers makes it easy to write the formulas of compounds formed by the transfer of electrons. However, not all compounds are formed by the actual transfer of electrons from one element to another. A compound may also be formed when some atoms *share* common electrons. Nonmetals usually bond in this manner. *When atoms share one or more pairs of electrons, a* **covalent bond** *is formed.*

Water is a good example of a molecule formed by the sharing of electrons. From Table 3-1, notice that oxygen needs two electrons to complete the **L** shell. Hydrogen

3-14 (left) Diagram showing two atoms of hydrogen and one atom of oxygen. (right) Diagram showing the formation of a molecule of water. Common electron pairs are shared, completing the shells of both elements. What kind of bonding has taken place?

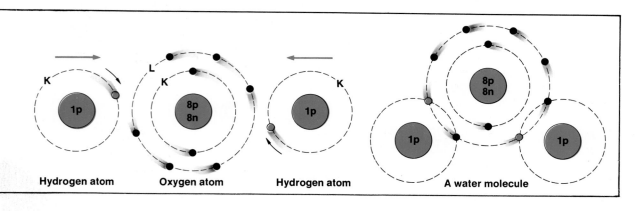

Hydrogen atom   Oxygen atom   Hydrogen atom   A water molecule

unit 1   matter and energy

needs one electron to complete the **K** shell. The electron from each of two hydrogen atoms, and two electrons from a single oxygen atom, are shared by the **L** shell of oxygen and the **K** shell of hydrogen. Thus, a molecule of water, H_2O, is formed. See Fig. 3-14. A laboratory model of a water molecule is shown in Fig. 3-15.

Molecules of common gases such as oxygen, hydrogen, chlorine, and nitrogen are also formed by the sharing of electrons. The hydrogen atom, for example, has only one electron. Since the first shell can hold a maximum of two electrons, two hydrogen atoms share their electrons with each other to form a hydrogen molecule. Chemists write the formula for this molecule as H_2 ("H two"). Molecules of other common gases are written as O_2 for oxygen, Cl_2 for chlorine, and N_2 for nitrogen. These molecules are called *diatomic* molecules. See Fig. 3-16.

Chemical energy is stored in a chemical bond. This stored energy is called *potential energy*. You will study more about potential energy later in the text. Energy is required to break chemical bonds. In most cases, when compounds are formed from elements, energy (in the form of heat, sound, or light) is given off. The reverse occurs when the compound is broken down, at which time the same amount of energy is absorbed.

Some groups of atoms bond as a unit. The behavior of some atom groups resembles that of a single atom. That is, certain groups of atoms stay united during a chemical reaction. These groups are called *radicals*. A radical acts just as though it were a single atom with a definite valence number. Two common radicals are the hydroxide group **(OH)** and the sulfate group **(SO_4)**.

The symbol for a radical may be written in parentheses: ( ). This is needed when a radical is used more than once in writing a formula. For instance, sodium hydroxide is **NaOH,** but calcium hydroxide is **$Ca(OH)_2$,** and aluminum hydroxide is **$Al(OH)_3$**. The valence numbers of the common radicals are shown in Table 3-5.

Compounds are named from their formulas. Notice that in writing chemical formulas for compounds, the element (or radical) with the positive valence number is written first. This is followed by the element (or radical) with the negative valence number. In naming almost all com-

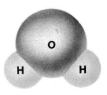

3-15 A laboratory model of a water molecule (H_2O). Can you explain why the oxygen (O) atom is shown larger than the hydrogen (H) atom?

3-16 Laboratory models of diatomic molecules of hydrogen (H_2), chlorine (Cl_2), and oxygen (O_2) gases. (Remember that atoms and molecules are *not* solid spheres.) How are these molecules formed?

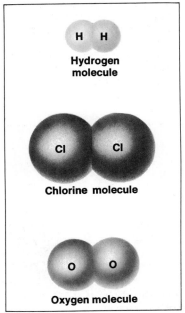

H   H

Hydrogen molecule

Cl   Cl

Chlorine molecule

O   O

Oxygen molecule

Table 3-5:

Valence Numbers of Common Radicals		
Name	Formula	Valence Number
Ammonium	NH_4	+1
Acetate	$C_2H_3O_2$	−1
Bicarbonate	HCO_3	−1
Carbonate	CO_3	−2
Chlorate	ClO_3	−1
Hydroxide	OH	−1
Nitrate	NO_3	−1
Phosphate	PO_4	−3
Sulfate	SO_4	−2
Sulfite	SO_3	−2

pounds containing two elements, the name of the last element ends in *ide*. For example, **NaCl** is sodium chlor*ide*, **KI** is potassium iod*ide*, and **CaO** is calcium ox*ide*.

The ending *ate* indicates that the compound contains oxygen in addition to the other elements mentioned in the name. For example, **NaNO$_3$** is sodium nitr*ate*, **CaCO$_3$** is calcium carbon*ate*, **CuSO$_4$** is copper(II) sulf*ate*, and **KClO$_3$** is potassium chlor*ate*. How would you write the formula for sodium bicarbon*ate*? ammonium chlor*ide*? hydrogen sulf*ate*?

Dmitri Mendeleyev and the Periodic Table

In 1864, an Englishman, John Newlands, arranged all the known elements in the order of their atomic masses. Newlands divided the elements into series of seven elements each. The eighth element was found to have chemical properties similar to the first element of the preceding series. For that reason, Newlands made this eighth element the first in a second series. This arrangement was called the *law of octaves*. Later, it was found that this system only worked for the lighter elements.

The periodic table in use today is based largely on the

unit 1 matter and energy

pioneering work of a Russian chemist, Dmitri Mendeleyev (men-deh-*lay*-eff) (1834–1907). When Mendeleyev first prepared a table of elements, he assumed that all the elements had not yet been discovered. He left gaps in the table where the undiscovered elements seemed to fit. Mendeleyev then predicted the properties that these undiscovered elements should have. When the elements were later discovered, they fit into the gaps just as Mendeleyev had predicted. Mendeleyev was a man of great genius and was regarded as the most outstanding teacher of his time.

Some 45 years later, a brilliant young English scientist, Henry Moseley, used X-ray patterns as a basis for listing the elements in the order of their atomic numbers. Today, scientists know that atomic number, rather than atomic mass, is the proper basis for the order of the elements in the periodic table.

∫ummary

1. All atoms are made of protons, neutrons, and electrons.
2. The atomic number of an element is equal to the number of protons in its nucleus.
3. The atomic number of an element identifies the element in the periodic table.
4. In a neutral atom, the number of protons (+) in the nucleus equals the number of electrons (−) in shells (energy levels) around the nucleus.
5. The mass number is equal to the total number of protons and neutrons in the nucleus.
6. An element's ability to combine with other elements to form compounds is shown by its valence number.
7. Ionic bonds are formed by the transfer of electrons from the outer shell of one element (usually a metal) to the outer shell of another element (usually a nonmetal).
8. Covalent bonds are formed when elements (usually nonmetals) share outer shell electrons with each other.

review

Match each item in the left column with the best response in the right column. *Do not write in this book.*

1. proton
2. atomic number
3. radical
4. ionic bond
5. covalent bond
6. mass number

a. neutron
b. equal to number of neutrons plus protons
c. formed by sharing of electrons
d. sulfate group, SO_4
e. positively charged particle
f. formed by electron transfer
g. equal to number of protons
h. valence number

questions

Group A

1. Name the three basic particles of an atom. What charge does each have?
2. What is a scientific model?
3. Where in the atom is most of the mass found? Explain.
4. How does an atom of iron differ in structure from an atom of gold?
5. What is the maximum number of electrons in each of the first three shells of an atom?
6. Set up a table of ten elements, listing the following information about each: (a) the number of protons in the nucleus, (b) the number of neutrons in the nucleus, and (c) the total number of electrons outside the nucleus.
7. Where in the periodic table are the most active elements located?
8. Where in the periodic table would you find the metals? Do metals have positive or negative valence numbers?
9. What is the valence number of Group I elements?
10. What are the formulas for molecules of oxygen, nitrogen, and chlorine?
11. Write the formula for each of the following compounds: (a) calcium hydroxide, (b) ammonium sulfate, (c) hydrogen phosphate, and (d) copper(II) chloride.
12. Name the compound represented by each of the following formulas: (a) $KClO_3$, (b) NH_4OH, (c) Na_2CO_3, (d) $Pb(C_2H_3O_2)_2$, and (e) K_2S.

Group B

13. Write a brief summary of the modern atomic theory of matter.
14. Sketch a diagram of the following atoms showing the number of protons and neutrons in the nucleus and electrons in the **K, L,** and **M** shells: (a) carbon, (b) magnesium, (c) nitrogen, and (d) sulfur.

15. Describe two ways in which atoms combine to form compounds. Give several examples of each.
16. Explain the difference between the atomic mass of an atom and its mass number.
17. From your knowledge of the periodic table and the fact that sodium reacts rapidly with water, predict how potassium will react with water.
18. Name five elements that are included in the "nitrogen" group of the periodic table.
19. Explain why some elements have positive and other elements have negative valence numbers.
20. How has the periodic table been of value in advancing the knowledge of chemistry?
21. What type of chemical bonding would you expect to take place in the making of lithium fluoride? Explain.
22. What evidence would you look for to tell that a chemical reaction has taken place?

problems

1. Write the chemical formulas for each of the following compounds: (a) barium chloride, (b) calcium iodide, (c) copper(II) oxide, (d) lead(II) nitrate, (e) aluminum hydroxide.
2. Write the names of each of the following compounds: (a) $MgSO_4$, (b) Na_2O, (c) $Ba(OH)_2$, (d) Li_2SO_4, (e) NH_4Cl, (f) $PbCO_3$.
3. Give the valence number of the following *underlined* elements (or radicals): (a) $\underline{Ca}Cl_2$, (b) $\underline{Ag}_2S$, (c) $K_2\underline{SO}_4$, (d) $\underline{Cu}_2O$, (e) $Al_2\underline{S}_3$, (f) $\underline{Fe}_2O_3$.

further reading

Asimov, Isaac, *The Beginning and the End*. Garden City, N.Y.: Doubleday, 1977.

Asimov, Isaac, *Asimov on Chemistry*. Garden City, N.Y.: Doubleday, 1975.

Palder, Edward L., *Magic with Chemistry*. New York: Grosset & Dunlap, 1976.

Trefil, J. S., *From Atoms to Quarks*. New York: Scribners, 1980.

Wilford, J. N., ed., *Scientists at Work: The Creative Process of Scientific Research*. New York: Dodd, Mead & Co., 1979.

4

the kinetic theory of matter

In this chapter, you will study the behavior of very small particles of matter, including atoms and molecules. The kinetic theory is a model for the behavior of matter. This model will help you to explain the properties of gases, liquids, and solids in terms of the forces between particles and the energy of the particles.

objectives:

☐ To explain the kinetic theory of matter.

☐ To identify the three phases of matter as they relate to the forces between particles and the energy these particles have.

☐ To identify kinds of solution and their properties.

☐ To describe the nature of electrolytes and the formation of ions and their properties.

molecules are always in motion

The kinetic theory is a useful model. If ammonia water is placed in an open dish in the front of the room, you will smell the odor in a very short time. You will smell this odor even if there are no air currents in the room. How fast these molecules move depends upon the energy they have. *The energy of motion of objects, including molecules, is called* **kinetic energy.** The higher the temperature, the greater the speed of the molecules, and so the greater their kinetic energy also.

The statement that matter is made up of particles (atoms, ions, or molecules) that are in constant motion is part of *the kinetic theory of matter*. This theory has been used to explain how matter behaves and to predict new facts about matter.

activity

Motion of molecules. Fill a large beaker with cold water. Fill another beaker with hot water. Wait several minutes and add a drop of food coloring to each beaker at the same time. Do not touch or stir the water. Observe what happens. How can you explain the difference in the behavior of the food coloring in the hot and the cold water?

Diffusion shows molecules are in random motion. A few drops of hydrochloric acid (HCl) were placed on a wad of paper in the bottom of a cylinder. Then some strong ammonia water was placed on a wad of paper in the bottom of another cylinder. When the open ends of the two cylinders were brought together, as shown in Fig. 4-1, a white smoke was formed. This smoke is the compound ammonium chloride (NH_4Cl). Ammonium chloride was formed as a result of a chemical reaction between the vapors of hydrochloric acid and ammonia. *The mixing of molecules of two or more substances as a result of their motion is called* **diffusion.** Diffusion takes place in liquids and solids as well as in gases. Gases diffuse through porous solids. "Porous" means having very small openings. For example, a sponge is porous. Oxygen gas diffuses into cell membranes while waste gas diffuses out of plants and animals, including humans.

Oxygen gas from underwater plants is diffused into the water. This water then passes over the gills of fish, where it is diffused again through the gill walls and finally enters the bloodstreams of fish.

There are three phases of matter. You can recognize many kinds of matter, such as wood, water, air, concrete, and

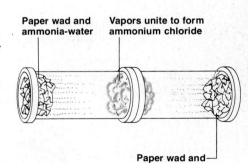

Paper wad and ammonia-water

Vapors unite to form ammonium chloride

Paper wad and hydrochloric acid

4-1 A demonstration of diffusion of gas particles. How do you know a chemical reaction has taken place? Describe the formation of ammonium chloride in terms of the kinetic theory of matter.

gasoline, by the way they look, feel, or smell. Matter can be divided into three familiar groups: solids, liquids, and gases. These groups are called *phases or states of matter*. Wood and steel are examples of matter in a solid phase. Water and gasoline are examples of matter in a liquid phase. The air you breathe is an example of matter in a gas phase.

In a solid phase, matter takes up a definite volume and has a definite shape. The atoms of a solid move slowly, usually vibrating in place. There is little space between the particles, and strong forces hold the solid particles together. Solids cannot be compressed; they have a high density and their rate of diffusion is very slow.

Matter in a liquid phase takes up a definite volume but has no definite shape. Liquids flow freely. The particles of a liquid move faster than the particles of most solids, and they have greater kinetic energy. There are greater distances between the particles of a liquid since the bonds are much weaker than those of a solid. Liquids cannot be compressed and may evaporate into a gas.

Matter in the form of a gas does not have a definite volume or shape. A gas takes on the shape and size of its container. Air in a toy balloon is a good example of how a gas fills its container. The spaces between gas molecules are large. Gas molecules have no fixed bonds. Gases have a low density and diffuse very rapidly. See Fig. 4-2.

Scientists also speak of a fourth state of matter called *plasma*. This form of matter, somewhat like a gas, is made up of charged particles and exists only at very high temperatures. Plasma is rare on earth but quite common throughout the universe. It is the major state of matter in the stars. The northern lights (aurora borealis) in the upper atmosphere and solar flares on the sun are other examples of matter in a plasma state. Scientists have estimated that over 99% of all matter in the universe is plasma.

Another unusual form of matter is that of neutron stars. This form of matter is so compressed that electrons are squeezed into protons to form neutrons. The density of a neutron star is many billions of kilograms per cubic centimeter. If the earth were compressed in a similar manner, it would measure about 200 m across.

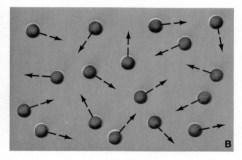

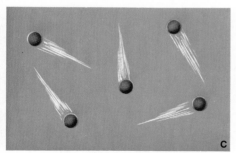

4-2 (a) Particles of a solid vibrate about fixed positions. (b) Weak forces of attraction between particles of a liquid allow for greater movement. (c) Particles of a gas are widely separated and move farther away.

unit 1 matter and energy

Changes in phase involve changes in energy. All forms of matter can be changed from one phase to another by changing the temperature, or the pressure, or both. Water is a well-known example. When a piece of ice (solid) is heated, it changes to water (liquid). Adding heat energy to ice increases the kinetic energy of the molecules. It also weakens the forces holding the ice molecules together. If enough heat is added to a solid, these forces will be weakened until they cannot hold the solid in a definite shape. The solid will then *melt* into a liquid.

As the water continues to be heated, it turns to steam (gas). The forces are not able to hold the rapidly moving molecules *within* the liquid together. The molecules will spread apart into a gas. The gas molecules will have more energy than the liquid and will rise to the surface. This process is known as *boiling*. Chemical bonds, however, do not break when a solid melts or when a liquid boils.

A substance can also change from a liquid to a gas through *evaporation*. Evaporation is the escape of some of the faster-moving molecules from the *surface* of a liquid. This change of phase can take place whether the liquid is hot or cold. Boiling, however, takes place only at a definite temperature. Water evaporates because some of its molecules at the surface have enough energy to break away from the rest of the molecules and become a gas. Energy is removed by these molecules, and the remaining liquid is cooled. Blow your breath on a wet finger and on a dry one at the same time. Can you tell which finger is losing energy? How? See Fig. 4-3.

When heat is removed from molecules of steam, the molecules will slow down until the forces of attraction are once again strong enough to hold them together to form water. The process of going from a gas to a liquid is called *condensing*. As more heat is removed, the liquid will finally *freeze* into a solid whose molecules have very strong forces of attraction. The total amount of heat energy removed is equal to the total amount of heat energy absorbed when the process is reversed.

Changes in phase can also be brought about by changing the pressure. Pressure applied to ice, for instance, will cause it to melt. *Melting* is the process of changing a substance from a solid to a liquid. You have often

4-3 Water evaporates because some molecules receive enough energy to leave the surface. Some water molecules return to the surface after hitting molecules of gases in the air. What happens to the evaporation process if the temperature of the water is increased?

seen snow stick to the soles of your shoes. Your weight provides the pressure to melt the snow. As you lift your foot, the pressure is removed and the water refreezes on the sole of your shoe in the form of ice. The next step melts more snow and adds more ice to your shoe.

Sometimes matter changes from a solid to a gas without passing through the liquid phase. This process is called *sublimation* (sub-lih-*may*-shun). Dry ice (carbon dioxide in the solid phase) is an example of matter having no liquid phase at ordinary pressures. At normal air pressure, dry ice sublimes directly into a gas. Because of its low temperature, dry ice is often used to keep ice cream from melting.

structure of crystals

Many solids have a definite internal structure. If a common salt solution is allowed to stand in an open dish until the water evaporates, salt will remain in the dish. If you look closely at the particles of salt, you will see a regular shape, with flat surfaces and straight edges. *Solids having an orderly internal pattern of their atoms are called* **crystals.** Common table salt, for instance, is made up of tiny, cube-shaped crystals. See Fig. 4-4.

The size of crystals depends upon the time it takes them to form. The longer it takes, the larger the crystals grow. Each crystal of a given substance has the same shape and distinct pattern as the other crystals of the same substance. Crystals are regular geometric structures. The atoms and molecules of which they are formed are arranged in regular, repeating patterns. Many other solids, such as plastic, gum, and even glass are not crystals. Their molecules do not form a regular, repeating structure. See Fig. 4-5.

4-4 Magnified crystals of common table salt (NaCl). Describe the shape of these crystals.

activity

Crystals can be grown from seed. Crystals of copper(II) sulfate ($CuSO_4$) are blue, six-sided solids whose opposite sides are parallel. To produce these crystals, a hot concentrated solution of copper sulfate is allowed to cool. When the solution is poured into another container, crystals of copper sulfate remain.

To make large copper sulfate crystals, allow the cool solution to evaporate slowly for a week or two. Then drop a very small, nearly perfect crystal into this solution to serve as a "seed." The copper sulfate collects on the seed as the crystal grows. It is possible to grow crystals having a mass of as much as 1 kg by this method. CAUTION: *Care should be taken in handling these crystals since copper sulfate is poisonous.*

The main use of copper sulfate, also called blue vitriol, is in copper plating. It is also used in pesticides. Sometimes small amounts of copper sulfate are added to city water supplies to destroy tiny plants that give the water a bad taste.

4-5 A garnet crystal (left) and a group of quartz crystals (right) show how crystal structure follows a definite pattern.

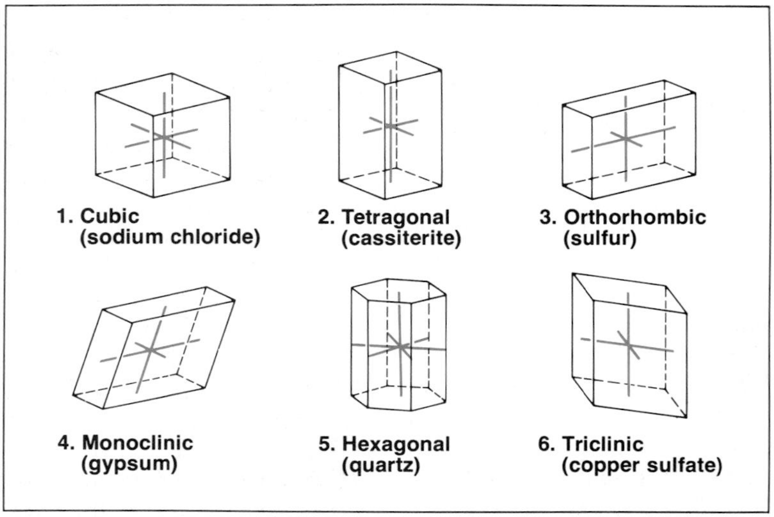

4-6 Diagrams of the six basic crystal systems. (An example of a substance is given for each system.) How would you describe the position of the axes in the tetragonal system?

1. Cubic (sodium chloride)
2. Tetragonal (cassiterite)
3. Orthorhombic (sulfur)
4. Monoclinic (gypsum)
5. Hexagonal (quartz)
6. Triclinic (copper sulfate)

Crystal patterns can help identify a substance. The shape of a crystal depends on the pattern in which the atoms or molecules of the solid are arranged. Crystal patterns are studied by scientists with the use of X rays. It was found that each crystal has its own distinct type of pattern on an X-ray picture. By measuring such patterns, scientists can tell how far apart the particles in a crystal are, and how these particles are arranged.

Crystals are grouped into six basic crystal systems. The simplest, the *cubic* system, is made up of three axes at right angles to each other, as in a cube. All of the axes are of equal length. See Fig. 4-6. Lead sulfide (PbS), sodium chloride (NaCl), and potassium chloride (KCl) are examples of compounds in the cubic system. Diagrams of other systems of crystals are also shown in Fig. 4-6.

solutions

Solutions are uniform mixtures. If you place a cube of sugar in a liter of water and stir, the sugar will seem to disappear. The molecules of sugar will spread evenly through the water. The sugar has *dissolved* in the water. No chemical reaction has taken place. In this case, the liquid (water) in which the sugar dissolves is called a *solvent*. The solid (sugar) that is dissolved is called the

unit 1 matter and energy

solute (sol-*yoot*). The liquid formed by dissolving a solute in a solvent is called a **solution.** *A solution is a uniform mixture of a solute and a solvent.* Solutions do not settle out on standing, and the solute cannot be removed from the liquid solvent by a filter. The solute, however, can be removed by evaporation of the liquid. Water can dissolve a gas, such as air, or a liquid, such as alcohol.

Some substances, such as oils, gums, and resins, will not dissolve in water. These substances are said to be *insoluble* in water. They will, however, dissolve in other liquids such as alcohol, gasoline, or turpentine. Different solvents are used to meet various needs. For instance, cleaning fluids, such as benzine, will remove grease spots. Turpentine is used to dissolve certain paints.

There are several kinds of solutions. Different words are used to describe solutions that contain varying amounts of solute. A spoonful of sugar dissolved in a liter of water is an example of a *dilute* solution. A cupful of sugar dissolved in the same amount of water is said to be *concentrated.*

In the above cases, the terms dilute and concentrated are not very definite. They mean a "little" or a "lot" of solute in the solvent. In the chemistry lab, you may see bottles labeled "Concentrated Sulfuric Acid" or "Concentrated H_2SO_4." This means that there is not much water (solvent) present in the acid.

If you were to keep adding sugar to a concentrated sugar solution, stirring it constantly, you would finally reach a point where no more sugar would dissolve at that temperature. The sugar would then settle to the bottom of the container. This solution would now be *saturated.* Such a solution contains a definite percent of solute, which is always the same for a given temperature and solute. In this example, only a fixed maximum amount of sugar would dissolve in a liter of water at some given temperature. See Fig. 4-7.

4-7 The amount of solute that will dissolve in a saturated solution varies according to the solute used. The water temperature is 60°C. Which solute is most soluble?

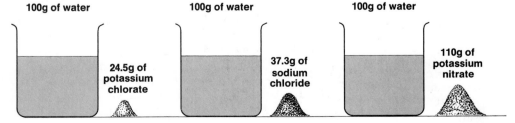

100g of water   100g of water   100g of water

24.5g of potassium chlorate

37.3g of sodium chloride

110g of potassium nitrate

If you heat a saturated solution of sugar and water, you will find that more sugar can be dissolved. In fact, a hot saturated solution contains a much greater amount of solute than the same solution cooled down. *Most solutes are more soluble in hot solvents than in cold ones.*

Kinetic theory explains the speed at which a solute dissolves. Not only is a solid generally more soluble in hot liquids than in cold ones, but the speed at which a solid dissolves is also greater as the liquid warms up.

When heat is applied, the particles of the solute have extra kinetic energy and they tend to move apart faster. The particles of the solvent move faster also. Therefore, the speed at which the solute goes into solution generally increases.

The speed of dissolving can also be increased by stirring or shaking. After adding sugar to a glass of lemonade, you stir it vigorously with a spoon. This stirring action brings fresh parts of the liquid in contact with the undissolved portions of the solid. The solid dissolves more quickly.

Grinding the solid material into a fine powder will also increase the speed of dissolving. Grinding greatly increases the surface area of the solid that is exposed to the liquid. You can see how grinding increases the surface area of a substance by taking a simple example: In Fig. 4-8, a 2-cm cube has a surface area of 24 cm² (2 cm × 2 cm/side × 6 sides = 24 cm²). Now imagine the 2-cm cube is cut into 1-cm cubes as shown by the dashed lines. The eight new cubes now have a total surface area of 48 cm² (1 cm × 1 cm/side × 6 sides/cube × 8 cubes = 48 cm²). You can see that the surface area exposed to a liquid has been doubled in this example! Thus, finely powdered solids will dissolve more rapidly in the solvent than large lumps of the same substance.

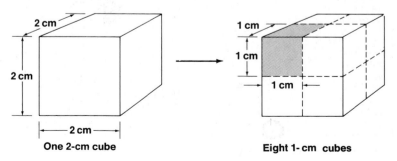

4-8 Dividing the cube into smaller parts increases the surface area. Does the volume change? the mass?

One 2-cm cube

Eight 1-cm cubes

unit 1   matter and energy

Gases are soluble in liquids. You have seen that a solid is generally more soluble in hot liquids than in cold ones. With gases, the reverse is true. *Gases are more soluble in cold liquids than in hot ones.* Very cold water has a small amount of air already dissolved in it. If you let a glass of cold water stand in a warm room, you will soon see gas bubbles on the side of the glass. As the water warms up, it loses some of its ability to hold as much air in solution. See Fig. 4-9.

Larger amounts of gas can be dissolved in a liquid if pressure is applied. Soda pop, for example, is water in which carbon dioxide gas has been dissolved under pressure. When the cap is removed, the pressure within the bottle will be less and the gas will escape from the liquid.

Some liquids are soluble in other liquids. Liquids such as alcohol and water mix so completely with each other that no boundary between them can be seen. Since both liquids are soluble in each other, they are said to be *miscible* (mis-ee-bil). A different condition exists when oil and water are combined. A definite boundary can be seen between the oil and water. When two liquids such as oil and water are mixed, they tend to separate on standing. Liquids that are insoluble in each other are said to be *immiscible.*

Immiscible liquids can form a temporary mixture, called an *emulsion* (ee-*mul*-shun), when shaken together. An emulsion consists of tiny droplets of one liquid scattered

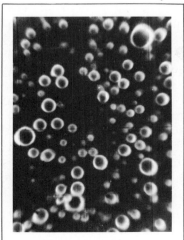 

4-9 As cold water warms up, bubbles of air collect on the sides of the glass. What general statement can you make about the solubility of gases in liquids?

4-10 Fat particles of raw milk (left) and homogenized milk (right). Which would take longer to settle out?

throughout another. When oil and water are mixed by shaking the two liquids in a container, the suspended oil particles will come to the surface when you stop shaking.

Whole raw milk is another example of an emulsion. On standing, the tiny fat particles in the milk come together and rise to the top as cream. When the milk is *homogenized* (ho-*mog*-ee-nized), the fat particles are broken down into still smaller particles by special machines. This forms a more lasting emulsion that does not settle out too quickly. Why do you shake a mixture of oil and vinegar salad dressing? See Fig. 4-10.

Kerosene and water do not mix well. If a special substance is added to the liquids, a more lasting mixture will result. Usually, a substance that is partly soluble in both the liquids is added. If a soap solution is added to a mixture of kerosene and water, the mixture forms an emulsion that stays mixed for a long period of time. Mayonnaise is a common emulsion formed by the addition of egg yolk to a mixture of olive oil and vinegar. Egg yolk is the substance that is partly soluble in both olive oil and vinegar.

electrolytes

Some solutions conduct an electric current. A setup like Fig. 4-11 shows which solutions conduct an electric current and which do not.

To begin the test, fill the beaker with pure water and close the switch of the electric circuit. Note that the lamp does not light. This means that pure water does not provide a path to complete the electric circuit.

Now add a few drops of sulfuric acid to the water. When you do this the lamp glows brightly. Now the solution completes the electric path. In other words, the solution conducts an electric current.

Try other liquids, such as solutions of common table salt (NaCl), hydrochloric acid (HCl), and sodium hydroxide (NaOH). You will see that the lamp also glows brightly when these solutions are used. *Substances that conduct an electric current when dissolved or melted are called* **electrolytes** (eh-*lek*-tro-lites).

Now try a solution of sugar, alcohol, or glycerin. You will see that these substances do not conduct an electric

4-11 What causes the bulb to light when the switch is closed?

Switch

Source of current

Electrodes

Electrolytic solution

current. *Substances that do not conduct an electric current are called* **nonelectrolytes.**

Electrolytes are solutions containing ions. You saw in Chapter 3 that the formula for sodium chloride can be written as $Na^{+1}Cl^{-1}$. Sodium chloride is a crystal made up of rows of *positive ions* and *negative ions. Ions are atoms, or groups of atoms, that carry an electric charge.* An atom that loses an electron becomes positively charged. An atom that gains an electron becomes negatively charged. Therefore, sodium chloride is an ionic compound. See Fig. 4-12.

When salt is dissolved in water, it separates into two kinds of ions. The Na^+ and Cl^- ions are released from their fixed places in the crystal pattern and become free ions in the water. This breaking down of sodium chloride can be written as follows:

$$Na^+Cl^- \longrightarrow Na^+ + Cl^-$$
$$\text{(solid crystal)} \qquad \text{(water solution)}$$

When substances react with water in this manner to form ions, they are said to be *ionized.* The ions in the water are the "charge carriers" that conduct an electric current. The Na^+ ions will then be attracted to the negative pole of the battery, while the Cl^- ions will be attracted to the positive pole. Ionization forms a complete path for electric current to flow. Generally, when metals react with non-metals, ionic compounds are formed. Would you consider calcium chloride an ionic compound? silver nitrate? Explain.

A few covalent compounds, such as HCl gas, also react with water to form ions:

$$HCl \longrightarrow H^+ + Cl^-$$
$$\text{(gas molecule)} \qquad \text{(ions in solution)}$$

Most covalent compounds on the other hand, do not form ions in water. Sugar, for instance, remains in solution as separate sugar molecules ($C_{12}H_{22}O_{11}$). Sugar is a nonelectrolyte. The particles of a nonelectrolyte do not ionize. Therefore, nonelectrolytes cannot conduct an electric current. Acids, bases, and soluble salts are an important group of electrolytes. You will study these compounds in Chapter 5.

The study of electrolytes is of such importance in chemistry that the following summary will help to better understand the next chapter.

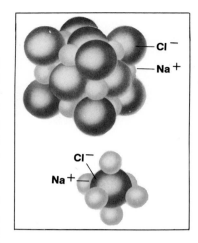

4-12 A crystal of sodium chloride (top) is composed of equal numbers of sodium ions (Na^+) and chloride ions (Cl^-). Within the crystal (bottom), how many sodium ions surround a chloride ion?

chapter 4 the kinetic theory of matter 63

1. Electrolytes in water exist in the form of ions.
2. An ion is an atom or a group of atoms that carries an electric charge.
3. The ions are free to move in the solution and serve to conduct an electric current.
4. The total charge of the positive ions equals the total charge of the negative ions.
5. Water plays an important part in the formation of ions.

Science Teaching

Science teachers can be divided mostly into two large groups, elementary and secondary. They can be further divided into public, private, and parochial (usually religious) teachers. By far, the greatest number of science teachers are found in the public schools.

Teachers in the elementary grades teach science as part of their work in reading, writing, and arithmetic. Elementary school teachers work on basic skills with their students in order to aid their mental, physical, social, and cultural development in later grades.

In the upper grades, the secondary level, science teachers perform a number of tasks. They help the student learn about the general areas of life science, earth science, and physical science (mostly chemistry and physics). Teachers prepare lesson plans and keep records of student progress. They also hold conferences with parents and/or students and attend teacher meetings. In addition, science teachers set up demonstrations, order lab supplies, conduct field trips, and act as advisors to science clubs.

All states require teachers to be certified by the education department of the state in which they teach. Nearly all states require at least a bachelor's degree from a college or university. For secondary teachers, a major field of study must be completed in biology, chemistry, physics, or general science.

Science teachers are required to do a great deal of reading and studying after getting their degrees. This reading helps them to keep up with new developments in science as well as with improved teaching methods.

summary

1. The kinetic theory of matter describes the rapid, zig-zag motion of molecules.
2. The higher the temperature of a substance, the greater the speed of the molecules, and the greater the kinetic energy.
3. Matter exists in three ordinary phases: solid, liquid, and gas.
4. Crystals have an orderly arrangement of atoms, molecules, or ions.
5. A solution is a uniform mixture of a solute (usually a solid) and a solvent (usually a liquid).
6. Solutions can be dilute, concentrated, or saturated.
7. Generally, an increase in temperature will increase the amount of solute that will dissolve in a solvent.
8. The speed at which a solute dissolves depends on the temperature, the size of the particles, and the amount of stirring of the mixture.
9. Substances that conduct an electric current when dissolved or melted are called electrolytes.
10. Ions are atoms, or groups of atoms, that are charged positively or negatively.

review

Match each item in the left column with the best response in the right column. *Do not write in this book.*

1. ion
2. electrolyte
3. water
4. sublimation
5. gas

a. phase of matter
b. common solvent
c. phase change from solid to gas
d. emulsion
e. saturated solution
f. charged particle
g. substance that conducts an electric current when dissolved or melted

questions

Group A
1. What is the kinetic theory of matter? Why is it an important idea?
2. Give several examples of diffusion. What causes diffusion?
3. What are the three ordinary phases of matter? Give several examples of each.
4. How can one phase of matter be changed into another phase?

5. How does the process of boiling differ from evaporation?
6. Describe a method for growing crystals.
7. Why are crystals studied in science?
8. What are the differences between a solute, a solvent, and a solution?
9. Name four common solvents.
10. List several properties of a solution.
11. What is an emulsion? Give several examples.
12. How does temperature affect the solubility of most solids? of gases?
13. What is the difference between a dilute solution and a concentrated one? What is a saturated solution?
14. What is an electrolyte? Name three electrolytes and three nonelectrolytes.
15. What is an ion? How is it formed?

Group B
16. Describe how the movement of atoms and molecules in solids, liquids, and gases is affected by a temperature increase.
17. How is the speed of molecules in matter related to kinetic energy?
18. Using the kinetic theory of matter, describe how melting of a solid takes place; the boiling of a liquid.
19. Describe three ways of increasing the rate at which a solid dissolves in a liquid.
20. What two factors help to dissolve more gas in a liquid?
21. Why is milk homogenized? How is it done?
22. Explain why some solutions conduct an electric current and others do not.

problems

The maximum amount of sodium chloride (NaCl) that can be dissolved in 100 mL of water at a given temperature is 35 g. At the same temperature, what is the minimum amount of water that will dissolve 175 g of NaCl? What is the name for this type of solution?

further reading

Hamlet, Peter, *Introducing Chemistry: A New View*. Boston: D. C. Heath, 1975.

Holden, Alan, *Crystals and Crystal Growing*. Science Study Series, 1961.

Rutherford, F. James, et al., *Project Physics*. New York: Holt, Rinehart and Winston, 1981.

Stone, G. K., *More Science Projects You Can Do*. Englewood Cliffs, N. J.: Prentice-Hall, 1970.

Williams, J. E., et al., *Modern Physics*. New York: Holt, Rinehart and Winston, 1980.

unit 1   matter and energy

acids, bases, and salts

5

You have learned that matter can be classified into elements, compounds, and mixtures. Elements can then be divided into metals and nonmetals. Matter can also be divided into solids, liquids, and gases. In this chapter, you are going to study special compounds, called electrolytes. Electrolytes can be divided into acids, bases, and salts. An important property of the group of compounds called electrolytes is that their solutions conduct an electric current.

objectives:

☐ To identify the properties of acids, bases, and salts.

☐ To describe the making and uses of several common acids, bases, and salts.

☐ To demonstrate skill in balancing chemical equations.

☐ To gain an understanding of the process of neutralization.

☐ To demonstrate knowledge of four major types of chemical reactions.

☐ To understand how chemists use the solubility of salts to separate and identify ions.

acids

Acids have certain properties. In a dilute solution, acids have a sour or sharp taste. You know how sour lemon juice tastes! This sour taste is due to the presence of *citric* (*sit*-rik) acid in the juice. Vinegar tastes sour because of the *acetic* (ah-*see*-tik) acid that it contains. Many foods turn sour when they spoil because the starches and sugars they contain break down into acids. When milk turns sour, for instance, some of the milk sugar is changed to *lactic* (*lak*-tik) acid.

CAUTION: *You should not taste unknown liquids in the lab to find out if they are acids.* Some chemicals will burn your tongue, and others are very poisonous. There is a better way of identifying acids.

When a piece of blue litmus paper is dipped into acetic acid, the paper turns red. Other acids affect blue litmus in the same way. Litmus is a plant dye that changes color when it comes into contact with acids. *Substances, such as litmus, that change color in the presence of certain ions are called* **indicators.**

Acids have properties that are opposite to those of another group of compounds, called bases. Acids and bases react with each other to neutralize each other's properties. Bases will be discussed later in the chapter.

Acids contain hydronium ions. If hydrogen chloride (HCl) gas is bubbled into water, a solution that turns blue litmus paper red is obtained. This solution is called hydrochloric acid (HCl). The following reaction takes place:

$$\text{HCl} \quad + \quad \text{H}_2\text{O} \quad \rightarrow \quad \text{H}_2\text{O}\cdot\text{H}^+ \quad + \quad \text{Cl}^-$$

| hydrogen chloride gas | water | water united with hydrogen ion | chloride ion |

Notice that the hydrogen chloride gas reacts with water in a special way, forming $\text{H}_2\text{O}\cdot\text{H}^+$ ions and chloride ions. Chemists sometimes write the $\text{H}_2\text{O}\cdot\text{H}^+$ ion as H_3O^+ and call it the *hydronium ion. An* **acid** *is a substance that gives hydrogen ions to water to form hydronium ions.* The hydronium ion turns blue litmus paper red.

To make the notation simpler, the term hydrogen ion (H^+) will be used throughout the rest of the text in place of the hydronium ion (H_3O^+).

Acid bottles in the lab are labeled with formulas like those shown in Table 5-1. Notice that all of these acids contain hydrogen.

Table 5-1:

Names and Formulas of Some Common Acids	
Acid	Formula
Sulfuric	H_2SO_4
Hydrochloric	HCl
Nitric	HNO_3
Carbonic	H_2CO_3
Boric	H_3BO_3
Phosphoric	H_3PO_4
Acetic	$HC_2H_3O_2$
Tartaric	$H_2C_4H_4O_6$

Sour taste like lemon

Corrode active metals

The symbol for hydrogen is always written first in the formulas. In a water solution, it is this hydrogen that becomes hydrogen ions. Certain acids, acetic and tartaric for example, contain added hydrogen (as part of a radical) that does not go into solution as hydrogen ions.

Not all substances that contain hydrogen are acids. For instance, there is hydrogen in water (H_2O), methane (CH_4), cane sugar ($C_{12}H_{22}O_{11}$), and ammonia (NH_3). These compounds are not acids.

Acids that produce large numbers of H^+ ions in water are called *strong* acids. Hydrochloric, nitric, and sulfuric acids are examples of common strong acids. Strong acids ionize completely, or nearly so, in water solutions and so provide a great many hydrogen ions.

Acetic, citric, and carbonic acids are common examples of *weak* acids. Since they are only slightly ionized, weak acids supply few hydrogen ions. Do not confuse the terms strong and weak with the terms concentrated and dilute. It is possible to have a dilute solution of a strong acid; dilute hydrochloric acid, for example.

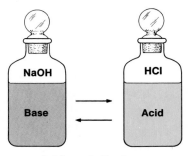

Acids neutralize bases

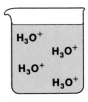

Form hydronium ions in solution

Acids corrode metals. Place a few clean iron nails in a beaker with some dilute sulfuric acid (H_2SO_4). See Fig. 5-1. You will see that a chemical reaction is taking place, as shown by gas bubbles that are given off. The gas given off can be identified as hydrogen (H_2). The nails become thinner as the iron atoms react with the acid. This experiment shows that acids are *corrosive*, meaning that they attack metals. After some time, if there is enough

Turns litmus red

5-1 Common properties of acids. Why is it good practice not to taste materials in the laboratory?

chapter 5 acids, bases, and salts

acid, the nails disappear. A green solution remains. This solution is iron(II) sulfate ($FeSO_4$). The equation for this reaction is

$$Fe + H_2SO_4 \rightarrow FeSO_4 + H_2 \uparrow$$

The arrow ($\uparrow$) shows that a gas is given off. Notice that Fe has replaced the H_2 in the sulfuric acid to form $FeSO_4$. This type of reaction is called **single replacement.** *A single replacement reaction is one in which one element replaces another in a compound.*

Acids attract other metals, such as magnesium and zinc, in the same manner. Look at the equation for the action of magnesium and hydrochloric acid. The word equation for this reaction is

magnesium + hydrochloric acid →
magnesium chloride + hydrogen

Using symbols and formulas, this chemical equation is written as

$$Mg + HCl \rightarrow MgCl_2 + H_2 \uparrow \qquad \text{(not balanced)}$$

Notice that on the left side of the equation there is one atom of hydrogen and one atom of chlorine. On the right there are two atoms of hydrogen and two atoms of chlorine. You will recall that according to the law of conservation of matter, chemical changes cannot create or destroy atoms. Is there something wrong with this equation? Have additional atoms of hydrogen and of chlorine been created? If you could put the substances taking part in the reaction on a pan balance, you would find that there is no gain or loss of mass after the reaction. Actually, it takes two parts of HCl to provide the additional atoms. This is shown by the equation

$$Mg + 2\,HCl \rightarrow MgCl_2 + H_2 \uparrow \qquad \text{(balanced)}$$

Now there are the same numbers and same kinds of atoms on both sides of the equation. The equation is balanced. Of course, in any chemical change, huge numbers of atoms and molecules take part. These atoms and molecules always combine in the proportion shown by the whole numbers written in *front* of the symbols and formulas.

Thus, to write a balanced chemical equation, numbers may be needed in front of a formula. These numbers insure that the numbers of atoms of each element are the

         unit 1   matter and energy

same on both sides of the equation. This process is called balancing an equation. The smallest whole numbers are used to balance equations.

In writing chemical equations, three conditions must be satisfied:

1. Chemical equations must represent what is observed in the lab.
2. The formulas of compounds or the symbols of elements involved must be written correctly.
3. The equation must be balanced. This means that the number of atoms of each element must be the same on both sides of the equation.

There are two things that a chemical equation does not tell: (1) the *rate* (speed) at which a reaction takes place, and (2) the *conditions* under which a chemical reaction takes place.

Sulfuric acid is a vital chemical. The amount of sulfuric acid that a nation uses shows the size of that nation's industry. Sulfuric acid is one of the most widely used chemicals. For the most part, the raw materials used to make this acid are water, sulfur, and the oxygen of the air. These substances are very common. See Fig. 5-2.

The first step in making sulfuric acid is to burn sulfur in air to form sulfur dioxide (SO_2):

$$S + O_2 \rightarrow SO_2$$

Then the sulfur dioxide is changed to sulfur trioxide (SO_3):

$$2\ SO_2 + O_2 \rightarrow 2\ SO_3$$

Finally, the sulfur trioxide is added to water to produce sulfuric acid:

$$SO_3 + H_2O \rightarrow H_2SO_4$$

These three equations shown above are examples of another type of chemical reaction, called **synthesis** (sin-the-sis). *A synthesis reaction is one in which two or more substances combine to form a more complex substance.*

Sulfuric acid has many uses. One of the major uses of sulfuric acid is in the making of fertilizers. The mineral called rock phosphate (calcium phosphate) does not dissolve readily in moist soil. However, when calcium phosphate is treated with sulfuric acid, it dissolves with

5-2 Sulfur is melted underground, pumped to the surface, and placed in vats to harden. Can you suggest a method for melting the underground sulfur deposits?

5-3 A steel strip being drawn from an acid pickling tank. Why is pickling necessary?

ease. This process releases phosphates to the soil that are needed for growing plants.

Metals are "pickled" in dilute sulfuric acid to remove any oxide (rust) coating before they are plated or enameled. Sulfuric acid is used in making chromium-plated bumpers for cars and enameled sinks for our homes. See Fig. 5-3.

Sulfuric acid is also used in making camera film, as the electrolyte in car storage batteries, and in the making of many products from oil, including gasoline. Sulfuric acid is needed to make paints, plastics, rayon, explosives, and many other products.

When concentrated sulfuric acid is left in an open dish, its volume increases because the acid absorbs moisture from the air and dilutes itself. Sulfuric acid also removes hydrogen and oxygen from table sugar ($C_{12}H_{22}O_{11}$) in the proportion needed to form water. The carbon is left as a residue. See Fig. 5-4. Sulfuric acid will react in the same way with material containing only carbon. A piece of wood or paper turns black when placed in sulfuric acid. Substances that remove water from other substances are called *drying agents.* Sometimes, even a trace of moisture will spoil a chemical reaction. This is one reason for the wide use of sulfuric acid as a drying agent.

Always handle acids with care. In fact, all chemicals must be handled with caution. For instance, if you add concentrated sulfuric acid to water, much heat will be given off. To avoid danger, slowly add acid to the water, while stirring. CAUTION: *Never add water to acid.* Be sure to wear goggles to shield your eyes as the mixture is likely to spatter. This care should be taken when using all strong, concentrated chemicals. Be sure to use a container that can withstand the large amount of heat that is produced. Concentrated sulfuric acid will burn your skin. Even dilute sulfuric acid can make holes in your clothes.

Hydrochloric acid can be made in the lab. If you add concentrated sulfuric acid to sodium chloride and heat the mixture, a choking gas is produced. This gas is hydrogen chloride. Hydrogen chloride gas is very soluble in water, forming hydrochloric acid. Figure 5-5 shows the lab setup for making small amounts of hydrochloric acid. The equation for this reaction is

$$2\ NaCl + H_2SO_4 \rightarrow Na_2SO_4 + 2\ HCl \uparrow$$

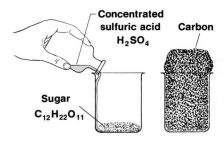

5-4 Concentrated sulfuric acid removes hydrogen and oxygen atoms from sugar, leaving a black carbon residue. What name is given to a material that removes water from other materials?

5-5 Heating sodium chloride with sulfuric acid produces hydrogen chloride gas. Describe what happens when the gas is bubbled into water. CAUTION: Remove the outlet tube from the water before the reaction stops.

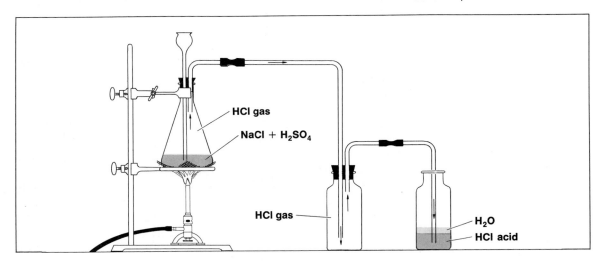

chapter 5   acids, bases, and salts   73

This reaction is a good example of still another type of chemical change. It is called a **double replacement reaction.** *A double replacement reaction is one in which substances change places in a chemical reaction.* It is just like two couples changing partners on a dance floor!

Hydrochloric acid has many uses. Galvanized iron is made by coating sheet iron with melted zinc. Before the metal is dipped in the zinc, hydrochloric and sulfuric acids are used to clean the metal. Hydrochloric acid is also used in cleaning excess mortar from stone and brick.

You may be surprised to learn that there is acid in your stomach. Gastric juices in your stomach contain very dilute hydrochloric acid that helps digest foods.

Nitric acid has many uses. Small amounts of nitric acid can be prepared in the lab by heating a mixture of sodium nitrate ($NaNO_3$) and concentrated sulfuric acid. Nitric acid gas (HNO_3) is produced and then condensed to a liquid. Can you write a balanced equation for this reaction?

About 75% of the nitric acid produced in the United States is used in making fertilizers. Nitric acid is also used in making dyes and plastics. A mixture of nitric and hydrochloric acids, called *aqua regia,* will dissolve gold.

Nitric acid is used in making all common explosives. Among these are smokeless gunpowder and dynamite. Another explosive made from nitric acid is TNT, a high explosive used in artillery shells. The molecules of all these explosives are unstable; that is, they break down quickly, forming gases that expand with great force.

Care must be used in handling nitric acid to avoid staining the skin yellow. Nitric acid produces this yellow stain on any protein and so is used as a test for protein. For instance, a drop of nitric acid added to a slice of hard-boiled egg white will turn the egg white yellow. This color change shows the presence of protein. The yellow color deepens to a bright orange when ammonia water is placed on the sample. A pure solution of nitric acid is colorless, but it may turn brown as its molecules break down.

bases

Bases can be identified by certain properties. Several substances that are found in many homes have long been known as bases. For example, household ammonia, a

unit 1 matter and energy

common cleaning agent, is a base. So is milk of magnesia, a mixture of magnesium hydroxide and water that is used as a medicine. Lye, a commercial grade of sodium hydroxide, is a base used for cleaning clogged sink drains.

Just as acids in dilute solutions have a sour taste, bases have a bitter taste. (CAUTION: *Never use the "taste test" in the lab.*) Solutions of bases change litmus paper from red to blue. Here is an easy way to remember these facts about bases:

B⟨
　Base
　Blue
　Bitter

If you were to rub a drop of dilute sodium hydroxide solution between your fingers, it would feel slippery. Sodium hydroxide is *caustic,* attacking the skin and other animal tissues. In addition to caustic effects on the skin, strong bases destroy hair and wool. Bases are very harmful to the eyes. See Fig. 5–6.

Bases contain hydroxide ions. A **base** is any soluble substance that can neutralize the properties of an acid when they are mixed. The most common bases are soluble hydroxides. *A base is a substance that will form hydroxide ions (OH^-) when dissolved in water.* The word *alkaline* (*al*-ka-line) is often used to describe substances with basic properties.

Each base contains one or more hydroxide radicals, or OH^- groups. In water, soluble bases will ionize as shown by the following examples:

$$NaOH \rightarrow Na^+ + OH^-$$
$$Ca(OH)_2 \rightarrow Ca^{++} + 2\ OH^-$$

In naming bases, the metal ion, such as sodium (Na^+) or calcium (Ca^{++}) in the above equations, is named first, followed by the word "hydroxide." Thus, NaOH is called sodium hydroxide and $Ca(OH)_2$ is called calcium hydroxide. The symbol for the metal is written first in the formula. Ammonia water, a common base often called ammonium hydroxide (NH_4OH), does not contain a metal. In this base, the ammonium radical (NH_4^+) acts like a metal.

Base solutions feel slippery

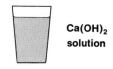

Ca(OH)$_2$ solution

Bases taste bitter

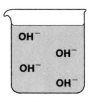

Form hydroxide ions in solution

Caustic action on wool

Turn litmus blue

5-6 Five common properties of bases. Why should the "taste test" not be used in the laboratory?

chapter 5　acids, bases, and salts　　　　　75

activity

Caustic effect of strong bases. Fill a beaker half-full of a 5% solution of sodium hydroxide. (Dissolve 5 g of solid NaOH in enough water to make 100 mL of solution. CAUTION: *Do not touch the solid; wear goggles for eye protection.*) Heat the solution gently until it boils. Then add a piece of woolen cloth and a piece of cotton cloth, each about 5 cm square. Stir with a glass rod. What happens to the woolen cloth? to the cotton cloth? Can you suggest a practical use of this experiment?

Not all substances having OH^- groups are bases. For example, the formula for methyl alcohol, commonly known as wood alcohol, is written CH_3OH. Since this compound does not break down in water to form OH^- ions, it is not a base.

Table 5-2:

Some Common Bases and Their Formulas	
Base	Formula
Sodium hydroxide	NaOH
Potassium hydroxide	KOH
Calcium hydroxide	$Ca(OH)_2$
Magnesium hydroxide	$Mg(OH)_2$
Aluminum hydroxide	$Al(OH)_3$
Ammonia water	NH_4OH
Iron(III) hydroxide	$Fe(OH)_3$

Sodium hydroxide and potassium hydroxide are very soluble in water. Their solutions are strongly basic because of the high numbers of OH^- ions. These bases are said to be *strong*. The word *alkali* (*al*-ka-lie) is often used in reference to strong bases.

Ammonia water is an example of a base that is *weak* because it is only slightly ionized. Calcium hydroxide, even though it ionizes completely in water, is moderately basic because it is only slightly soluble in water.

unit 1   matter and energy

Sodium hydroxide is made by electrolysis. One way to make sodium hydroxide is to pass an electric current through a water solution of sodium chloride. The equation for the chemical change that takes place is

$$2\ NaCl + 2\ H_2O \rightarrow 2\ NaOH + H_2 \uparrow + Cl_2 \uparrow$$

Sodium hydroxide is made in large factories located in regions close to an abundant source of electric energy. One advantage of this method is that it gives off hydrogen gas and chlorine gas as valuable by-products. Both of these by-products can also be sold for consumer use.

Sodium hydroxide is a white crystal solid that is sold under the name of lye. Sodium hydroxide reacts with grease to form soap. For this reason, sodium hydroxide is used to clean plumbing drains clogged with grease. Large amounts of sodium hydroxide are used in making rayon and cellophane, and in refining gasoline and other petroleum products.

Ammonia water is a common weak base. Ammonia (NH_3) is a colorless gas with a sharp, penetrating odor. It is very soluble in water. Some of the ammonia gas reacts with the water, forming ammonium ions (NH_4^+) and hydroxide ions as shown by the following equation:

$$NH_3 + HOH \rightleftarrows NH_4^+ + OH^-$$

The double arrow ($\rightleftarrows$) in the equation shows that this is a *reversible* reaction. In other words, ammonia and water react to form ammonium and hydroxide ions. These ions, in turn, react with each other to form ammonia and water. Both of these reactions take place at the same time. Many chemical reactions are reversible. Nearly all particles in ammonia water solution are covalent molecules of NH_3.

Ammonia water is one of the most common and useful lab bases. It has a special use as a household cleaning agent because it removes grease and dirt.

Calcium hydroxide is a moderately strong base. Limestone, or calcium carbonate, is the raw material used in the making of calcium hydroxide. Limestone is abundant in many sections of the country.

Calcium carbonate ($CaCO_3$) is first heated in a large furnace. The heat converts the $CaCO_3$ into calcium oxide (CaO) and carbon dioxide (CO_2) as shown by the following equation:

$$CaCO_3 \xrightarrow{\text{heat}} CaO + CO_2 \uparrow$$

chapter 5 acids, bases, and salts

This type of chemical change is called a *decomposition reaction*, the reverse of the synthesis reaction studied earlier in the chapter. A decomposition reaction is one in which a complex substance, like $CaCO_3$, is broken down into two or more simpler substances, like CaO and CO_2.

Calcium oxide, usually called lime or quicklime, is a white solid that reacts violently with water. In the next step in making calcium hydroxide, water is added to quicklime. The lime will give off a great deal of heat and will act as though it were "alive." This is why CaO is called quicklime. "Quick" is an old-fashioned word meaning alive. The reaction of CaO with water is the final step in making calcium hydroxide:

$$CaO + H_2O \rightarrow Ca(OH)_2$$

Calcium hydroxide is used in large amounts for making mortar and plaster. It is also used to treat acid soils and to soften hard water. A water solution of calcium hydroxide, called *limewater,* is used to detect the presence of carbon dioxide gas. As a moderately strong base, calcium hydroxide is much less caustic than sodium or potassium hydroxide. Yet it is strong enough to remove hair from hides that are to be made into leather.

5-7 The pH scale. What does a pH of 7 indicate in terms of the numbers of hydrogen and hydroxide ions in a solution?

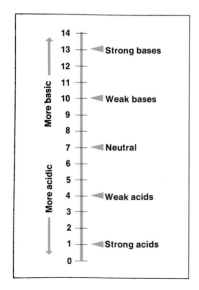

The pH of a solution tells how acidic or basic the solution is. Scientists find it helpful to use a pH scale to show just how acidic or basic a solution is. This scale ranges from 0 to 14. A pH of 1, for example, shows a strong acid solution. This means a very high number of hydrogen ions present in the solution. A pH of 14 shows a strong basic solution having a very high level of hydroxide ions. Pure water is neutral and has a pH of 7. See Fig. 5-7.

Any solution with a pH less than 7 is acid, having an excess of hydrogen ions. With a pH greater than 7, a solution is basic, having an excess of hydroxide ions. A solution with a pH of 2 is more acidic than one having a pH of 3. A solution with a pH of 12 is more basic than one having a pH of 10.

In the lab, the pH of a solution can be found by using Hydrion paper as shown in Fig. 5-8. When the paper is moistened with the solution the paper changes color. The color of the test paper is then compared with a color scale mounted on the case. When the colors match, the pH number can be read directly from the scale.

unit 1   matter and energy

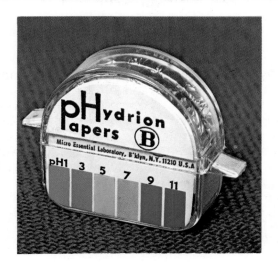

5-8 Easy-to-use pH Hydrion paper for finding the pH of solutions. Describe the accuracy of this method.

Table 5-3 shows the pH values of several common substances.

Table 5-3:

Approximate pH Values of Several Common Liquids	
Liquid	pH
Lemon juice	2.3
Vinegar	2.8
Soft drinks	3.0
Orange juice	3.5
Milk	6.5
Pure water	7.0
Human blood	7.4
Sea water	8.5
Milk of magnesia	10.5

salts

A salt is one of the products of an acid-base reaction. When proper amounts of an acid and a base are mixed, the properties of the two solutions cancel each other. The products formed are a **salt** and water. The following is a word equation of the reaction that takes place:

$$\text{acid} + \text{base} \rightarrow \text{salt} + \text{water}$$

When solutions of hydrochloric acid and sodium hydroxide are mixed, for instance, the reaction is written as

$$HCl + NaOH \rightarrow NaCl + HOH$$
$$\text{acid} \quad\quad \text{base} \quad\quad \text{salt} \quad\quad \text{water}$$

Since the acid, base, and salt are highly ionized in a water solution, a better way to write the above equation is in an *ionic* form:

$$H^+ + Cl^- + Na^+ + OH^- \rightarrow Na^+ + Cl^- + HOH$$
$$\text{hydrochloric} \quad\quad \text{sodium} \quad\quad \text{sodium} \quad\quad \text{water}$$
$$\text{acid} \quad\quad\quad \text{hydroxide} \quad\quad \text{chloride}$$

If the solution is heated to evaporate the water, white crystals of sodium chloride (table salt) remain in the container.

Of course, the reaction of any acid with any base will produce a salt. In this case, sodium chloride was produced by the reaction of hydrochloric acid and sodium hydroxide. *A salt is defined as a compound made up of the positive ions of a base and the negative ions of an acid.* Sodium chloride is only one example of the many thousands of salts used in chemistry.

Notice that, in the above ionic equation, sodium ions (Na^+) and chloride ions (Cl^-) appear on both sides of the equation. These ions take no part in the reaction. They are called *spectator ions*, because they remain evenly mixed throughout the solution. Thus, the last equation can be written more simply as

$$H^+ + OH^- \rightarrow HOH$$

This equation shows that the hydrogen ions of the acid and the hydroxide ions of the base joined together to form water. In other words, the H^+ ions and the OH^- ions lose their effect in the solution by forming a nonionized compound, water. This type of reaction is called a *neutralization* reaction. Of course, if there were more of the H^+ ions than OH^- ions, the solution would be acidic. If, on the other hand, there were more OH^- ions than H^+ ions, the solution would be basic. Therefore, an acid and a base completely neutralize each other only if the same number of H^+ ions and OH^- ions are present. No matter how small the amount of an acid and a base used, many billions of ions will be involved in the reaction. You cannot count billions of ions. However, you can use indicators, like litmus, to tell you when proper amounts of

unit 1 matter and energy

the acid or base have been added to give a neutral solution.

You will find that salts as a group do not have any special properties. Some taste salty, some bitter. Many dissolve well in water, others slightly, and still others not at all. Some salts form crystals and some do not. Salts vary in color, although most are white.

Salts are compounds that are formed by the transfer of electrons, forming ionic bonds. Those salts that are soluble in water break down freely as ions and conduct an electric current. Table 5-4 lists the solubility of common salts in a water solution. Knowing these properties of salts helps chemists find ways to identify ions or to separate one ion from another.

Table 5-4:

Solubility of Salts
1. Common sodium, potassium, and ammonium compounds are soluble in water.
2. Common nitrates, acetates, and chlorates are soluble.
3. Common chlorides are soluble except silver, mercury(I), and lead. Lead(II) chloride is soluble in hot water.
4. Common sulfates are soluble except calcium, barium, strontium, and lead.
5. Common carbonates, phosphates, and silicates are insoluble except sodium, potassium, and ammonium.
6. Common sulfides are insoluble except calcium, barium, strontium, magnesium, sodium, potassium, and ammonium.

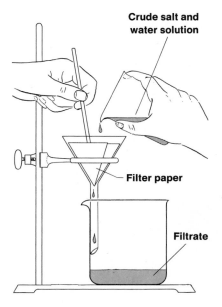

Crude salt and water solution

Filter paper

Filtrate

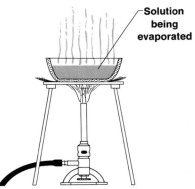

Solution being evaporated

5-9

activity

Purifying salt. Small samples of salt can be readily purified in the lab. Dissolve about 100 g of rock salt in a large beaker of water. Filter the solution as shown in Fig. 5-9 (top). Then boil away the water until the salt begins to separate out. See Fig. 5-9 (bottom). Now let the solution stand until it cools down to room temperature. Pour off the liquid and observe the salt. Why is filtering needed?

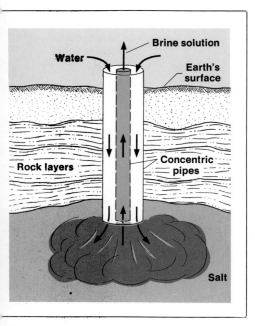

5-10 Water is forced down a salt well to dissolve underground salt. The salt solution is brought to the surface. How is the salt recovered from the crude salt solution?

Common table salt is abundant. Sodium chloride (table salt) is found all over the world. Much of it is dissolved in the oceans. Salt from sea water is sometimes obtained by letting water dry up in shallow bays or inlets. This method has been used since people first noticed salt crystals appearing in trapped pools of sea water. The sun's heat is used to evaporate the water.

Underground deposits of rock salt are mined in the same way as coal. Rock salt is a grayish substance. It is used for freezing ice cream, curing animal hides, and melting snow and ice on roads.

Most table salt, however, comes from salt wells. To get the salt, a large pipe is driven down to the center of an underground salt deposit. Water is pumped down through the pipe to dissolve the salt. The salt solution is brought to the surface through a smaller pipe placed inside the larger one. See Fig. 5-10. The salt solution, or brine, that comes from the salt wells is filtered and purified before it is ready for the dinner table.

Do you know that pure sodium chloride does not get sticky in damp weather? Very small amounts of other salts in table salt cause the trouble. The moisture they pick up from the air makes the sodium chloride sticky so that it will not pour easily.

Common table salt has many uses. At one time or another, many nations have used table salt as we use money. Salt was so valuable at one time that it was used to pay Roman soldiers. This practice led to the common expression "worth his/her salt" and the word "salary."

Salt is vital to our diet. It is found in solution in body fluids such as blood, urine, and perspiration.

If you look at a box of salt, you may see the words "iodized salt" on the label. This means that small amounts of sodium or potassium iodide have been added to the salt. The iodide protects people from the disease called *goiter*. A goiter is an enlargement of the thyroid gland, found in the neck, often caused by a shortage of iodine in the diet.

Salt is important in the making of many other chemicals. Lye, chlorine, hydrochloric acid, washing soda, and baking soda are some of the well-known chemicals made from sodium chloride. Salt is also used to prevent decay in foods. Before the invention of the refrigerator, much food had to be preserved in salt. See Table 5-5.

unit 1 matter and energy

Table 5-5:

Some Important Salts and Their Uses		
Name	**Formula**	**Uses**
Ammonium chloride (Sal ammoniac)	NH_4Cl	Dry cell batteries, medicine
Potassium nitrate (saltpeter)	KNO_3	Meat preservative; making of fertilizers; fireworks.
Silver nitrate	$AgNO_3$	Medicine, photography
Potassium carbonate (Potash)	K_2CO_3	Making of glass and soap
Sodium bicarbonate	$NaHCO_3$	Making of baking powder; source of carbon dioxide gas for fire extinguishers
Copper(II) sulfate	$CuSO_4$	Copper plating; making of fungicides
Aluminum sulfate	$Al_2(SO_4)_3$	Water purification; deodorant
Calcium sulfate (plaster of paris)	$CaSO_4$	Molds and casts
Sodium hypochlorite	$NaClO$	Bleaching solution

Svante Arrhenius (1859–1927)

At the age of 28, Svante Arrhenius, a Swedish chemist, first published his theory of how ions are formed. His theory, which served for many years, has been changed slightly as a result of new findings by modern-day chemists. For his ion theory, Arrhenius was awarded the Nobel Prize in chemistry in 1903.

Arrhenius believed that ions were formed by breaking apart molecules of certain substances in water solutions. Arrhenius thought the ions were electrical in nature. When the molecules were ionized, they formed equal numbers of positive and negative charges, which made the solution a good conductor of electric current. Arrhenius believed that the breakdown of molecules into ions was complete in dilute solutions.

summary

1. A substance that gives up hydrogen ions to water to form hydronium ions is called an acid.
2. Acids have a sour or sharp taste, turn blue litmus to red, are corrosive, and neutralize bases.
3. Substances that change color in the presence of certain ions are called indicators.
4. A balanced chemical equation has the same number of atoms of each kind on both sides.
5. A substance that will form hydroxide ions when dissolved in water is called a base.
6. Bases have a bitter taste, turn red litmus blue, are caustic, and feel slippery.
7. A salt is a compound made up of the positive ions of a base and the negative ions of an acid.
8. Neutralization is the reaction between the hydronium (hydrogen) ions of an acid and the hydroxide ions of a base to form water.
9. Four general types of chemical reactions are (a) synthesis, (b) single replacement, (c) double replacement, and (d) decomposition.

review

Match each item in the left column with the best response in the right column. *Do not write in this book.*

1. litmus paper
2. weak acid
3. iodized salt
4. synthesis
5. NaCl
6. ammonia
7. water
8. strong acid
9. hydronium ion

a. common table salt
b. pH = 7
c. H_2SO_4
d. metal
e. NH_3
f. one type of chemical reaction
g. H_3O^+
h. vinegar
i. an example of an indicator
j. helps to prevent goiter
k. .an example of a double replacement reaction

questions

Group A
1. What is an acid? List four properties of acids.
2. How does the hydronium ion differ in structure from the hydrogen ion?

unit 1 matter and energy

3. What two substances are formed when dilute sulfuric acid reacts with iron? Write the balanced equation for this reaction.
4. How is hydrochloric acid prepared in the laboratory?
5. What is a base? List four properties of strong bases.
6. What is the difference between a strong base and a weak one?
7. What substances would you use to make sodium hydroxide? Name two by-products.
8. Describe how you would prove that a piece of cloth was all wool.
9. The pH of a solution is found to be 4.7. What does this mean? Compare its number of hydrogen ions with a solution having a pH of 3.2.
10. What are salts?
11. What is meant by neutralization?
12. Do salts have general properties like acids and bases? Explain.
13. What is iodized salt? Why is it iodized?
14. Describe how table salt is obtained from salt wells.

Group B
15. Explain the statement, "All acids contain hydrogen, but not all compounds containing hydrogen are acids."
16. Describe a safe method of diluting concentrated sulfuric acid.
17. Describe what happens when concentrated H_2SO_4 is used as a drying agent.
18. How is sulfuric acid prepared for commercial use?
19. Describe how you would test a food for protein.
20. What important uses does calcium hydroxide have? sodium hydroxide?
21. A dish of gasoline and a dish of concentrated sulfuric acid are weighed and left uncovered for several hours. Then they are weighed again. What change in weight, if any, would you expect to occur in each substance? Explain.
22. Write a balanced ionic equation for the reaction between magnesium and hydrochloric acid.
23. Why is a solution of ammonia gas in water more accurately called ammonia water rather than ammonium hydroxide, a name commonly used?
24. Describe one method for finding the pH of a solution in the lab.
25. What is meant by the statement, "Acids neutralize bases?"
26. Write the chemical formulas for the following salts, and use Table 5-5 to predict their solubility. (Refer to Tables 3-4 and 3-5 for valence numbers of elements and radicals.)
   a. lead(II) nitrate   c. sodium sulfide   e. lead(II) sulfate
   b. barium carbonate   d. calcium chloride   f. ammonium chloride
27. Write balanced chemical equations for the following reactions. Using Table 5-5, predict if an insoluble product (precipitate) is formed.
   a. sodium sulfate + barium chloride
   b. magnesium sulfate + ammonium chloride
   c. zinc sulfate + barium sulfide
28. Copy the chemical equations below on a separate sheet of paper and supply the lowest numbers before the symbols and formulas that will balance the equations.

Which general type of chemical reaction is shown by each equation? *(Do not change the formulas.)*

a.   Na   +   HOH   →   NaOH   + $H_2 \uparrow$
b. H_2O → $O_2 \uparrow$ + $H_2 \uparrow$
c. H_2 + Cl_2 → $HCl \uparrow$
d.   Mg   +   HOH   → $Mg(OH)_2$ + $H_2 \uparrow$
e.   Na   + Cl_2 →   NaCl
f.   NaCl   + $AgNO_3$ →   AgCl   + $NaNO_3$

29. For each of the following word equations, write a balanced chemical equation on a separate sheet of paper. Refer to Table 3-4 and Table 3-5 for valence numbers of elements and radicals.
   a. zinc + hydrochloric acid → zinc chloride + hydrogen
   b. calcium oxide + water → calcium hydroxide
   c. iron(II) sulfide + hydrochloric acid → iron chloride + hydrogen sulfide
   d. barium chloride + sulfuric acid → barium sulfate + hydrochloric acid
   e. potassium chloride + silver nitrate → potassium nitrate + silver chloride

problems

1. A sample of sea water contains 4% mineral salts. If you evaporated 5 kg of the sample, how much mineral salts would you have?
2. The mass of an empty beaker is 100 g. Salt solution is added until the total mass reaches 500 g. The water is evaporated and the salt is left. The beaker and salt now have a mass of 116 g. (a) What is the mass of the salt solution? (b) What is the mass of salt obtained? (c) What percent of the solution is salt?

further reading

Asimov, Isaac, *Adding a Dimension*. New York: Avon Books, 1975.

Asimov, Isaac, *Opus 100*. Boston: Houghton Mifflin, 1969.

Choppin, G. R., et al., *Chemistry*. Morristown, N.J.: Silver Burdett.

Coulson, A. E., et al., *Test Tubes and Beakers: Chemistry for Young Experimenters*. Garden City, N.Y.: Doubleday, 1970.

Dickson, T. R., *Introduction to Chemistry*. New York: John Wiley, 1975.

Metcalfe, H. C., et al., *Modern Chemistry*. New York: Holt, Rinehart and Winston, 1982.

Shalit, N., *Cup and Saucer Chemistry*. New York: Grosset & Dunlap, 1974.

nuclear reactions

6

You have seen that in chemical reactions the number and kinds of atoms or ions do not change. Rather, the substances are regrouped to form new substances. You have also seen that these changes are brought about by the transfer or sharing of electrons outside the nucleus. An ordinary chemical reaction does not involve any changes in the nucleus of the atom. A nuclear reaction is like a chemical reaction in that new substances with new properties are formed. However, in a nuclear reaction, new elements are formed by changes in the nucleus of the atom. In this chapter, you will study these nuclear changes.

objectives:

☐ To develop an understanding of the nature, properties, types, and uses of nuclear radiations.

☐ To understand and use nuclear equations.

☐ To describe how nuclear fission and fusion may be used as a source of energy.

☐ To become familiar with the uses and safety rules for radiations produced in nuclear reactors.

☐ To define the following: chain reaction, breeder reactor, moderator, control rod, and radioactive dating.

natural radioactivity

Radioactivity was discovered by accident. In 1896, a French scientist named Henri Becquerel (Bek-*rel*) made a startling discovery. He placed an unexposed photographic film near a sample of uranium ore. To his surprise, Becquerel found that when it was developed, the film darkened just as though it had been exposed to light. After further study, Becquerel found that uranium ore gave out strange high-energy radiations. *These uncontrolled changes in the nucleus of certain atoms, resulting in the giving off of particles and rays, are called* **radioactivity.**

Uranium and all its compounds constantly send out these highly penetrating radiations. Nothing scientists can do to uranium affects the radioactivity in any way. We can toss uranium into a hot furnace or into frigid cold with no effect on the radiations. The radioactivity of an element is not even affected by its chemical combination with other elements.

Radioactive elements have common properties. All of the elements with atomic numbers greater than 83 are radioactive. A few radioactive elements with atomic numbers smaller than 83 are also known. All radioactive elements are alike in the following ways:

1. *They darken a photographic film.* Even though photo film is wrapped in heavy black paper and kept in the dark, radiations can pass through the wrapping and affect the film. The radiations pass through paper, wood, flesh, and even thin sheets of metal. See Fig. 6-1.
2. *They break down into simpler atoms.* The atoms of all radioactive elements eventually break down into atoms of lower atomic numbers at the time they give off radiations.
3. *They produce ions in gas molecules.* Radiations knock outer electrons from gas molecules of the air. The gas molecules become positive ions.
4. *They have damaging effects on living tissue.* Large doses of radiation are fatal to plant seeds, bacteria, and even large animals. Radiation can produce flesh wounds that heal with great difficulty. An overdose of radiation can cause sickness and even death in human beings. The chief symptoms of radiation sickness are nausea, diarrhea, vomiting, internal bleeding, and a feeling of weakness. Even small amounts of radiation may cause damage to reproductive cells, which could later result in the birth of defective children. However, if the amount is controlled, radiation can be used to destroy diseased

6-1 This photograph was produced by radiations from a tiny piece of uranium ore. The radiations passed through the wrappings of the film.

unit 1   matter and energy

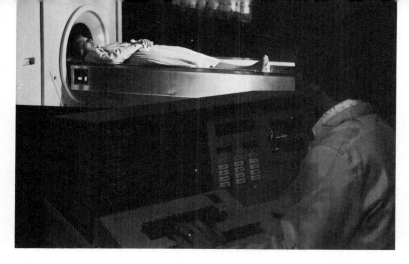

6-2 A device that uses radiation to help doctors locate brain ·tumors. Can you suggest any safety precautions in the use of this instrument?

tissue in the treatment of cancer and certain skin diseases. See Fig. 6-2.

Radioactive elements may emit three kinds of radiation. The nucleus of a radioactive atom is made up of particles held together by very powerful binding forces. When these forces are upset, energy and matter escape as *alpha, beta,* or *gamma* radiations. These radiations can be separated by means of an electrically charged set of plates as shown in Fig. 6-3.

1. Alpha (α) particles are positively charged helium nuclei, made up of two protons and two neutrons. Alpha particles lack great penetrating power. They can be stopped by a thin sheet of aluminum foil, by a few sheets of paper, or by a few centimeters of air.

6-3 Radium, a radioactive element, continuously gives off three kinds of radiation from its nucleus. Which of the three are particles?

2. Beta (β) particles are high-speed electrons traveling at nearly the speed of light. Beta particles have about 100 times the penetrating power of alpha particles. They can pass through wood nearly 3 cm thick.
3. Gamma rays (γ) are similar to visible light but have a much higher energy. Gamma rays are the most penetrating form of radiation. Even thick layers of lead or concrete will not completely stop all gamma rays. They are similar to X rays, which you will study later in the text.

Alpha and beta particles are seldom, if ever, given off at the same time from the same nucleus. Gamma rays, however, are frequently produced when alpha or beta particles are given off.

Each radioactive element has a half-life. The rate at which the nucleus will break down varies with different

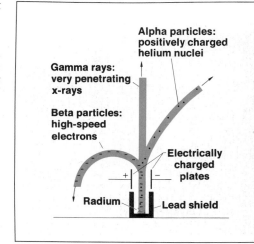

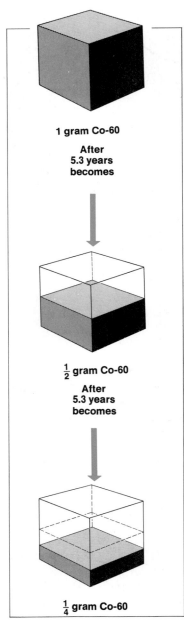

1 gram Co-60

After
5.3 years
becomes

$\frac{1}{2}$ gram Co-60

After
5.3 years
becomes

$\frac{1}{4}$ gram Co-60

6-4 The half-life of radioactive cobalt-60 is 5.3 years. Cobalt-60 gives off ß and γ rays to become nickel-60. How much cobalt-60 remains after the third breakdown? How much nickel-60 is formed after the third breakdown?

elements. This rate is called the element's **half-life.** *The half-life of a radioactive element is the time needed for half of the atoms of a given sample to break down.* For example, the half-life of radium is about 1600 years. No matter how many atoms of radium there are to begin with, in 1600 years only half of them will remain unchanged. During the next 1600 years, half of the remainder will break down, and so on. Another radioactive element called cobalt-60 has a half-life of 5.3 years. See Fig. 6-4.

Atoms do not disappear during the breakdown of the nucleus. They change to atoms of another kind of element. For some radioactive elements, the half-life is less than 0.0000001 sec. For others, it is billions of years.

A large amount of heat energy is set free by radioactivity. For instance, the heat given off every hour by 1 g of radium would be more than enough to warm up 1 g of water from freezing to boiling. This may not seem like much heat, but remember that this process has been going on for billions of years.

Radioactive dating provides a good time clock. When radioactive uranium-238 breaks down, it finally reaches the stable element lead as one of its products. By finding the amount of lead mixed in with the uranium-bearing ore, scientists can compute how long ago the material was formed. Radioactive minerals from the earth help scientists estimate the age of the earth and of the ancient remains of plants and animals.

One method used to obtain a record of early human history is through the breakdown of radioactive carbon-14. Carbon-14 is formed by high-energy particles (cosmic rays) from outer space striking nitrogen atoms of the air. Plants and animals contain carbon dioxide made up of some carbon-14 atoms along with ordinary carbon atoms. All living matter, therefore, contains some carbon-14. The intake of carbon-14 stops when living matter dies. However, the breakdown of carbon-14 continues. The approximate age of early plant or animal life can be found by comparing the amount of carbon-14 in the test sample with the amount in a similar living sample. Since the half-life of carbon-14 is 5770 years, a fossil with only 1/4 as much carbon-14 as a similar present-day plant or animal must have been alive about 11,540 years ago (2×5770).

Carbon-14 dating was devised by Dr. Willard F. Libby, who was awarded the Nobel Prize in chemistry for his work. Carbon-14 is useful in making age estimates as far back as about 30,000 years. The Dead Sea Scrolls, a group of religious documents of the Old Testament, were discovered in 1947 in a cave near the Dead Sea. The Scrolls, which are about 2000 years old, were dated by counting the number of beta particles given off per unit of time by the carbon-14 they now contain. See Fig. 6-5.

6-5 Fragments of the Dead Sea Scrolls. Can you describe the radioactive dating procedure as it applies to this discovery?

artificial radioactivity

Rutherford was the first to change an atom. The study of radioactive elements in their natural state led scientists to believe that they could make other elements in the laboratory. Scientists could do this by finding a way to add particles to the nucleus. In 1919, Ernest Rutherford, a British scientist, found a way to change one element into another. Rutherford found that by shooting atoms with very high-speed atomic "bullets," he was able to hit the nucleus. This direct hit caused a change in the nucleus, that is, a *nuclear reaction.*

Rutherford's first successful reaction was changing nitrogen atoms into oxygen atoms. The nucleus of a nitrogen atom was struck by alpha particles given off by radium. One proton was released from the nucleus and oxygen was formed.

Remember that an alpha particle (α) is the nucleus of a helium atom (He), and that a proton is a hydrogen nucleus. Therefore, the word equation for this reaction can be written as

helium + nitrogen → oxygen + hydrogen

A helium nucleus, traveling at high speed, strikes the nitrogen nucleus and is absorbed by it. The resulting nucleus then becomes unstable and breaks apart, giving off a high-speed hydrogen nucleus (proton). The result of the breakdown is the formation of an oxygen nucleus. See Fig. 6-6.

Scientists use *nuclear equations* that look much like chemical equations. The word equation above can be written using symbols as follows:

$$\underset{\substack{\text{helium}\\\text{nucleus}}}{^4_2\text{He}} + \underset{\substack{\text{nitrogen}\\\text{nucleus}}}{^{14}_7\text{N}} \rightarrow \underset{\substack{\text{oxygen}\\\text{nucleus}}}{^{17}_8\text{O}} + \underset{\substack{\text{hydrogen}\\\text{nucleus}}}{^1_1\text{H}}$$

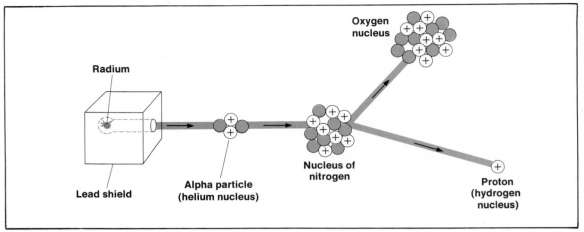

Oxygen
nucleus

Radium

Lead shield

Alpha particle
(helium nucleus)

Nucleus of
nitrogen

Proton
(hydrogen
nucleus)

6-6 This diagram shows the first human-made nuclear reaction: the making of oxygen from nitrogen. From the diagram, what is the atomic number of each of the two elements? the atomic mass?

The number at the lower left of each symbol is the atomic number of the element, that is, the number of protons in its nucleus. Notice that the number of protons adds up to nine on each side of the equation.

The number in the upper left indicates the mass number of the element. The mass number represents the total number of protons and neutrons in the nucleus. The sum of the mass numbers on each side of the equation must also be equal. In this case, the mass numbers on each side of the equation equal 18.

Rutherford's experiments were important because they showed that scientists could break down the atomic nucleus. However, the final key to unlocking nuclear energy had not yet been discovered. Although the resulting hydrogen nuclei had more energy than the helium "bullet," there were so few hits on nitrogen nuclei that energy was actually lost in the total process. Not until scientists split the uranium atom were great amounts of energy released. This process will be discussed in the next section on nuclear fission.

Keep in mind that the loss of an alpha particle ($^4_2$He) from the nucleus of an atom *decreases* its atomic number by two and its atomic mass by four. The loss of a beta particle (electron, $^0_{-1}$e) from the nucleus *increases* the atomic number by one, with no change in mass number. A helpful model that describes what happens is to think of a neutron in the nucleus as being made up of a proton (+) and an electron (−).

Table 6-1 shows some properties of subatomic particles and rays.

unit 1 matter and energy

Table 6-1:

Subatomic Particles and Rays			
Symbol	Name	Charge	Mass Number
^1_0n	Neutron (n)	0	1
^1_1H	Proton (p) (hydrogen nucleus)	+1	1
$^0_{-1}\text{e}$	Electron (beta particle, β)	−1	0
$^0_{+1}\text{e}$	Positron (positive electron)	+1	0
^4_2He	Alpha particle (α) (helium nucleus)	+2	4
γ	Gamma ray (very high-energy form of "light")	0	0

sample problem

1. The radioactive element thorium ($^{234}_{90}\text{Th}$) loses a beta particle ($^0_{-1}\text{e}$). What is the new element formed?
2. The new element formed is also radioactive and loses an alpha particle (^4_2He). What element is then formed?

solution:

1. The loss of a beta particle ($^0_{-1}\text{e}$) results in an added positive unit for the new nucleus. Thus, the atomic number is *increased* by 1 to 91 (90 + 1). In the periodic table in Fig. 3-6, the element with atomic number 91 is protactinium (Pa). Since there is no change in mass (except for the loss of an electron), the mass number is the same. The symbol for the new element is $^{234}_{91}\text{Pa}$.
2. The loss of an alpha particle (^4_2He) *reduces* the atomic number by 2 to 89 (91 − 2), and the mass number by 4 to 230 (234 − 4). From the periodic table, the element with atomic number 89 is actinium ($^{230}_{89}\text{Ac}$).

Isotopes. If the nucleus of an atom has eight protons, it is an oxygen atom. The common oxygen atom also has eight neutrons, for a total mass number of 16 ($^{16}_8\text{O}$ or oxygen-16). However, some oxygen atoms contain an

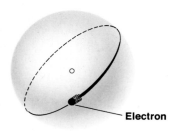

Hydrogen, $_1^1$H

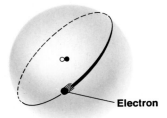

Deuterium, $_1^2$H

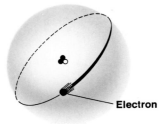

Tritium, $_1^3$H

6-7 The three isotopes of hydrogen. Describe the structural difference between them.

extra neutron, giving them a mass number of 17 ($_8^{17}$O) or oxygen-17). For example, the oxygen atom in Rutherford's equation has nine neutrons instead of the usual eight. *Atoms of the same element (same atomic number) but with different masses (different number of neutrons) are called* **isotopes.** Most elements have several natural isotopes.

Chlorine was found to have some atoms with a mass number of 35 ($_{17}^{35}$Cl or chlorine-35) and others with a mass number of 37 ($_{17}^{37}$Cl or chlorine-37). In nature, chlorine always consists of the same mixture of isotopes. The *average* mass of all chlorine atoms is 35.453 atomic mass units. This is the value given in the periodic table (Fig. 3-6). Thus, there must be more of the chlorine isotope with a mass of 35 than of 37 since the average mass is closer to 35.

For every 5000 ordinary hydrogen ($_1^1$H) atoms, there is one atom of an isotope called *heavy hydrogen,* or *deuterium* ($_1^2$H) (dew-*teer*-ee-um). As shown in Fig. 6-7, the nucleus of deuterium contains one neutron in addition to the proton of ordinary hydrogen. Hydrogen also contains a trace of the isotope *tritium* ($_1^3$H) (*tri*-ti-um), having two neutrons in its nucleus.

Since isotopes are alike chemically, no chemical method can be used to separate them. Instead, a method that depends on their differences in mass (a physical property) is used.

Radioisotopes. Scientists have now made *radioisotopes* of all elements by using neutrons and other particles as bullets. *A* **radioisotope** *is an isotope of an element that is radioactive.*

Radioisotopes have many important uses in modern life. With the aid of radioisotopes, scientists are learning more about the laws of nuclear structure. Radioisotopes can also be used as "tagged atoms" or *tracers.* By detecting the radioactivity of the few "tagged atoms," the behavior of the rest of the atoms can be inferred. For example, chemists can follow the path of tagged atoms during an experiment and thus learn what is happening in complex chemical reactions. Tracers are used in biological and medical research as well. See Fig. 6-8.

Radioisotopes are often used in the diagnosis and the treatment of cancer. Radioactive iodine is used in the diagnosis of cancer of the thyroid gland. Radiophos-

unit 1   matter and energy

phorus is used to treat certain types of bone cancer. Some examples of radioisotopes used in medicine and other areas are shown in Table 6-2.

Table 6-2:

Some Common Radioisotopes			
Radioiso-tope	Type of Radiation	Half-life	Use
Carbon-14	β	5770 years	Treating tumors; measuring age of fossils
Iron-59	β, γ	46 days	Blood studies
Phosphorus-32	β	14 days	Studying use of fertilizers by plants; treating bone cancer
Sodium-24	β, γ	15 hours	Study of circulatory diseases
Strontium-90	β	27 years	Treating lesions

6-8 A photograph made after a plant had absorbed a specific radioactive isotope that was added to the soil. Can you suggest a use for such a picture?

Another isotope, cobalt-60 ($^{60}_{27}$Co), is formed when a neutron enters the nucleus of an atom of ordinary cobalt ($^{59}_{27}$Co). The powerful gamma rays given off by cobalt-60 are used not only for medical treatments but also in industry, as a substitute for X rays, to test large metal castings for flaws. (See Fig. 6-9.)

nuclear fission

Accelerators shoot particles into target atoms. Once Rutherford succeeded in altering the structure of the nucleus of the atom, scientists began to build bigger and more powerful machines. They applied greater voltages to speed up charged nuclear particles. The higher the voltage, the faster the particles traveled and the harder they smashed into the target atoms.

One such machine in use today is the *linear accelerator.* This machine applies successive bursts of voltage. These pulses pull the charged particles through a long tube

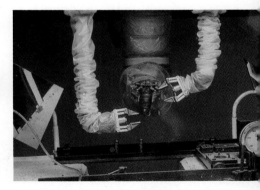

6-9 Highly radioactive materials are handled by mechanical hands. The technician is standing behind a glass shield that is about 1 m thick. Why are these precautions necessary?

at ever-increasing speeds. Some accelerators are over 1.5 km long. See Fig. 6-10.

Scientists use a variety of particles to set off nuclear reactions. Most often the nuclei of the lighter elements, such as hydrogen, deuterium, and lithium, are used.

Neutrons make good atomic bullets. Beams of neutrons cannot be speeded up in high-voltage machines. The reason for this is that neutrons, having no net plus or minus charge, are not affected by electric charges. Even so, neutrons do make very good bullets for atom smashing. Since neutrons are uncharged, they are neither attracted nor repelled by the positive charge on the target nucleus. Thus, the neutron bullets are more likely to score a hit and be captured by the target atom. The target nucleus would then become unstable and break apart. See Fig. 6-11.

It is not easy to obtain neutrons for use as bullets. Neutrons are tightly bound in atomic nuclei and can be freed only by using other bullets. Scientists cannot store a supply of free neutrons for several reasons. First, neutrons would pass through the walls of the container. Another reason is that neutrons, when removed from the nuclei of atoms, break down into protons and electrons. The average life of a neutron before it breaks apart is only about 17 minutes. Thus, scientists must produce neutrons when and where they are needed. One way to produce neutrons is to shoot helium nuclei ($^{4}_{2}He$) at a beryllium ($^{9}_{4}Be$) target. The nuclear equation for the production of neutrons ($^{1}_{0}n$) is as follows:

$$^{9}_{4}Be \quad + \quad ^{4}_{2}He \quad \rightarrow \quad ^{12}_{6}C \quad + \quad ^{1}_{0}n \quad + \text{ energy}$$

beryllium	alpha particles	carbon-12	neutron

Notice that energy, in the form of heat, is given off in the above reaction. When these experiments were first performed, scientists knew that the nucleus was a great storehouse of energy. Nuclear reactions seemed to support Einstein's theory that energy and mass are related. That is, mass can be changed into energy.

During nuclear reactions, some mass disappears and an equivalent amount of energy is given off in its place. This energy is mostly in the form of heat.

6-10 A linear accelerator used in nuclear research. How is the speed of charged particles increased in this device?

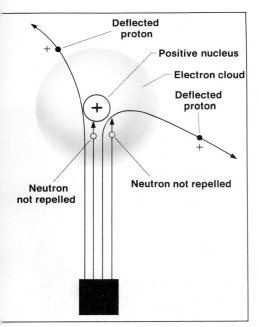

Deflected proton

Positive nucleus

Electron cloud

Deflected proton

Neutron not repelled

Neutron not repelled

6-11 From this diagram, can you explain why neutrons make good bullets for striking target atoms?

unit 1 matter and energy

The relation between mass and energy is given by Einstein's equation as follows:

$$E = mc^2$$

In this equation, **E** stands for the energy obtained, **m** for the mass that disappears, and **c**^{2} for the square of the speed of light.

Using this equation, we find that when 1 g of mass is completely destroyed, an amount of energy equal to the burning of about 3000 metric tons of coal is obtained. At the present time, no way has been found to change all of a given mass into energy. Only a small portion of any mass can be changed to energy.

Splitting the atomic nucleus gives off a large amount of energy. By 1938, several scientists, including Enrico Fermi, Lise Meitner, and Otto Hahn, were using neutrons to hit different kinds of atomic targets. When they used uranium-235 ($^{235}_{92}U$) as a target, something quite surprising happened.

When uranium-235 was used as a target, the nucleus was split into two smaller nuclei and gave off a small number of fast neutrons. In addition, large amounts of energy were released. This process is called **nuclear fission.** *Fission is the splitting of a heavy nucleus into two or more lighter nuclei.* See Fig. 6-12.

After fission has taken place, the total mass of all the pieces is slightly *less* than that of the original uranium nucleus. The mass that disappears is changed to energy as predicted by Einstein's equation, $E = mc^2$.

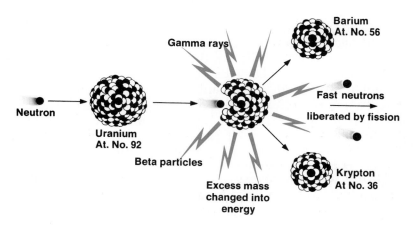

6-12 When a neutron bullet hits a $^{235}_{92}U$ nucleus, fission results. Many kinds of radioactive fission products are produced, depending on how the nucleus divides. From the diagram, describe what takes place in this particular reaction.

There are two main reasons why the discovery of fission excited the scientific world. First, large amounts of energy were set free. Second, it was believed that the fission process could lead to a *chain reaction*. That is, the neutrons set free by fission of one uranium nucleus might, in turn, cause fission in nearby nuclei. Thus, the reaction would spread and huge amounts of energy could be set free. The fission process was discovered just as World War II was starting. The possibility of using fission energy in a bomb gave great urgency to nuclear research.

Uranium-235 ($^{235}_{92}U$) will produce a chain reaction only if the amount available is large enough and pure enough. If the mass is too small, many neutrons that could produce fission escape before they can strike other nuclei.

If, however, the amount of uranium is large enough, most of the neutrons set free by one fission will strike other nuclei before they can escape. *The smallest mass of material needed for a chain reaction to occur is called the* **critical mass.** The fissionable material of an atomic bomb is stored in small parts. The explosion is triggered by bringing these small masses together suddenly. See Fig. 6-13.

Nuclear fission can be controlled. Most peacetime uses of nuclear energy require a slowed-down or controlled chain reaction. Enrico Fermi and other scientists first built a *nuclear reactor* for this purpose at the University of Chicago. The first controlled nuclear fission reaction was set off on December 2, 1942.

Only slow neutrons can start and maintain a chain reaction in uranium. The fast neutrons released by the fission of $^{235}_{92}U$ must be slowed down. *The material used to slow down the neutrons is called a* **moderator.** The atoms of a moderator slow down the neutrons in much the same way that a billiard ball slows down each time it strikes another ball. Water and graphite, a form of carbon, are good moderators.

The first nuclear reactor was a room-sized pile of graphite blocks and small rods of uranium. This kind of construction led to the name "atomic pile" for a nuclear reactor. Cadmium and boron steel are good *absorbers* of neutrons. Therefore, rods of these materials are used to control the reaction. When the control rods are lowered into the pile, the chain reaction stops.

6-13 The change of mass into energy is demonstrated by the explosion of this nuclear device. What is the relationship between energy and mass?

unit 1 matter and energy

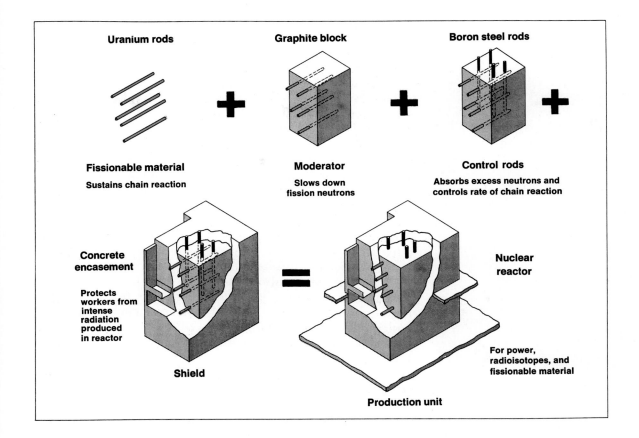

Uranium rods **+** Graphite block **+** Boron steel rods **+**

Fissionable material
Sustains chain reaction

Moderator
Slows down
fission neutrons

Control rods
Absorbs excess neutrons and
controls rate of chain reaction

**Concrete
encasement**
Protects
workers from
intense
radiation
produced
in reactor

=

**Nuclear
reactor**

For power,
radioisotopes, and
fissionable material

Shield

Production unit

More neutrons are absorbed by the control rods than are set free by the fission of $^{235}_{92}$U. Thus, the reactions taking place in the pile can be slowed down or speeded up by moving the control rods in or out. See Fig. 6-14.

Since the first nuclear reactor was built, many other types of reactors have been developed. If a reactor contains the right mixture of $^{235}_{92}$U and $^{238}_{92}$U, it can act as a **breeder reactor.** *A breeder reactor will produce new fissionable material at a greater rate than the fuel can be used up.* In this case, $^{238}_{92}$U is a material that will not fission directly. Using a breeder reactor, it is possible to change almost the entire supply of natural uranium into a material ($^{239}_{94}$Pu, plutonium-239) that will undergo fission. See Fig. 6-15.

Nuclear reactors have special uses. There are several basic uses for nuclear reactors. First, they produce great amounts of heat that can be used as a source of energy. See Fig. 6-16. Second, the radiations that

6-14 Diagram showing the parts that make up an atomic pile, or nuclear reactor. Why is the construction so massive?

are set free can be used to produce useful radioisotopes. Third, the nuclear reactor is an excellent source of neutrons, which are used in research.

The United States was the first country to use nuclear energy to power ships and submarines. When a standard submarine is underwater, it cannot run on its diesel engines because the engines quickly use up all the oxygen. Thus, electric motors, drawing on energy from storage batteries, are used when the submarine is underwater. However, the batteries soon run down and the submarine must surface and use diesel power to recharge its batteries.

Because nuclear reactors do not use oxygen, a nuclear-powered submarine can stay underwater for a long time. In early tests, one submarine cruised underwater for over two months without surfacing. Unlike conventional submarines, nuclear submarines have been able to make scientific studies under the ice of the Arctic Ocean. The fissionable material in the submarines' reactors must be replaced every few years.

Nuclear energy is also being used to run surface ships. The first of these ships, N. S. *Savannah,* sailed over 430,000 km after launching in 1961 before being refueled in 1968. This is equal to a distance of 12 times around the world.

Nuclear reactors have many safeguards. Reactors must be surrounded by thick metal or concrete, or water, or be buried underground. For example, the core of a typical nuclear power reactor near New York City is located inside a vessel with steel walls that are 18 cm thick. The vessel itself is in a water-filled steel tank, surrounded by a thick concrete wall. The whole unit, in turn, is enclosed in a steel sphere.

Many scientists believe that nuclear reactors can be used safely, even in populated areas. However, there is a possibility that radioactive pollution of the environment may occur. Because of this possibility, some people are unwilling to have nuclear power plants located in their communities.

6-15 Bundles of uranium fuel rods being inspected before placement in a nuclear reactor. Which isotope of uranium is most commonly used in a breeder reactor?

6-16 A nuclear power plant differs from other power plants only in the way in which the steam to run the turbines is made. What safety measures are needed in a nuclear power plant?

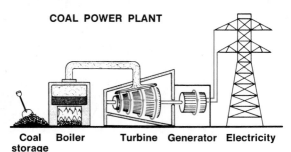

COAL POWER PLANT

Coal storage   Boiler   Turbine   Generator   Electricity

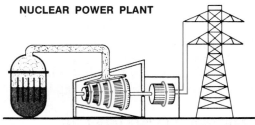

NUCLEAR POWER PLANT

Reactor   Turbine   Generator   Electricity

Fossil-fuel power plants cannot continue to supply growing demands for energy in the near future. Greater numbers of nuclear power plants may be built unless other sources of energy can be developed to meet growing energy needs. At the present time, about 13% of our electricity is generated by nuclear energy in 75 nuclear power plants.

nuclear fusion

An atomic nucleus can be built up. For many years, scientists did not understand how the sun has been able to pour out huge amounts of energy for billions of years, without cooling off. Scientists now know that nuclear reactions are the source of the sun's energy. Just as large nuclei can be split into smaller ones (fission), it is also possible to combine lighter nuclei to form heavier ones (fusion). One important example of a *nuclear fusion reaction* is the fusion of four hydrogen nuclei to form one helium nucleus. The energy of the sun and certain other stars is made by this process.

Four hydrogen nuclei (^1_1H) react to form a helium nucleus (^4_2He) and two *positrons* ($^0_{+1}e$). A positron is an electron with a positive instead of a negative charge. This fusion reaction is shown by the following equation:

$$4^1_1H \rightarrow\ ^4_2He\ +\ 2^0_{+1}e\ +\ \text{energy}$$

4 hydrogen nuclei	1 helium nucleus	2 positrons	

It is interesting to compare the total mass on the left side of the above equation with that on the right. This difference is the mass that is converted into energy.

An isotope of carbon (carbon-12) is defined as having exactly 12 atomic mass units. On this scale, a helium nucleus has a mass of 4.0015; a hydrogen nucleus, a mass of 1.0073; and a positron, a mass of 0.0005. The calculation for the ^1_1H fusion reaction is shown in Table 6-3.

The difference in mass shown in Table 6-3 is called the *nuclear mass defect*. It is the mass that is changed to energy according to the equation $E = mc^2$, when smaller nuclei join to form a larger one.

The above nuclear reaction will only take place at temperatures in the millions of degrees. For this reason, it is known as a *thermonuclear reaction*.

Table 6-3:

Mass Calculation for Fusion Reaction
Left Side:
4^1_1H @ 1.0073 = 4.0292
Total 4.0292
Right Side:
^4_2He @ 4.0015 = 4.0015
$2^0_{+1}e$ @ 0.0005 = 0.0010
Total 4.0025
Difference = 0.0267

Controlled fusion may provide for future energy needs.
One way of making a large amount of energy by the fusion process is by heating a compound of lithium and heavy hydrogen called lithium hydride ($^6_3Li^2_1H$) to a very high temperature. The nuclear reaction is

$$^6_3Li^2_1H \rightarrow 2^4_2He + energy$$

For a given mass of material, the energy given off by a fusion reaction is much greater than that for a fission reaction. However, as stated, fusion reactions can take place only at a temperature of millions of degrees. This temperature is so high that it is hard to produce and maintain even in scientific laboratories. At such a temperature, all materials become completely ionized gases, called *plasma*. This plasma cannot be contained in any known material, since the container would also turn into plasma. One way to hold the plasma together is by means of a very strong magnet, called a *magnetic bottle*. Even with so many problems, fusion reactors show promise for providing energy in the future. See Fig. 6-17.

The fusion process is preferable to the fission process for the following reasons:

1. The oceans of the world can be used as a source of heavy hydrogen (^2_1H), used as a fusion fuel. Another fusion fuel can be obtained from lithium ore.
2. A smaller amount of radioactive waste per unit of energy is produced.
3. There is less heat loss to the environment.
4. Fusion energy is expected to cost less than present-day fossil and nuclear-fission fuels.

Fusion energy is not likely to be practical until after the year 2000. But when used with other forms of energy, controlled fusion holds great promise for hundreds of years.

Antiparticles are a form of antimatter. For every kind of nuclear particle, there exists a sort of "left-handed" twin, called an *antiparticle*. Many kinds of antiparticles have already been observed by scientists. When a particle meets its antiparticle, they both disappear. In their place, an equal amount of energy appears. Antiprotons are now being made routinely in large amounts from atomic nuclei in particle accelerators. Antiprotons are the

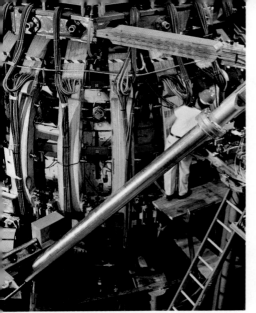

6-17 Very high temperatures (60,000,000°C) required for nuclear fusion are produced by the Princeton Large Torus. What is another requirement for obtaining power from nuclear fusion?

unit 1 matter and energy

same as ordinary protons except that they have a negative charge instead of a positive one.

The opposite, or antiparticle, of an electron is a positron. When electrons meet positrons, they destroy each other to form gamma rays.

Scientists feel sure that antimatter exists elsewhere in the universe. With the use of high-altitude balloons, scientists have observed antiprotons in the upper atmosphere. Atoms of antimatter would have a nucleus of antiprotons and antineutrons, with positrons instead of electrons outside the nucleus. See Fig. 6-18. What would happen if a small amount of antimatter hit the same amount of ordinary matter? The total mass would instantly change into a huge amount of energy as predicted by Einstein's energy-mass equation, $E = mc^2$.

If a world of antimatter actually existed, what would it look like? Dr. Emilo Segre, one of the physicists who first created the antiproton, said, "As far as physics is concerned, the antiworld would be identical with our world. An antiegg would taste like an ordinary egg if you, too, were antiman."

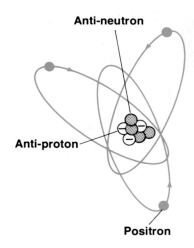

6-18 Model of an antilithium atom. What is the charge on the positrons outside the nucleus?

Marie Curie

Maria Sklodowska Curie was born in Warsaw, Poland, in 1867. Curie knew when she first started school that she wanted to become a scientist.

During the day, Curie worked for a private family, teaching and training their children. After she had earned enough money, she went to Paris at the age of 24 to enter the university.

Curie had very little money, so she suffered from hunger and cold. However, she did not mind facing many hardships in order to continue her studies. After much hard work, she graduated with the highest grade in her class and was awarded her degree in physics.

In her experiments, Curie found that pitchblende, the ore from which uranium is removed, was far more radioactive than the uranium it contained. This finding suggested the possible presence of a new element. At this point, Pierre, her scientist husband, joined Curie in further research. After working together for four years, the Curies announced the discovery of the elements polonium and radium in 1899. For this work, the Curies

shared the 1903 Nobel Prize in physics with Henri Becquerel, the discoverer of radioactivity.

After the death of her husband, Curie continued her research alone and was awarded a second Nobel Prize in chemistry in 1911. She died in 1934.

summary

1. Radioactive elements give off alpha particles (helium nuclei), beta particles (high-speed electrons), and gamma rays (high-energy form of light).
2. Radioactive elements darken film, ionize molecules, break down into simpler elements, and have a damaging effect on living things.
3. The half-life of a radioactive element is the time needed for half the atoms of a given sample to break down.
4. One element can be changed into another by bombarding the target nucleus with fast-moving charged particles.
5. Atoms of the same element with a different mass number are called isotopes.
6. Einstein's equation states that energy is equal to the mass destroyed times the square of the speed of light ($E = mc^2$).
7. Nuclear fission is the splitting of a heavy element into two or more lighter elements with the release of energy and neutrons.
8. Nuclear fusion is the joining of lighter elements to form heavier ones, with the release of huge amounts of energy.

review

Match each item in the left column with the best response in the right column. *Do not write in this book.*

1. alpha particle
2. ^3_1H
3. beta particles
4. gamma rays
5. carbon-14
6. neutron
7. fission
8. water
9. mass defect

a. the most penetrating radiation
b. element used for radioactive dating
c. deuterium
d. process of splitting heavy elements into lighter ones
e. high-speed electrons from atomic nuclei
f. good moderator for nuclear reactors
g. helium atom minus two outer electrons
h. amount of matter changed into energy
i. uncharged particle of mass number 1
j. tritium
k. controls rate of chain reaction in nuclear reactor

unit 1   matter and energy

questions

Group A

1. Which elements are naturally radioactive? List four common properties of these elements.
2. Describe the three kinds of radiation emitted naturally by the nucleus of a radioactive atom.
3. What is meant by the half-life of a radioactive element?
4. What is the common name for the isotope called deuterium? Describe the structure of its nucleus.
5. How does the nucleus of oxygen-16 differ from that of oxygen-17 and oxygen-18?
6. Why is it difficult to separate the isotopes of an element?
7. What happens to the mass number of an atom of a radioactive element when an alpha particle is given off from its nucleus?
8. What safeguards are necessary in handling radioisotopes?
9. What is Einstein's energy-mass equation? Define the symbols used.
10. Why is it impossible to speed up neutrons in a high-voltage accelerator?
11. Why are neutrons so effective as bullets for smashing atoms?
12. What causes nuclear fission?
13. What use does a moderator have in a nuclear reactor? What two materials are widely used as moderators?
14. Describe the use of control rods in nuclear reactors.
15. What is the chief hazard in the operation of nuclear reactors?

Group B

16. Compare the penetrating power of the three types of radiation from a radioactive element.
17. Why are most atomic masses not whole numbers?
18. Suggest a means of storing a radioisotope that gives off only beta particles.
19. In what way is the breakdown of radioactive carbon-14 an aid to the study of early human history?
20. Outline the chain reaction of the $^{235}_{92}U$ fission process.
21. Discuss the advantages of nuclear submarines over conventional submarines.
22. Where does the energy from nuclear fusion come from?
23. Describe an antihelium atom in terms of antimatter particles.
24. What are some of the problems in developing nuclear fusion as a practical source of energy?

problems

1. Fill in the missing spaces in the chart below. The spaces for phosphorus have already been filled. Use a separate sheet of paper for this exercise. *Do not write in this book.*

Atom	Protons	Neutrons	Electrons	Atomic Number	Mass Number
Phosphorus	15	16	15	15	31
Hydrogen	1	–	–	–	1
Deuterium	1	–	–	–	2
Tritium	1	2	–	–	–
Helium	–	2	–	–	4
Carbon	–	6	–	–	12
Oxygen	–	–	–	8	16
Uranium	–	–	92	–	238

2. How long would it take for 1 g of radium to break down so that only 1/8 of the sample was still radium? (Use 1600 years as the half-life of radium.)
3. Thorium ($^{230}_{90}\text{Th}$) gives off an alpha particle and becomes radium ($^{226}_{88}\text{Ra}$). Write the nuclear equation for this reaction.
4. When bismuth ($^{210}_{83}\text{Bi}$) breaks down, it loses a beta particle. (a) What new radioactive element is formed? The new element gives off an alpha particle. (b) What is the symbol for the final nucleus? (c) Write the nuclear equation for this reaction.
5. When neutrons are fired at $^{238}_{92}\text{U}$ target atoms, the atoms capture some of the neutrons and temporarily become $^{239}_{92}\text{U}$. Write the nuclear equation for this reaction.

further reading

Asimov, Isaac, *Left Hand of the Electron*. Garden City, N.Y.: Doubleday, 1972.

Enghadl, S., et al., *The Subnucleus Zoo: New Discoveries in High-energy Physics*. New York: Atheneum, 1977.

Hunt, S. E., *Fission, Fusion and the Energy Crisis*. Elmsford, N.Y.: Pergamon Press, 1974.

Libby, L. M., *The Uranium People*. New York: Scribners, 1979.

unit 1 matter and energy

unit 2
chemistry in our world

7

common gases of
the atmosphere

The atmosphere is an ocean of air around the earth. "Air" is a common word for the mixture of gases that make up the atmosphere. Air is needed to live. You can get along without food or water for days, but you cannot live without air for more than a few minutes.

objectives:

☐ To describe the laboratory and industrial preparations of oxygen, nitrogen, and hydrogen.

☐ To explain some of the uses of oxygen, nitrogen, carbon dioxide, ammonia, and hydrogen.

☐ To list some of the physical and chemical properties of the common gases of the atmosphere.

☐ To define catalyst, slow oxidation, rapid oxidation, spontaneous combustion, and kindling temperature.

oxygen: the breath of life

Air is a mixture. Clear air is a colorless, tasteless, and odorless gas. It is a mixture mostly of nitrogen and oxygen. Other gases such as carbon dioxide, water vapor, and a number of noble gases are also present in very small amounts.

Look at Table 7-1. Notice that there is more nitrogen in the air than oxygen. However, oxygen is more plentiful in the earth as a whole. It is found combined with other elements to form many compounds in the earth's

unit 2 chemistry in our world

crust. Oxygen compounds such as water, sand, clay, and most minerals make up about 50% of the earth's crust. All plant and animal tissues contain oxygen. Oxygen also makes up 89% of the mass of water.

Priestley discovered oxygen. Joseph Priestley, an English scientist, is given credit for the discovery of oxygen. Priestley was the first scientist to publish the results of his experiments with this gas. As recently as 200 years ago, people thought that the air was made up of only one gas.

The nature of burning, rusting, and other processes involving oxygen was a mystery that challenged many scientists of that day, including Priestley. In 1774, while studying the behavior of mercury(II) oxide, Priestley heated the oxide by focusing the rays of the sun on it with a lens.

Priestley saw that the red oxide of mercury turned black as it was heated. Then a small drop of mercury metal formed. Priestley stirred the mixture with a wooden stick. Much to his surprise, instead of smoldering, the stick burst into flame. Next, Priestley heated some of the powdered oxide in a container and collected the gas that was given off. He found that a lighted candle burned much more brightly than normal in this gas. See Fig. 7-1.

In another experiment, Priestley placed a mouse in the gas and saw it become very active. He breathed some of the gas himself and found that his breath felt strangely "light" and "easy" for some time later. Priestley described the gas as "good air." He did not know that this gas (oxygen) is present in the air.

Lavoisier advanced the knowledge of oxygen. A few years later, in the late 1700's, Antoine Lavoisier (la-*vwah*-zee-ay), a French scientist, read of Priestley's experiments. Lavoisier thought that Priestley's new gas might also be present in the air, so he conducted a 12-day experiment to test his theory.

Using a glass retort, Lavoisier heated some mercury in the presence of air as shown in Fig. 7-2. A reddish powder began to gather on the surface of the mercury in the retort. After 12 days, no more red powder was formed and about one-fifth of the air in the bell jar was gone. Mercury rose in the bell jar to take the place of the

Table 7-1:

Gases of the Air	
Nitrogen (N_2)	78%
Oxygen (O_2)	21
Noble gases	0.94
Carbon dioxide (CO_2)	0.04
Water vapor (H_2O) (% by volume)	varies

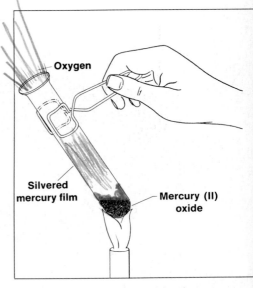

7-1 When mercury(II) oxide is heated, it breaks down into mercury (Hg) and oxygen (O_2). CAUTION: Mercury vapors are poisonous. Do not perform this experiment yourself.

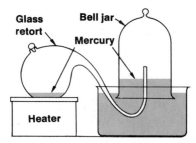

Glass retort **Bell jar** **Mercury**

Heater

7-2 Lavoisier heated mercury (Hg) in a glass retort, producing mercury(II) oxide (HgO). Upon heating, the level of mercury in the bell jar rises. Why?

air, as shown in the diagram. The oxygen of the air had combined with the heated mercury, forming mercury(II) oxide (HgO).

In another experiment, Lavoisier heated the oxide to a higher temperature in a smaller retort. The mercury(II) oxide changed back to mercury and a colorless gas was given off. When he tested this gas, Lavoisier found that substances burned brightly, as Priestley had seen. Here, then, was proof that about one-fifth of the air is an active gas that unites chemically with other substances. Lavoisier named the gas *oxygen*. See Fig. 7-3.

activity

Find the percent of oxygen in the air. Wash a piece of new steel wool with a detergent and rinse. Place the wet steel wool in a bottle so that it will remain firmly in place when the bottle is inverted in a pan of water, as shown in Fig. 7-4.

After a day or two, carefully put a glass plate over the mouth of the bottle while it is still underwater without letting any water leak out. Lift the bottle from the water and quickly turn it right side up. With a graduated cylinder, measure the volume of water in the bottle. This volume is equal to the volume of oxygen from the air that combined with the steel wool. Next, fill the bottle full of water and again measure its volume.

From these two measured volumes, find the percent of oxygen that is in the air sample. Compare your value with that given in Table 7-1.

(1) Do you think a chemical reaction has taken place? Why? (2) Why was the steel wool washed with a detergent? (3) Why was the steel wool placed in the bottle while wet? (4) What facts did you assume in finding the percent of oxygen? (5) How would your results be affected if any of your assumed facts were wrong?

Oxygen can be made in many ways. When mercury(II) oxide is heated, it decomposes (breaks down) into mercury (Hg) and molecular oxygen (O_2), as shown by the following equation:

$$2\,HgO \stackrel{heat}{\rightarrow} 2\,Hg + O_2 \uparrow$$

Small amounts of oxygen can be made in another way.

A chemist might mix and heat equal parts of potassium chlorate ($KClO_3$) and manganese dioxide (MnO_2), as shown in Fig. 7-5. The heat will break down the potassium chlorate into potassium chloride (KCl) and oxygen. The equation for this reaction is

$$2\ KClO_3 \xrightarrow{\text{heat}} 2\ KCl + 3\ O_2 \uparrow$$

You may wonder why MnO_2 has not been included in this equation, even though it was added to $KClO_3$ at the very start of the reaction. The reason is that MnO_2 helps $KClO_3$ give up its oxygen more quickly and at a lower temperature. The manganese dioxide is *not* permanently changed in this reaction. *Substances that change the rate of a chemical reaction but are not themselves changed are called* **catalysts** (*kat*-uh-lists). In this case, a catalyst (MnO_2) is used to speed up a chemical change. Sometimes catalysts are used to slow down chemical reactions. Catalysts are widely used in chemistry.

Heating a mixture of potassium chlorate and manganese dioxide produces oxygen. Does the oxygen come from $KClO_3$ or MnO_2? Heating these compounds one at a time in a test tube will give the answer. When only the catalyst, manganese dioxide, is heated, oxygen is *not* given off. However, at high temperatures, potassium chlorate melts when heated and gives off a gas that tests positively for oxygen. When potassium chlorate is heated without the catalyst, it must be heated to a higher

7-3 Antoine Lavoisier, a French scientist, is shown performing his historic experiment in his laboratory. Describe his famous 12-day experiment.

7-4 Diagram of a setup for finding the percent of oxygen in a sample of air. Why does the water level rise in the collecting bottle?

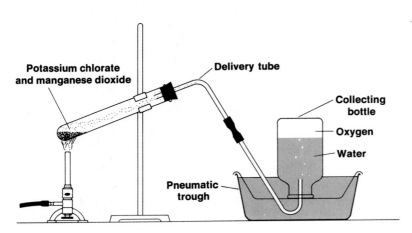

7-5 The apparatus for making and collecting oxygen by displacement of water. Can you explain why you must remove the delivery tube before removing the burner? CAUTION: Do not attempt this experiment without your teacher's approval and supervision.

temperature before oxygen is given off. Again, the manganese dioxide catalyst allows the breakdown of potassium chlorate to take place quickly at a lower temperature.

CAUTION: *Because of the potentially volatile reaction of potassium chlorate and manganese dioxide, making oxygen in the lab must be done with extreme care. The heated mixture will explode if there are any burnable impurities in the mixture. Therefore, this way of making oxygen is not recommended for use in school labs.*

The methods of making oxygen that have been described are not suitable for making large amounts of oxygen for commercial uses. Oxygen for consumer use is usually made in large amounts either by electrolysis of water or by evaporation of liquid air.

Earlier in the text (see Fig. 2-4), you saw that an electric current can be used to break down water into its elements, oxygen and hydrogen. This process is called electrolysis. In making oxygen by this method, hydrogen gas is an important by-product.

Air can be made into a liquid if it is highly compressed and, at the same time, cooled to a very low temperature (−200°C). ("C" refers to Celsius, a scale used to measure temperature. You will learn about this scale in Chapter 20.) This process brings the air molecules closer together to form stronger bonds as the air becomes a liquid. Liquid air is made up of mostly nitrogen and oxygen. Liquid nitrogen has a boiling point (−196°C) that is about 13° lower than that of liquid oxygen (−183°C). Therefore, as the liquid air warms on standing, the nitrogen will boil away first. Mostly liquid oxygen is left behind. The nitrogen produced is also recovered for consumer use.

Oxygen has many uses. Since pure oxygen is very important to sick people who have breathing problems, hospitals and ambulances always have a supply of compressed oxygen. See Fig. 7-6.

Additional oxygen is also necessary to provide enough oxygen for normal breathing at high altitudes where the air is too thin. To solve this problem, modern planes have pressurized cabins. Air from the outside of the plane is compressed so that it has a higher pressure inside the cabin.

Firefighters and rescue workers often need gas masks and tanks of oxygen when entering smoke-filled build-

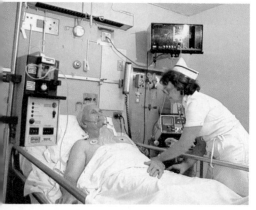

7-6 Oxygen is shown being used in a hospital room. For what purpose is the oxygen being used?

ings. Deep-sea divers and skin divers use tanks of compressed gases so that they will have enough oxygen to breathe while underwater.

An important property of oxygen is that it is slightly soluble in water. Fish and other marine life use this dissolved oxygen for *respiration*. Respiration is the exchange of oxygen and carbon dioxide in living things and results in the making of energy.

Oxygen is used with another gas, acetylene (uh-*set*-ih-leen), in a torch that makes a very hot flame. An oxyacetylene torch is hot enough to cut through metals or to weld pieces of metal together. See Fig. 7-7.

Today, large amounts of oxygen are needed in industry. All modern steel mills, for example, use many metric tons of oxygen daily. Blast furnaces use oxygen-enriched air to make iron from iron ore. The ore can be changed to iron more quickly by burning out the impurities with oxygen instead of air.

The space program also needs great amounts of oxygen. Huge liquid-fuel rockets that launch satellites and space vehicles use *lox* (liquid oxygen) to burn the fuel because there is no oxygen in space.

Oxygen can be identified by a simple test. A wood splint burns in air, but it burns more brightly in pure oxygen. Therefore, if a glowing splint is lowered into a bottle of pure oxygen, it will burst into flame at once. This behavior is commonly used as a test to identify the presence of oxygen. Oxygen is the only common, odorless gas that supports the burning of a wood splint. See Fig. 7-8.

Oxidation is a common chemical change. Modern chemistry really began when scientists discovered what an important part oxygen plays in the chemical changes taking place all around us. Lavoisier is often called the father of modern chemistry because chemistry and all other sciences advanced rapidly after his famous experiments.

When a substance such as coal, wood, or gasoline burns, it undergoes *oxidation*. The atoms of carbon and hydrogen that make up these substances join with oxygen in the air to form new compounds. **Oxidation** *is the chemical combination of oxygen with other substances.*

Oxidation that takes place quickly and gives off heat and light is called rapid oxidation, or combustion. Most often,

7-7 The oxyacetylene flame cuts through slabs of steel quickly and efficiently. Why do the workers wear shields over their heads?

7-8 If a glowing wooden splint is thrust into a test tube of oxygen, the splint will burst into flame. What does this observation indicate?

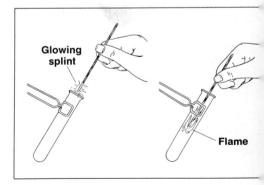

Glowing splint

Flame

air supplies the oxygen needed for burning. Can you give several examples of rapid oxidation (combustion)?

Many serious explosions have taken place in grain elevators, coal mines, starch mills, and other places where the air was filled with burnable dust. These blasts were caused by a spark or flame. The rapid burning of the dust particles gives off heat and so warms the nearby particles. Why is combustion so fast? Consider that a wooden log burns more quickly when split into kindling. The kindling burns more quickly when made into still smaller particles and blown or scattered over a fire. The increased speed of burning is due to an increase in the surface area of the particles in contact with the oxygen of the air. Kindling burns quickly; small particles burn explosively. The burning spreads from particle to particle very quickly, causing a violent explosion. Coal mines and many industrial plants are careful to keep the air clear of burnable dust. See Fig. 7-9.

It may surprise you to know that partially decayed wood does not give off as much heat as fresh wood. The reason for this is that in decayed wood, oxidation has already used up some of the wood.

The decay of dead plants and animals is another example of oxidation. Unlike burning, the process of decay is a very slow chemical change. *Decay is called slow oxidation.* Slow oxidation, like rapid oxidation, gives off heat. The decay process is so slow, and takes place over such a long time, that the heat is not usually noticeable.

Slow oxidation, which breaks down dead leaves and other plant and animal material, is useful. The decay

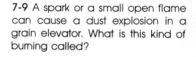
7-9 A spark or a small open flame can cause a dust explosion in a grain elevator. What is this kind of burning called?

unit 2   chemistry in our world

process produces substances that nourish growing plants. However, slow oxidation may also cause undesirable changes. For example, when an iron barbecue grill is left outdoors, the iron unites with the oxygen of the air to form an iron oxide, commonly called *rust*.

Because rusting of iron causes millions of dollars worth of damage every year, a strong effort is made to prevent it. Steel tools are coated with a thin film of oil to keep the oxygen from coming in direct contact with the metal. Steel bridges and water tanks are usually painted with a mixture of red lead and linseed oil, and often with a coat of aluminum paint. Some steel products are plated with metals such as tin, nickel, zinc, or chromium. These metals also oxidize, but the resulting oxide forms a thin coating that protects the iron or steel below it from further oxidation.

If you leave a pile of oily rags in a closed place, you will see another undesirable example of the effects of slow oxidation. The rags may heat up and then burst into flame. Such a fire is caused by *spontaneous combustion*. The word "spontaneous" means "by itself."

The fire starts because the oily rags slowly oxidize, building up heat. If the heat cannot escape, the temperature of the rags will rise. Finally, the rags get so hot that they burst into flame. *The lowest temperature at which a material begins to burn is known as its* **kindling temperature.** See Fig. 7-10.

Most industries have a strict rule that oily rags be placed in covered metal containers. It is wise to follow the same practice at home. For instance, rags that have been used to polish furniture or to wipe oil and paint should be properly stored in an airtight container, or dried outside and disposed of.

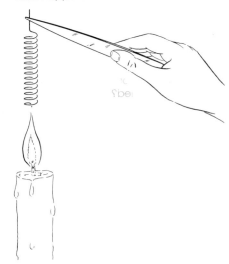

7-10 A copper coil lowered into a candle flame conducts heat away so quickly that the wax is cooled below its kindling temperature. What happens to the flame?

other gases of the air

Nitrogen is the most common gas of the air. In contrast to oxygen, nitrogen is a rather inactive element. It does not combine readily with other elements. Like oxygen, nitrogen has no odor, no taste, and no color. Nitrogen does not dissolve readily in water and will not burn. When nitrogen does react with another element, the union is seldom stable and the compound breaks down easily.

7-11 The nodules on these pea roots contain nitrogen-fixing bacteria. Describe the functions of these bacteria.

Oxygen atoms

Carbon atom

7-12 A model of a carbon dioxide molecule (CO_2) showing its straight-line structure. What are some properties of carbon dioxide?

Common explosives such as dynamite, gunpowder, TNT (tri-nitro-toluene), and nitroglycerin are compounds of nitrogen. The destructive power of these explosives is due to the rapid breakdown of their molecules. This breakdown, in turn, gives off heat and causes gases to expand quickly, producing rapid combustion.

All living things and all proteins contain compounds of nitrogen. While most crops take nitrogen compounds from the soil, certain other crops such as peas, beans, clover, and alfalfa put nitrogen compounds back into the soil. Bacteria that grow on the roots of these plants change the nitrogen of the air into nitrogen compounds. This process is called *nitrogen fixation.* See Fig. 7-11. In this way, the nitrogen compounds are returned to the soil for further use by other crops. Another way of adding nitrogen compounds to the soil is through the use of fertilizers.

Carbon dioxide is needed for plants. Carbon dioxide and water vapor are compounds found in the air. Although the amount of carbon dioxide in the air is very small, this gas is needed for life on earth. Green plants must have carbon dioxide to grow.

The food-making process in green plants is called *photosynthesis* (fo-toh-*sin*-theh-sis). In the presence of the green chemical *chlorophyll* (*kloh*-roh-fil) and sunlight, plants use water from the soil and carbon dioxide from the air to make sugars and starches. These sugars and starches are used by the plants for growth and development. Oxygen is given off as a waste product.

Animals, including humans, depend on plants for their food and oxygen. Animals breathe the oxygen that plants put into the air and give off carbon dioxide as one of the products of food oxidation. Most of the carbon dioxide of the air comes not from animals, however, but from burning fuels in homes, cars, and factories. Plants remove some of the carbon dioxide from the air, but most of it is absorbed by the oceans of the world. Thus, the total amount of carbon dioxide in the air stays about the same. A model of the carbon dioxide (CO_2) molecule is shown in Fig. 7-12.

In the laboratory, carbon dioxide can be made by adding hydrochloric acid (HCl) to chips of marble. Marble is impure calcium carbonate ($CaCO_3$). The setup for this reaction is shown in Fig. 7-13. The equation is

$$2 \, HCl + CaCO_3 \rightarrow CaCl_2 + H_2O + CO_2 \uparrow$$

If you collect a bottle of CO_2 and thrust a burning splint into the bottle, what happens? Suppose you "pour" a bottle of the invisible CO_2 gas over a burning candle. What happens to the candle flame? How can you see that CO_2 is denser than air? Carbon dioxide does not burn or support the burning process. It reacts with water to form carbonic acid (H_2CO_3), or soda water.

If limewater [$Ca(OH)_2$] is added to a bottle of CO_2 gas and shaken, the limewater turns white. Calcium carbonate ($CaCO_3$) is formed as follows:

$$Ca(OH)_2 + CO_2 \rightarrow CaCO_3 \downarrow + H_2O$$

The arrow ($\downarrow$) means that an insoluble product, or *precipitate*, is formed. The formation of this white precipitate is a common test used to detect CO_2.

Ammonia results from protein decay. Ammonia gas is one of the most important of all chemical compounds. A very small amount of ammonia gas is found in the air. Ammonia is formed when proteins in dead plants and animals are broken down in the decay process. An odor of ammonia can be noticed around barns and stables where farm animals are kept. Bacteria break down the nitrogen compounds in manure to form ammonia gas.

Ammonia is a colorless gas, less dense than air, with a choking odor. Ammonia is poisonous and is extremely soluble in water. One liter of water can dissolve about 700 L of ammonia gas at room temperature.

In the laboratory, ammonia (NH_3) can be made by

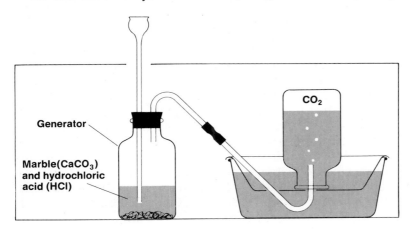

Generator

Marble($CaCO_3$) and hydrochloric acid (HCl)

CO_2

7-13 A common laboratory setup for making carbon dioxide. Which of the reacting substances produces the carbon dioxide?

chapter 7   common gases of the atmosphere   117

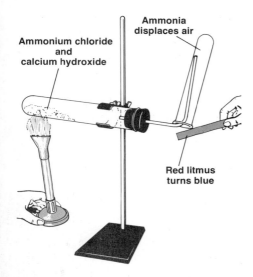

Ammonia displaces air

Ammonium chloride and calcium hydroxide

Red litmus turns blue

7-14 A typical setup for making small amounts of ammonia gas (NH_3). Why does the wet litmus paper change color as shown above? Write the equation for this reaction.

7-15 Liquid ammonia is applied directly to the soil as a fertilizer. What are some other uses for ammonia?

heating a mixture of calcium hydroxide [$Ca(OH)_2$] and ammonium chloride (NH_4Cl) as shown in Fig. 7-14. The chemical change that takes place is described by the following equation:

$$Ca(OH)_2 + 2\,NH_4Cl \overset{heat}{\rightarrow} CaCl_2 + 2\,H_2O + 2\,NH_3 \uparrow$$

Large amounts of ammonia are made for the commercial market by reacting nitrogen (N_2) with hydrogen (H_2), using a catalyst to speed up the process. This reaction takes place at about 600°C and at a pressure of about 1000 times normal air pressure. The equation for this reaction is

$$N_2 + 3\,H_2 \rightarrow 2\,NH_3 \uparrow$$

The nitrogen for this process is produced by the evaporation of liquid air, described earlier. Hydrogen is obtained from the electrolysis of water, a process also described earlier.

Because ammonia contains about 82% nitrogen (by weight), large amounts of it are used in making fertilizers. Ammonia is also used in making nitric acid and ammonia water, in ice-making, and in frozen food plants. See Fig. 7-15.

Hydrogen is the least dense gas. Hydrogen gas is very scarce in the atmosphere. However, analysis of light given off by stars shows that the stars are mostly composed of hydrogen. As far as scientists have been able to tell, hydrogen atoms make up about 90% of all atoms in the entire universe. (The whole universe is mostly empty space.) Hydrogen, in combined form, is found in all living things, in fuels, and in water.

Most active metals will replace the hydrogen in an acid, thereby setting the hydrogen free. In the lab, hydrogen can be made by the reaction of zinc (Zn) with dilute sulfuric acid (H_2SO_4). The single replacement reaction is described by the following equation:

$$Zn + H_2SO_4 \rightarrow ZnSO_4 + H_2 \uparrow$$

The zinc sulfate ($ZnSO_4$) that is formed remains in the bottle as shown in Fig. 7-16. Hydrogen gas bubbles off and drives the water out of the collecting bottle.

Hydrogen has the lowest density of all the elements. This property is one reason why it was once used to fill balloons. However, hydrogen can explode and is so

unit 2 chemistry in our world

Thistle tube

Dilute sulfuric acid

Zinc

Hydrogen

Water

7-16 A typical setup for the lab preparation of hydrogen gas (H_2). Why is the thistle tube placed below the surface of the liquid?

7-17 A photo of the *Hindenburg* disaster in 1937. In the presence of a flame or electric spark, the hydrogen in the dirigible combined with the oxygen of the air with explosive force. What product is formed in this chemical reaction?

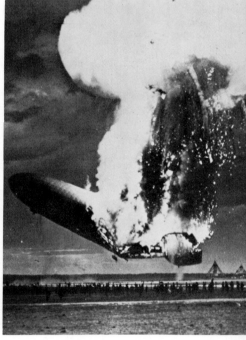

highly dangerous that it is no longer used. Helium gas, the next lightest gas and a totally inactive element, is now used in place of hydrogen. See Fig. 7-17.

Hydrogen is not very active at room temperature, but it reacts with many substances when heated. For example, suppose a lighted candle is held carefully to the mouth of a small open (plastic) container of hydrogen. If the bottle also contains some air, there will be a "pop" as the hydrogen burns rapidly. If an open bottle of hydrogen is allowed to stand for a minute or two and a burning candle is brought near the bottle, nothing will happen. This is because hydrogen, which is less dense than air, escapes from the open bottle.

If a flaming splint is placed inside an inverted bottle of hydrogen, the flame will go out, but the hydrogen will continue to burn at the mouth of the bottle. See Fig. 7-18. This is a rough test for the presence of hydrogen gas. While the hydrogen gas itself burns, it cannot support the burning of other substances. See Fig. 7-19.

activity

The action of magnesium with an acid. Place a 5-cm length of magnesium (Mg) ribbon in a test tube containing a small amount of *dilute* hydrochloric acid (HCl). (1) What do you observe? (2) Identify the gas given off. (3) Write the balanced chemical equation for this reaction.

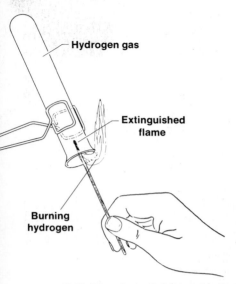

Hydrogen gas

Extinguished flame

Burning hydrogen

7-18 A burning splint is used to test for hydrogen gas. Pure hydrogen burns with an almost colorless flame. A loud "pop" is heard when a mixture of hydrogen and air is burned. Why is the test tube held upside down? CAUTION: Do not try this test for hydrogen in the lab.

7-19 In the process of burning, hydrogen combines with oxygen of the air to produce water. How many atoms of hydrogen and oxygen are there on each side of the equation?

When hydrogen is added to cottonseed, coconut, or other food oils, it combines chemically to form solid cooking fats. This process is called *hydrogenation* (hy-*droj*-en-nay-shun). Fine grains of nickel (Ni) are used as a catalyst to speed up the process. Peanut butter is hydrogenated to prevent the separation of the oils from the pulp.

Special welding and cutting devices use hydrogen to produce extremely hot flames. Hydrogen is also used to make high-grade gasoline from petroleum. Liquid hydrogen is a very efficient fuel for rockets.

Noble gases of the air. The air contains very small amounts of other gases and impurities, including the *noble gases*. Noble gases in the air include *argon, helium,* and *neon*. Argon means "lazy," a name that correctly describes the activity of this element. Helium gets its name from the Greek word for the sun. The element helium was found in the sun before it was discovered on the earth. The name of the element neon means "new."

Argon is used with nitrogen in electric light bulbs. The hot tungsten filament in the bulb evaporates, blackening the inner surface of the bulb. Argon slows the rate of evaporation so that the bulb will stay bright longer.

Helium is obtained from some of the natural gas wells in Kansas, Oklahoma, and Texas. Helium is useful for filling balloons because it is very light and yet will not burn. It is also used in welding metals. Deep-sea divers breathe a mixture of helium and oxygen because the nitrogen of the air may be harmful if it is inhaled under pressure. Large amounts of helium are used as a coolant in some nuclear reactors.

Neon is widely used in advertising signs and for air-

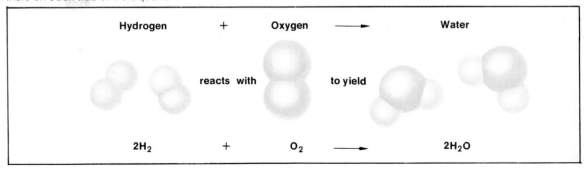

Hydrogen	+	Oxygen	$\longrightarrow$	Water
	reacts with		to yield	
$2H_2$	+	O_2	$\longrightarrow$	$2H_2O$

unit 2   chemistry in our world

plane beacon lights. A high-voltage charge applied across a neon-filled glass tube causes the gas to give off a familiar, glowing orange-red light. Various mixtures of gases are used to produce a variety of colored lights. For instance, argon mixed with mercury will produce a bright blue color; argon mixed with helium produces a yellow color. See Fig. 7-20.

The noble gases were once thought to be completely inert, or chemically inactive. This lack of chemical activity can be explained by their atomic structure. The outer shell of electrons of the noble gases is filled. Scientists thought that such elements would not combine with other elements because they do not tend to give, receive, or share electrons. However, scientists now know that it is possible to make a few compounds of the heavier noble gases (krypton, xenon, and radon) with such active elements as fluorine. The noble gases are sometimes called the *inert* or *rare gases*. They are found in Group VIII of the periodic table (Fig. 3-6).

7-20 So-called neon lights are used in advertising but can also be works of art, as in this abstract sculpture. What causes the various colors of light in these tubes?

George Washington Carver
Scientist and Inventor

George Washington Carver (1864–1943) was an American scientist who was born in Missouri during the Civil War. He enrolled at Simpson College in Iowa and later transferred to Iowa State College where he received his degree in agricultural science in 1896. Carver then joined the faculty at Tuskegee Institute in Alabama where he spent the rest of his life.

Carver devoted his life to developing better methods for growing crops and conserving the soil. He found that crop rotation would enrich the soil and increase crop yields. He persuaded farmers to plant peanuts and sweet potatoes after long periods of cotton planting had taken out much of the soil nutrients.

Carver made some 300 different products from peanuts, including such items as flour, dyes, insulating board, wood stains, and ink. From sweet potatoes, he made more than 100 different products, such as molasses, candy, vinegar, and flour. Carver received many awards during his lifetime for his methods of improving the science of farming.

summary

1. Air is a mixture of many gases, mostly nitrogen and oxygen.
2. Oxygen can be made in the lab by breaking down mercury(II) oxide or potassium chlorate.
3. Oxygen can be made commercially by electrolysis of water and by the evaporation of liquid air.
4. Oxygen is the only common, odorless gas that supports burning.
5. Oxidation is the chemical combination of oxygen with other substances.
6. Catalysts are substances that change the speed of a chemical reaction but are not themselves changed by the reaction.
7. Nitrogen compounds are needed for the proper growth of most plants.
8. Carbon dioxide and water are needed in photosynthesis.
9. Ammonia is a colorless gas that is lighter than air, has a strong choking odor, is poisonous, and is very soluble in water.
10. Hydrogen, the element with the lowest density, is found in combined form in all living things, in fuel, and in water.

review

Match each item in the left column with the best response in the right column. *Do not write in this book.*

1. hydrogen
2. burning
3. limewater
4. rusting of metals
5. oxygen
6. ammonia
7. neon
8. CO_2

a. noble gases
b. slow oxidation
c. gas used as a fertilizer
d. supports combustion
e. clear solution of $Ca(OH)_2$
f. mercury(II) oxide
g. calcium carbonate + acid
h. rapid oxidation
i. chemical equation
j. least dense element

questions

Group A
1. Describe a convenient way of making oxygen in the chemistry lab.
2. Define oxidation and kindling temperature.

3. Is the activity shown in Fig. 7-4 an example of slow oxidation? Justify your answer.
4. How can oily rags cause a fire by spontaneous combustion?
5. List several uses of nitrogen compounds.
6. How is carbon dioxide made in the laboratory? How is it tested?
7. List two conditions that are necessary for photosynthesis to take place.
8. How is ammonia gas made in the laboratory? How is it made commercially? Of what use is it?
9. How is hydrogen made in the laboratory? How is it made in industry?
10. What are five uses for hydrogen?
11. List three uses for helium.
12. Where in the periodic table (Fig. 3-6) are the noble gases found?

Group B
13. What evidence can you give to support the idea that air is a mixture?
14. Describe the action of manganese dioxide in the making of oxygen by the decomposition of potassium chlorate.
15. Describe how dust explosions are produced.
16. Why is argon used in making electric light bulbs? Suggest several other gases that might be used.
17. Why are the noble gases inactive compared with other elements?
18. In a table, list four gases found in the air. In separate columns, list some physical or chemical properties of each gas.

further reading

Asimov, Isaac, *Of Matters Great and Small.* New York: Ace Books, 1976.
Coulson, E. H., *Test Tubes and Beakers: Chemistry for Young Experimenters.* Garden City, N.Y.: Doubleday, 1971.
Davis, Kenneth, *The Cautionary Scientists: Priestley, Lavoisier and the Founding of Modern Chemistry.* New York: G P Putnam's Sons.
Erwin, N. H., et al., *Joseph Priestley: Scientist, Theologian, and Metaphysician.* Lewisburg, Penn.: Bucknell University Press, 1980.
Holmes, E., et al., *Great Men of Science.* New York: Franklin Watts, Inc., 1979.
Wilford, J. N., ed., *Scientists at Work: The Creative Process of Scientific Research.* New York: Dodd, Mead & Company, 1979.

8

the chemistry of water

If you look at a globe, you will see that oceans of water cover about three-fourths of the earth's surface. Large amounts of water are also found in the form of vapor, which is always present in the air, even over the deserts. About two-thirds of the mass of the human body is water. Vegetables and fruits, such as cucumbers, tomatoes, celery, and lettuce, as well as other foods, actually contain more than 90% water.

objectives:

☐ To explain why water is a good solvent and why it exists in a liquid phase at room temperature.

☐ To describe and explain the reason for each step in the process of purifying water.

☐ To explain the cause of hard water and to describe the processes used to make it soft.

☐ To describe the treatment and disposal of sewage.

☐ To define the following terms: hydrogen bond, suspension, potable water, distillation, freeze separation, reverse osmosis, and ion exchange.

water: an ever-present compound

Water has some remarkable properties. Pure water is a liquid that has no color, odor, or taste. You have seen that it exists in all three phases: solid, liquid, and gas.

Under normal conditions, water freezes at 0°C and boils at 100°C. See Fig. 8-1.

In many ways, water is a most remarkable compound. Almost all liquids shrink when they freeze. For instance, when a beaker full of melted wax is allowed to cool, a hollow will form in the center where the freezing liquid shrinks. See Fig. 8-2. Unlike wax, water expands when it freezes. When a tankful of water freezes, it exerts about 20,000 newtons (N) of force on each square centimeter of the tank. A tightly sealed container filled with water may burst when the water freezes, even if it is quite strong. Water pipes and car radiators are likely to burst in winter if the water in them freezes.

When water at 0°C is warmed, it shrinks until the temperature reaches 4°C. At 4°C, water reaches its greatest density. Then, as it is warmed further, the water slowly expands.

Chemically, water is a very stable compound. Thus, water does not break apart until it is heated to about 2700°C.

Water is a good solvent. Water is a covalent molecule that is formed when an oxygen atom shares a pair of electrons with each of two hydrogen atoms. This arrangement is shown in Fig. 3-14. A laboratory model shows that a water molecule has a bent structure. The angle of bonding, or bond angle, is 105°.

Like all molecules, water molecules are electrically neutral. However, you can see from Fig. 8-3 that the hydrogen ends are somewhat positive. The oxygen atom at the opposite end is somewhat negative. A lopsided molecule of this type is called a **polar** molecule. *The polar, covalent property of the water molecule is one of the main reasons why water is such a good solvent.* Because of its polar nature, the water molecule can attract other molecules or ions of a solute, surround them, and pull them into solution. A molecule with a straight-line structure, such as CO_2 (O—C—O), is a symmetrical, or *nonpolar*, covalent molecule. (See Fig. 7-12.)

The fact that water is a good solvent accounts for the saltiness of the oceans. For billions of years, rain has been falling on land areas of the earth and running off into streams, then into rivers, and finally into oceans. Rainwater dissolves solid matter from the land, mostly in the form of salts, and carries it to the sea. Thus, for

8-1 The change of phase from water to ice (freezing) or from ice to water (melting) takes place at a temperature of 0°C. At what temperature does the water boil under normal conditions?

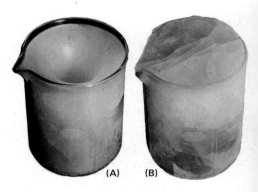

8-2 (a) Most substances, like wax, contract when they change from a liquid to a solid. (b) Water, however, expands when it freezes. Why do water pipes often break when the water in them freezes?

(A) (B)

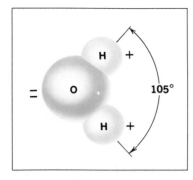

8-3 This diagram shows the polar nature of the water molecule. Can you explain why the hydrogen atoms are partially positive and the oxygen atom is partially negative?

billions of years, salts have been added to the oceans. The water that evaporates from the oceans returns to the land as rain, leaving behind the salts. Therefore, the ocean water that remains becomes more salty as time goes on.

The ocean also acts as a storehouse of raw materials. The waters that run from the land into the ocean carry with them not only common salt, but many dissolved solids as well. A wide range of chemical resources is found within this vast storehouse. From a single cubic meter of seawater, 21,000 g of sodium chloride, 100 g of magnesium, 400 g of potassium, and 60 g of bromine can be obtained. Many other substances are also present in smaller amounts.

The resources of the ocean are, for the most part, unused. It is difficult and expensive to remove these chemicals from the water. At the present time, only sodium chloride, magnesium, bromine, and iodine are taken from the sea in large amounts.

Water molecules are attracted to each other by weak forces. Water molecules are held to each other by the attraction of the positive hydrogen ends of each water molecule to the negative oxygen ends of other water molecules. Thus, a weak but effective *hydrogen bond* is formed. See Fig. 8-4. This diagram shows that water is not simply a group of separate H_2O molecules, but rather a large number of H_2O molecules linked together. Water is a liquid at room temperature because of the formation of molecular groups by hydrogen bonding. If hydrogen bonding did not exist, water would be a gas at room temperature.

Hydrogen bonds play an important part in fixing the melting and boiling points of many substances. These bonds cause the open structure of ice crystals. This

8-4 Hydrogen bonds are formed when the hydrogen atoms of the water molecule are weakly attracted to oxygen atoms of other water molecules. Can you explain why this attraction exists?

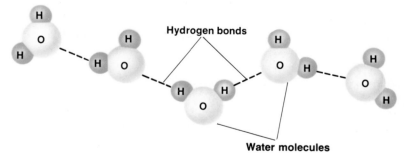

126

open structure accounts for the fact that ice has a lower density than water. See Fig. 8-5. Hydrogen also forms similar bonds with fluorine (HF) and with nitrogen (NH_3).

Pure water is not common. Even though there is so much water on the earth, pure water is not common. That is because water dissolves many substances that it touches. Water that contains impurities that are harmless to the human body may be suitable for drinking. However, water is almost never "pure" in a chemical sense. In fact, water tastes better if some minerals are allowed to remain in solution.

The purest water in nature is rainwater, but it, too, contains impurities. As rain falls from the clouds, rainwater dissolves oxygen and other gases of the air. Bits of dust and bacteria floating in the air are also picked up. Even so, rainwater is fairly free of dissolved solids, except in heavily polluted areas. In many areas of our country, mostly in the northeast, raindrops react with the oxides of sulfur and nitrogen pollutants in the air to form sulfuric and nitric acids. The result of this reaction is called **acid rain.** *Acid rain has a pH below 5.6.* Highly acid rain harms plant and animal life in our rivers and lakes.

As soon as rainwater touches the ground, it dissolves minerals and impurities from plant and animal sources. These impurities often make water unfit for drinking because of the presence of harmful bacteria. Dissolved minerals may even make water unsuitable for laundry purposes.

Water that is fit to drink is called **potable water.** Potable water is clear, colorless, pleasant-tasting, free of harmful bacteria, and fairly free of dissolved solids. A small amount of dissolved air gives the water a better taste.

Water contains matter in suspension and in solution. In addition to dissolved solids, flowing water holds material in suspension. *A* **suspension** *is a mixture formed by mixing a liquid with particles that are much larger than ions and molecules.* For example, chocolate powder mixed with milk is a suspension. Suspended matter will settle if the liquid is allowed to stand quietly. If you look at a glass container filled with river water, you can see suspended matter. The river may have carried this matter

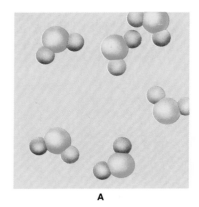

A

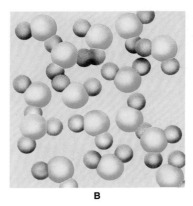

B

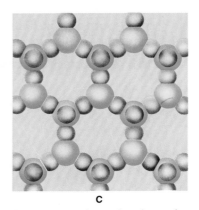

C

8-5 A comparison of water molecules in **(a)** water vapor (gas), **(b)** liquid, and **(c)** ice (solid). Describe what happens to the water molecules as water freezes.

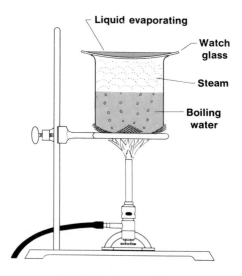

Liquid evaporating

Watch glass

Steam

Boiling water

8-6 After the liquid evaporates, what kind of materials remain on the watch glass?

for hundreds of kilometers. Some streams are very muddy because of the dirt and clay held in suspension.

A stream flowing through a limestone region will contain calcium compounds either in suspension or in solution. These materials affect the taste of the water. "Mineral water" is sometimes recommended as part of a healthy diet.

Inorganic, or nonliving, matter is most likely to consist of mineral compounds. The kinds of inorganic matter found in rivers and streams depend on the composition of soils and rock through which the water flows. Some of these materials either become dissolved in the water or are carried along by the river in suspension.

Animal and plant substances are classified as *organic* matter. Dissolved organic matter may be very harmful, depending on its nature and the amount contained in the water. What appear to be harmless bits of leaves or straw can be dangerous because bacteria grow rapidly on these materials.

activity

Solids in potable water. Place a small amount of tap water in a watch glass. Let the water evaporate by placing the glass over a beaker of boiling water as shown in Fig. 8-6. Look at the watch glass carefully after all the water has evaporated. What do you observe? Now repeat this experiment using a small sample of distilled water obtained from your teacher. Describe the differences, if any, in your observations.

Boiling and distillation purify water. Water that has been boiled for about 10 min is probably safe to drink. While boiling the water does not remove dissolved or suspended matter, it does kill harmful bacteria, making the water potable.

Chemists and pharmacists use distilled water in making solutions or suspensions. Distilled water is also used in car batteries and steam irons. **Distillation** *is the process of evaporating a liquid and then condensing the vapors in a separate container.*

Water is distilled by boiling it in a closed container until steam is formed. The steam is then led through a tube to another container that is cooled, usually by cold water. This cold container causes the steam to condense back into liquid water. Dissolved solids that are not changed into gases remain in the closed container. Water that is free of dissolved solids can be obtained by this process. Impurities that evaporate easily, however, are not removed. Can you explain why? Most often, the first sample of distilled water is thrown away. Can you give one reason why? (See Fig. 8-7.)

The distillation process purifies water. However, this method of making pure water is too costly when large amounts of pure water are needed.

8-7 (see below) The liquid in the flask vaporizes, leaving behind solid impurities. Why is the test tube placed in cold water?

activity

Distillation of water. Use the setup shown in Fig. 8-7 for distilling liquids in the lab. Dissolve about 5 g of powdered copper(II) sulfate in 100 mL of warm water. CAUTION: *Copper(II) sulfate is poisonous.* Add this solution to the flask. You will be starting with water that is not pure, as the blue color clearly shows.

Next, heat the solution to boiling and distill the liquid until a small amount of distilled water has been collected. Study the distilled sample and describe what you observe.

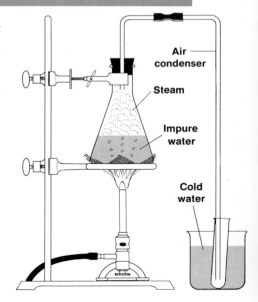

Air condenser

Steam

Impure water

Cold water

Drinking water comes from many sources. The source of your drinking water depends on where you live. In rural areas, water is pumped from wells, springs, or streams.

Wells should be dug as far as possible from places where organic materials might make the water unfit for human use. Many farm wells are a danger to health because they are poorly located or poorly built. Impure surface water may seep into the well.

Rivers and lakes are the major sources of drinking water in this country. They supply about 75% of the water used by cities and towns for drinking, and by farmers for irrigation of crops.

Most of the water used by New York City comes from the Catskill Mountains, about 160 km away. The water flows through large, underground pipes. Los Angeles obtains part of its water supply from the Sierra Nevada Mountains, 400 km away from the city. More water is

8-8 The settling basin is the first step in water purification. Why is this step necessary?

8-9 Diagram of a cross-section of a sand filter used in water purification. What is the purpose of the charcoal?

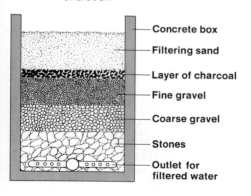

- Concrete box
- Filtering sand
- Layer of charcoal
- Fine gravel
- Coarse gravel
- Stones
- Outlet for filtered water

brought through long pipes from the more distant Colorado River. The needs of this rapidly growing area are so great that water is even brought from the rivers of northern California, more than 800 km away.

St. Louis and other large cities get their water from the Mississippi River. This supply must be filtered and treated with chemicals before it is used.

Cities along the Great Lakes have a supply of fresh water nearby. The water commonly enters an intake pipe about 15 m below the surface of the lake and about 5 km from the shore. The water is then treated to make it safe to drink.

Large cities use many millions of liters of potable water daily. In the United States, an average family uses over a million liters of water a year. It is believed that by the end of the decade, this country will need to double the present amount of available fresh water. To meet these needs, the United States government is testing a number of ways of getting fresh water from the ocean.

Cities and towns prepare water for drinking in many ways. Water obtained from a river or lake is seldom safe to drink. Such water, as you have seen, often contains organisms that may cause disease. Therefore, cities and towns treat water before it is used. To make water potable, sediment or suspended material is removed and harmful bacteria are killed. This process is usually carried out through the following five basic steps:

1. Water is taken from its source and allowed to stand in a large pool, or *settling basin*. The basin may cover an area of 5000 m^2, perhaps the size of a small pond. Here, most of the heavy suspended matter and debris settle to the bottom. See Fig. 8-8.

2. The water is then fed into a *second settling basin*. Since small particles suspended in water settle slowly, chemicals are used to speed up the settling process. Small amounts of aluminum sulfate [$Al_2(SO_4)_3$] and calcium hydroxide [$Ca(OH)_2$] are added to the water. These two chemicals react with each other to form sticky precipitates of aluminum hydroxide [$Al(OH)_3$] and calcium sulfate ($CaSO_4$). The double replacement reaction is

$$Al_2(SO_4)_3 + 3\ Ca(OH)_2 \rightarrow 2\ Al(OH)_3 \downarrow\ +\ 3\ CaSO_4 \downarrow$$

The fine particles that are coated with the precipitates become heavier and settle to the bottom. Bacteria also become caught in the precipitates. In this step, the second basin removes much more of the original suspended material and bacteria.

3. The water is next *filtered* through layers of sand and gravel in filters that may be larger than your classroom. These filters are composed of a layer of sand about 2 m thick that is placed above a layer of gravel. A layer of charcoal is often used between the sand and the gravel to take out coloring matter and foul-tasting substances. See Fig. 8-9.

4. The filtered water is often sprayed into the air, much like a large fountain. This process is called *aeration*. Oxygen of the air dissolves in the water and kills some types of bacteria. Aeration also improves the taste of water. See Fig. 8-10.

5. In the United States, chlorine gas is often used to make sure that all harmful bacteria are killed. The amount of chlorine used depends on the condition of the water and the season of the year. In the fall, more chlorine is used because there is more organic matter in the water. Can you explain why?

A diagram summarizing the basic steps usually undertaken in water purification in towns and cities is shown in Fig. 8-11.

Water supplies in many cities are lacking in the fluoride ion (F$^-$). Fluorine is a trace element found in the diet. To make up for this shortage, fluoride salts are often added to the water supply. The fluoride ion is

8-10 In the aeration process, oxygen of the air dissolves in the water. Explain why this step is desirable in making water potable.

8-11 A diagram of the basic steps in the purification of municipal drinking water. What important step is missing from the diagram below?

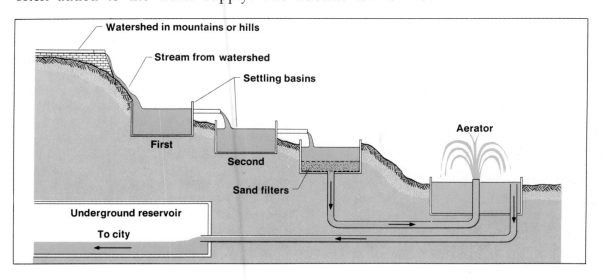

Watershed in mountains or hills
Stream from watershed
Settling basins
First
Second
Sand filters
Aerator
Underground reservoir
To city

8-12 A water chemist is shown testing drinking water in a laboratory.

known to have special value in protecting children's teeth against decay. Fluoride strengthens the enamel coating of the teeth, thus increasing the resistance to decay.

Many scientists believe that adding one part of a fluoride compound to one million parts of water is safe and helps reduce tooth decay up to 60%. Studies from many sources have shown that adding fluorides to the water supply has been very effective in reducing tooth decay. Fluoride compounds are also added to many toothpastes. (See Fig. 8-12.)

activity

Removal of coloring matter from water. Dissolve about 10 g of brown sugar in a 100-mL beaker of warm water. Filter a portion of this solution and observe the color of the liquid that passes through the filter paper. Then add about 10 g of boneblack (animal charcoal) to the remaining solution in the beaker. Stir and filter. Compare the color of the two liquids that passed through the filter paper. Explain what you observe.

8-13 This water plant in New Mexico makes millions of liters of potable water a day from brackish (salty) water by the distillation process. What is an important disadvantage of this method?

Potable water can be prepared by desalting seawater. The oceans have resources other than minerals. Perhaps the most abundant of these resources is potable water made from salt water. The making of drinking water from seawater is a very old process that dates back to at least 350 B.C. Today, several desalting plants using different systems are in operation. Among the methods used are distillation, ion exchange, reverse osmosis, and freeze separation.

You know that small amounts of water can be purified in the laboratory by *distillation*. On a large scale, however, this process can be very costly and is used only in special situations. The high cost of distillation is due to the large amounts of energy needed to heat the water. See Fig. 8-13.

Ion exchange is another method of making potable water. In this method, an ion of one kind of atom is exchanged with an ion of another kind of atom with the same charge. Ion exchange is also used to soften hard water. You will learn more about hard water in the next section.

132   unit 2 chemistry in our world

Osmosis is a process that occurs in nature. In osmosis, water passes through a thin membrane, or skin, from the dilute side (in this case, the fresh water) to the more concentrated solution (the salt water). *Reverse osmosis* is a simple way of reversing nature's normal process. Salt water is placed under pressure and water molecules are forced to pass through a thin membrane. Since only water is able to pass through the membrane, the salt (sodium and chloride ions) is left behind. This process also removes bacteria from the salt water. Reverse osmosis uses a small amount of energy and shows great promise for making large amounts of potable water from seawater. See Fig. 8-14.

Freeze separation is another method by which fresh water can be obtained from ocean water. When water freezes, it forms ice crystals that are almost free of salt. The ice is then melted to obtain fresh water. It requires less energy to freeze seawater than to distill it.

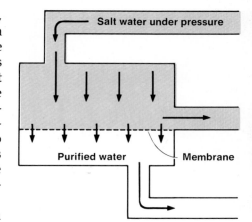

8-14 In reverse osmosis, pressure is applied to salt water to overcome natural osmotic pressure and force fresh water through a thin cellulose membrane. Can you give one advantage of this method?

hard and soft water

Water is either hard or soft. Making water potable is only one problem. Many communities face still another major task. If you live in a section of the country where the water is hard, then you are well aware of the trouble it causes. **Hard water** *is water that precipitates, or curdles, soap.*

The word "hard" means "it is hard to make suds with." Clothes cannot be properly cleaned with ordinary soap in hard water. Furthermore, the curd that forms leaves a greasy film on the fabric and gives it a dingy look. Distilled water is free of hardness and is called *soft water.*

When water falls as rain, it dissolves a small amount of the carbon dioxide in the air. When it rains, the water soaks through the ground, dissolving more carbon dioxide from dead roots and other plant matter. If this carbon dioxide solution flows over limestone ($CaCO_3$), a reaction takes place forming calcium bicarbonate [$Ca(HCO_3)_2$]. The calcium bicarbonate dissolves in the water, forming calcium ions (Ca^{+2}) and bicarbonate ions (HCO_3^-). The chemical equation for this reaction is

$$CO_2 + H_2O + CaCO_3 \rightarrow Ca(HCO_3)_2 \rightarrow Ca^{+2} + 2\ HCO_3^-$$

8-15 Equal amounts of soap powder were added to hard water (left) and to soft water (right). What is the cause of hardness in water?

Calcium ions in solution are the most common cause of hard water. Soluble magnesium and iron salts also produce ions that cause hardness in water.

When soap is added to hard water, a curd forms. See Fig. 8-15. This sticky material forms as a product of the reaction of the calcium ions in the water with the soap. As a result, an insoluble calcium compound is formed. You may have seen this curd as the greasy ring that forms around the washbowl or bathtub.

To get a lasting lather with soap and hard water, all of the curd-forming ions must first be precipitated out of solution. More soap is required. Obviously, this method of taking out the unwanted ions is wasteful, and better means of softening water have been developed.

Some water can be softened by boiling. Hard water containing calcium bicarbonate can be softened by boiling. Boiling causes the minerals to settle out when the calcium bicarbonate decomposes. This breakdown is shown by the following equation:

$$Ca(HCO_3)_2 \overset{heat}{\rightarrow} H_2O + CO_2 \uparrow + CaCO_3 \downarrow$$

The carbon dioxide is given off into the air and the calcium carbonate precipitate settles out. The action of the soap is not affected by the calcium carbonate because calcium carbonate is insoluble and does not break down into calcium ions. However, softening large amounts of water containing calcium bicarbonate by boiling is costly because it uses too much energy.

Most soft water is prepared chemically. The calcium, magnesium, or iron ions that cause hard water can be removed by adding soda ash, slaked lime, or borax. A typical reaction for the removal of the calcium ions using slaked lime $[Ca(OH)_2]$ is

$$Ca(HCO_3)_2 + Ca(OH)_2 \rightarrow 2 H_2O + 2 CaCO_3 \downarrow$$

The calcium ions in solution are precipitated as insoluble calcium carbonate and settle to the bottom of the container. Many cities and industries use slaked lime to soften the water at the same time it is made potable. This method is an effective way of softening large amounts of hard water. Soda ash and borax work just as well for softening water. They are convenient when small amounts of soft water are needed for home use.

Certain sodium compounds exchange their sodium ions for calcium ions. In this ion-exchange process, hard water containing calcium ions flows through a tank containing particles of a compound known as sodium zeolite. The calcium ions in the hard water replace the sodium ions in the zeolite, as described by the following reaction:

$$Ca^{+2} + 2 \text{ Na zeolite} \leftrightarrows Ca \text{ (zeolite)}_2 + 2 \text{ Na}^+$$

The calcium ions form insoluble calcium zeolite, which remains in the water-softening tank. The sodium ions that are set free have no effect on the hardness of water.

When all of the sodium zeolite in the tank is used up, a concentrated solution of sodium chloride is added to the tank. The chemical action is then reversed, changing the calcium zeolite back into sodium zeolite. The zeolite can be used again and again, any time salt is added. The tank is then ready to be used again to soften the water. The zeolite-type softener is used in both homes and industries.

8-16 A cross-section of a hot-water pipe showing the hard-water scale of magnesium and calcium salts. What will eventually happen to the pipe?

Soft water has many uses. Hard water may be no more than a nuisance at home, but it is a major problem in many industries. It is most troublesome in steam boilers and hot water pipes. Minerals deposited by the hard water clog the pipes. See Fig. 8-16.

These deposits insulate the boiler from the source of heat, thus reducing its efficiency. For this reason, water-softening methods are vital to industry. Steam-electric power plants must use carefully treated water in their boilers.

Paper mills need a supply of water that is free of iron compounds, which may stain the paper. Other compounds may keep the paper from having a smooth, glossy surface. It is easy to see why the removal of these compounds is a major concern of the paper industry. Soft water is also needed in textile mills to produce exact shades of color evenly spread through the yarn and cloth.

In homes, synthetic detergents have almost totally replaced soap for laundry use. These detergents work well in both hard and soft water. The major advantage of these cleaning agents is that they do not form a scum with the calcium ions in the hard water. Detergents will be discussed further in a later chapter.

sewage: treatment and disposal

Waste water must be treated. What happens when a large volume of rain or snow falls on an urban area? The water is drained from roofs, yards, and streets into *storm sewer* lines. These lines normally carry the water to nearby rivers or streams, or into other large bodies of water. Sewer water carries trash, organic matter, and chemicals. Pieces of rock, soil, metal, oil and grease, fertilizers, pesticides, and microorganisms may also be carried by the water.

To dispose of human waste and other organic and inorganic matter as well as detergents, another system of pipes, called *sanitary sewer* lines, is used. These organic wastes are treated in sewage disposal plants before the waste water is discharged into large bodies of water. In some cities, however, the sanitary and storm sewer lines are combined. This combination of lines causes many problems. When there is a rainfall, or when snow melts, the amount of water in the single sewer system increases greatly. This waste water can easily overload the sewage treatment plant. When such an overload occurs, the excess waste bypasses the treatment plant and takes the raw sewage with it. Thus, the river or lake that receives the overflow becomes polluted.

The waste water includes a rich source of phosphate ions (PO_4^{-3}) from detergents and fertilizers, and nitrate ions (NO_3^-) from organic matter. These nutrient ions, when dumped into bodies of water, speed the growth of algae and other water plants. When these plants die and decompose, the dissolved oxygen in the water is used up. Fish and other animal life may die from the lack of oxygen. Of course, plant growth and decay take place normally all the time, but at a much slower rate. Increased amounts of nutrient ions from untreated sewage can speed up the "death" of many lakes and streams. See Fig. 8-17. This procedure is called *eutrophication*.

Also, since untreated sewage contains bacteria and other material harmful to health and safety, it must be treated before it is discharged into rivers and lakes. The goal in sewage treatment is to make the waste water as much like potable water as possible.

8-17 University of Minnesota researchers are shown testing samples of water from a peat bog that is threatened by mining activities. Nutrient ions added to the water ecosystem could cause *eutrophication*. Explain what this means.

136   unit 2   chemistry in our world

In earlier years, most sewage was dumped into the waterways without treatment. As long as the amount of sewage was fairly low, the natural action of moving water purified the water. As the volume of sewage increased, however, nature was no longer able to do the job effectively. In recent years, people have insisted on full sewage treatment before discharge into the waterways.

There are three basic steps in sewage treatment. These steps are called *primary*, *secondary*, and *tertiary*.

Primary treatment of sewage is the first step. The primary treatment of waste water is little more than a screening and settling process. The waste is first fed through a *grit chamber*, where huge grates screen out the large, bulky matter. Then the liquid goes to *settling tanks*. There suspended material is allowed to settle. The insoluble material on the bottom of the tanks is then pumped into other large tanks, called *digesters*. There the bacteria in the sludge breaks down the organic matter in the waste water. This solid material is then dried and used as fertilizer. After primary treatment, the liquid may either be dumped into the waterways or passed on for secondary treatment. Chlorine may be added before discharge in order to kill most of the disease-causing bacteria and to reduce odors. Figure 8-18 shows a simplified flow chart of this filtering process.

Secondary treatment is the next step. The primary treatment of sewage by itself is not adequate. In secondary treatment, up to 90% of the organic material is removed from the sewage after it comes from the settling tanks.

One method used to remove the organic material is to allow the waste water to pass through gravel filters to an *aeration tank*, where it is mixed with oxygen of the air. This aeration allows bacteria to break down the

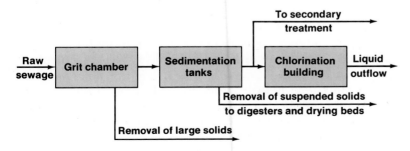

8-18 A flow chart showing solid waste removal in primary sewage treatment. What is the value of the chlorine treatment?

organic material further. After passing through another settling tank where more solids and suspended materials are removed, the liquid is then treated with chlorine to complete the secondary treatment. The liquid is then either discharged into the waterway, or passed to the third stage of treatment.

The third step is called tertiary treatment. Tertiary is the last stage of sewage treatment and the most advanced. This process is complex and costly. In this step, the phosphate and nitrate ions are finally taken out of the water. Reverse osmosis, ion exchange, and distillation are among the methods used in tertiary treatment.

In the ion-exchange method, the phosphate and nitrate ions are replaced with chloride (Cl^-) and carbonate (CO_3^{-2}) ions, which will not cause an increase in plant growth. When water is finally discharged from tertiary treatment, it is almost completely free of pollutants. The water is very close to drinking-water quality. In fact, scientists now know how to turn raw sewage into potable water. However, the treatment is very costly! See Fig. 8-19.

Sewage treatment produces useful by-products. Properly treated, the dried sludge from the sewage that is taken from the secondary and tertiary steps contains many nutrients that can be sold as fertilizers. Much of the solid, rocky material can be used in road building, for making building blocks, or for land fill.

8-19 A sewage treatment plant where tertiary treatment removes almost all pollutants. Describe what is done to the sewage in this stage of treatment.

Careers in health services

In recent years, the demand for health care services in this country has continued to grow at a rapid rate. The need for such services creates a large number of jobs for interested students. Over 4 million people are employed in the health care field, and the number grows yearly.

Many of these people work in hospitals, clinics, nursing homes, offices, mental health centers, and other health care centers. Many different health care people work together to provide proper services to patients.

unit 2 chemistry in our world

The table below contains a brief description of the duties of only a few types of careers.

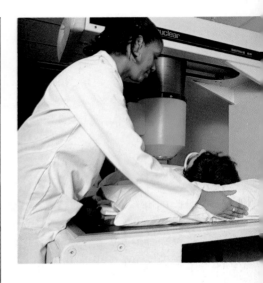

Job	Duties
Chiropractor	Treats illness relating to abnormal nerve function by adjusting parts of the body.
Dental assistant	Aids the dentist in treatment of patients.
Dental hygienist	Cleans and polishes teeth and advises patients on proper dental care.
Dentist	Diagnoses and treats disorders of teeth and gums. Prescribes and fits dentures.
Dietitian	Plans and supervises the preparation of food to assure balanced meals for schools, hospitals, and public and commercial food services.
Medical artist	Makes drawings, sketches, diagrams, and charts of medical information, surgical operations, and research data.
Medical librarian	Aids in selecting, processing, and issuing medical books, magazines, films, photographs, and other forms of recorded knowledge.
Medical office assistant	Assists physicians, dentists, and veterinarians in clerical and typing duties. May act as a receptionist and prepare patients for examination or treatment.
Medical social worker	Helps patients with social and emotional problems to better adjust to their illnesses.
Nurse's aide	Bathes and feeds patients, makes beds, and assists patients with their bodily needs.
Optometrist	Examines eyes to determine vision problems and disease. Prescribes corrective lenses for eye defects.
Osteopath	Diagnoses and treats human diseases by using physical therapy and adjustment of body parts.
Pharmacist	Prepares and gives out drugs and medications as prescribed by physicians, dentists, and veterinarians.
Physical therapist	Provides treatment, exercise, and massage programs for patients.

(continued)

Physician	Diagnoses and treats human diseases and injuries. Prescribes medications and performs surgery.
Podiatrist	Cares for human feet. Treats foot diseases, injuries, and other foot disorders.
Practical nurse	Works under the direction of physicians and registered nurses.
Registered nurse	Plans, assists, directs, and supervises the care of patients.

summary

1. A hydrogen bond is a weak attraction between the negative oxygen atom of a water molecule and the positive hydrogen atom of other water molecules.
2. Potable water is water that is fit to drink.
3. Distillation is a process of evaporation of a liquid followed by condensation of vapors in a separate container.
4. Community water supplies are made potable by settling, filtration, aeration, and by the use of chlorine and other chemicals.
5. Hard water is water containing ions such as calcium and magnesium, which form precipitates with soap.
6. Hard water containing calcium bicarbonate can be softened by boiling, or by adding soda ash, slaked lime, or borax.
7. Among the methods used in desalting seawater are distillation, ion exchange, reverse osmosis, and freeze separation.
8. The three basic stages of sewage treatment are (1) screening and settling, (2) removal of organic material by filtration and oxidation, and (3) removal of nutrient ions.

review

Match each item in the left column with the best response in the right column. *Do not write in this book.*

1. water
2. borax

a. home water softener
b. cause of hard water

3. hydrogen bond
4. PO_4^{-3}
5. suspension
6. $Al(OH)_3$
7. Ca^{+2}
8. potable

c. forms a sticky precipitate
d. fine solid material mixed with a liquid
e. helps explain some properties of water
f. boiler scale
g. polar molecule
h. nutrient ion in sewage
i. chlorination of drinking water
j. water that is safe to drink

questions

Group A
1. What is usually meant by "pure" water? Why is rainwater not entirely pure?
2. What makes the oceans salty?
3. What is the difference between organic and inorganic matter? Give examples of each.
4. What is acid rain? Why is it harmful?
5. Why is organic matter dissolved in water harmful to health?
6. How is water distilled?
7. List five steps commonly used for making city water potable. Describe the purpose of each step.
8. What chemicals are used to settle suspended particles in water? Describe their action.
9. Describe a sand filter. How is it used to purify water?
10. What is meant by hard water? Describe the difference between hard water and potable water.
11. What causes a deposit to form in steam boilers and hot-water pipes? Why are these deposits harmful?
12. Why is soft water needed in paper and textile plants?
13. What are some methods used to remove salt from ocean water?
14. Describe reverse osmosis as a method used to make water potable.
15. What does sewage consist of? Why should it be treated before being dumped into our waterways?
16. Distinguish between a storm sewer and a sanitary sewer system.
17. What nutrient ions that cause plant growth are contained in sewage? Where do they come from? Why should these ions be removed?
18. What does primary treatment of sewage usually consist of? Is this treatment sufficient for most purposes? Explain.

Group B
19. Explain why water is such a good solvent.
20. Suggest a reason why water is a liquid at room temperature.

21. What is the difference between a solution and a suspension? Which could you separate by filtering? Explain.
22. How would you set up an experiment to find the percentage of solid material in a sample of river water?
23. How do calcium ions get into the water supply? Do these calcium ions make water harmful for drinking?
24. How can water that contains calcium bicarbonate be softened by boiling? Write the equation for this reaction.
25. Explain how slaked lime softens water containing calcium bicarbonate. Write the equation for this reaction. Why is it necessary to remove the calcium ions from solution?
26. How does sodium zeolite soften water by the ion exchange method? How is a zeolite unit made reusable?
27. Why might distilled water be safer to drink than water obtained by ion exchange?
28. Describe how potable water may be obtained by freezing salt water.
29. Describe in order the three steps of sewage treatment.
30. What use can be made of the dried sewage sludge from secondary and tertiary treatment? What use can be made of the rocky, insoluble material?
31. Why is sewage treated with chlorine before it is discharged into rivers and streams?

further reading

Allen, Herbert E., and Kramer, James R., eds., *Nutrients in Natural Waters*. New York: Wiley, 1972.

Banik, Allan E., and Wade, Carlson, *Your Water and Your Health*. New Canaan, Conn.: Keats Publishing, 1974.

Besselievre, Edmond, and Schwartz, Max, *The Treatment of Industrial Waste*. New York: McGraw-Hill, 1976.

Chorley, Richard J., ed., *Water, Earth, and Man*. New York: Methuen, Inc., 1979.

Gilford, Henry, *Water: A Scarce Resource*. New York: Franklin Watts, 1978.

Keough, Carol, *Water Fit to Drink*. Emmaus, Penn.: Rodale Press, 1980.

Lehr, J. H., et al., *Domestic Water Conditioning*. New York: McGraw-Hill, 1979.

Smith, Norman, *Man and Water*. New York: Charles Scribner's Sons, 1975.

Steel, E. W., et al., *Water Supply and Sewage*. New York: McGraw-Hill, 1979.

Stevens, Leonard, *Clean Water*. New York: E. P. Dutton, 1974.

unit 2   chemistry in our world

environmental pollution

9

When astronauts go out into space, they are very careful not to pollute the air and water in their spaceship. Since they have only what they take with them, the air and water must be used carefully. Once used, the air and water are cleaned and used again.

In a sense, the entire planet earth is like a spaceship. Instead of just a few astronauts, however, there are about four billion "passengers" on earth. Just as in a spaceship, the earth's air and water go through a cleaning process so that they can be used again by human beings.

Nature's cleaning process worked well as long as people were scattered over a wide area of the globe. However, when people began to crowd together in cities, impurities were concentrated, too. As a result, nature's system of cleaning air and water was placed in danger. In recent years, people have become aware of that danger and have begun to control **pollutants.** *A pollutant is an impurity in water, air, or land caused by the activities of people.* The fight against pollution includes knowing its causes and how to control them.

objectives:

☐ To describe the causes and effects of water, air, and land pollution on the environment.

☐ To identify pollutants and how to control their effects.

☐ To explain the role of weather in scattering air pollutants.

9-1 This unsightly mess was caused by solid waste materials. What are other causes of water pollution?

☐ To describe the various methods used in solid waste disposal.

☐ To investigate radiation as a potential hazard and to study the disposal of radioactive waste.

water pollution

Water is polluted in many ways. Clean water is needed for good health, but it is becoming more and more difficult to find in many parts of the country. Many cities are finding it hard to get the pure water they need for daily use. In addition, large amounts of water are also used on farms to grow our food and by industries to produce our clothing and shelter. Pollution of water sources and the increasing demand for clean water are making current water shortages even more of a problem.

You have seen that phosphate and nitrate ions from untreated sewage shorten the life spans of lakes and streams. These ions speed the growth of water plants. When the plants die and decay, they use up much of the dissolved oxygen in the water. As a result, fish and other forms of animal life die from the lack of oxygen. A healthy body of water, therefore, must contain a good supply of dissolved oxygen.

A clean water supply involves much more than cleaning up sewage. The nation must also cope with the problems caused by industrial and farm wastes. These wastes include pesticides, chemicals, oils, greases, detergents, and organic matter. Some of the pollutants enter rivers and streams as surface runoff from land areas. Other materials are discharged directly by factories and industrial plants. Small amounts of some pollutants appear to have no harmful effects on human health. However, health officials are concerned by the effects that these materials may have on people over a period of years. See Fig. 9-1.

Oil pollutes water. Few pollutants are as visible as oil covering a white beach along a shoreline. The effect of oil on a beach is shocking in terms of dead birds and fish, fouled beach, and the likelihood of a resort area suffering great financial loss.

Most of this oil comes from the normal operation of ships. Seawater is pumped into empty fuel tanks to make ships more stable. However, the tanks are not completely empty. They hold "left-over" amounts of fuel oil. Then, when the ship is ready for refueling, the oily sea water is pumped out again, covering large patches of water with an oily film. Also, large amounts of crude oil are spilled, flushed, or leaked from ships. Laws against this practice exist, but they are hard to enforce on the high seas. Although the sinking of an oil tanker or an underwater oil leak may cause great damage, most oil spills come from the day-to-day operation of many thousands of ships.

The oily film that covers the water reduces the water's ability to absorb oxygen from the air. This, in turn, affects the health of fish and other sea life. Even if they are not killed by the lack of oxygen, fish caught in these oily waters cannot be eaten. Water birds, their feathers matted with oil, are unable to swim or fly, and soon die. See Fig. 9-2.

9-2 Fish and water birds can be killed as a result of water pollution. Here, volunteers are trying to clean the feathers of a bird trapped by an oil spill. Why do fish and birds die if they are covered with oil?

Mercury compounds can pollute water. Mercury and mercury compounds are used in some manufacturing processes. They are used to make chlorine and lye, plastics, and paper. Mercury compounds are also used by farmers as a coating for seeds and as a spray to kill fungus.

Mercury and mercury compounds find their way into rivers and streams through runoff and waste discharge. Because mercury is a dense metal, it sinks to the bottom of a stream. Bacteria in the water act on the mercury, changing it into a compound that is soluble and highly poisonous. The bacteria are then eaten by tiny forms of water life, which in turn are eaten by small fish. In this way, mercury compounds move up through the *food chain* (one animal feeding on another) into higher animal forms and finally into humans. Low levels of mercury compounds have been found in some seafoods and occasionally in human beings.

Waste heat may cause thermal pollution. The discharge of hot water from fossil fuel and nuclear power plants may cause *thermal pollution* in some lakes and rivers. Power plants use large amounts of heat to generate electricity. To get rid of the excess heat, water from a stream or lake is used to cool the main steam condensers of the power plants. This outside water is heated in the process and then returned to the stream or lake.

In the discharge areas, the added heat will drive some of the oxygen out of the water. This will alter the life functions of certain animals and change the ecology of the immediate area. Warming the water does not always create harmful effects, but it does change the environment. Cooling towers, air cooling, and spray ponds are often used to cool the heated water before it is released back into the main stream. See Fig. 9-3.

The heated water from fossil fuel and nuclear power plants could be used as a source of energy. Warmer pond or lake water can also cause fish to grow faster.

Controlling water pollution is costly. Because of public concern and demand, many industries are trying to stop or reduce water pollution. This process is expensive and may affect jobs and raise the cost of consumer goods. Many cities and towns reduce water pollution by building better sewage treatment plants. These new plants help keep the lakes and rivers clean.

Manufacturers have found ways to reduce the use of phosphates in detergents. Many other items, including

9-3 Cool water from the river condenses the exhaust steam in loop B. The heated water is returned to the river. Can you describe what happens in each loop?

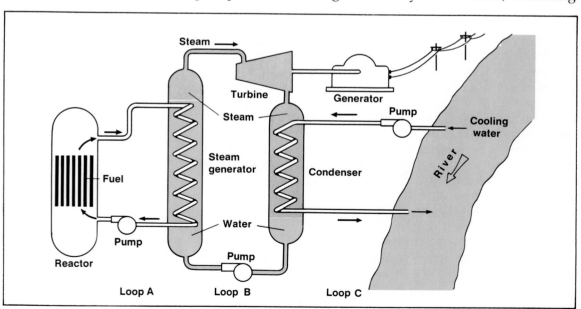

unit 2 chemistry in our world

containers, are now made of *biodegradable* materials. Biodegradable materials can be broken down into simpler compounds by bacteria or other natural processes.

When water drains from open mine areas, it is often highly acidic. Plants and animals cannot survive in acidic water. Many mining companies are now restoring open mines to safe and useful lands. See Fig. 9-4.

Up to a certain point, the water of flowing rivers and streams cleans itself by exposure to air and sunlight. The movement of water dilutes and scatters the pollutants. Oxygen from the air and from water plants dissolves in the water. The oxygen breaks down the organic matter into harmless compounds. The action of bacteria also plays an important part in cleaning the water.

However, when a stream is choked with sewage and other impurities, there is not enough oxygen to clean the water. Because of the lack of oxygen, the sewage produces harmful gases and other compounds. This water is not drinkable; the stream cannot even support its fish life.

9-4 Cattle now graze where an open aluminum-ore mine once stood. Who pays the cost of reclaiming strip-mined lands?

activity

Water pollutants. Obtain several samples of water from nearby streams or ponds. Examine the water samples under a microscope. Why is this water unsafe to drink? Next, evaporate several milliliters of each sample and observe the film that remains. Check with your local health department to find out what steps are taken to prevent pollution of the local water supply.

air pollution

Air is hard to keep clean. Water and air are perhaps our most vital natural resources. People need clean air and water to live.

The struggle for clean air is not new. For thousands of years, people have complained about smoke, soot, and odors. The Romans objected to the smell of burning coal and the deposit of soot it left on their clothes. During the Middle Ages, and even during the early days of this country, people were faced with these same problems.

In the past several hundred years, there has been a great increase in the amount of fuel burned in homes, factories, and cars. This increased burning of fuel has caused an increase in the amount of waste matter given off into the air. As a result, the air has become more polluted.

Pollutants spread out from the cities into the countryside, the mountains, and the beaches. Even areas where few people live have air that is not clean. However, impurities are at much higher levels near large cities and over industrial areas. Clouds of solids, liquids, and gases are constantly being added to the air.

In order to stop polluting the air, people must stop or greatly reduce those activities that cause pollution. Can you live without cars, heat, electricity, or food? Perhaps you cannot, but maybe you can change or reduce some of your needs.

Auto emissions pollute the air. The exhaust from cars and trucks is one of the major sources of air pollution. Motor vehicles account for an estimated 60% of the air pollutants in a large city. The waste products of motor vehicles come from the incomplete burning of fuel. These exhaust gases include carbon monoxide, oxides of nitrogen, and unburned hydrocarbons (compounds containing only hydrogen and carbon). See Fig. 9-5.

The word *smog* was once used to describe a mixture of smoke and fog. The word was later applied to pollu-

9-5 Major pollutants of the air in the United States. What are the sources of these pollutants?

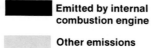

Emitted by internal combustion engine

Other emissions

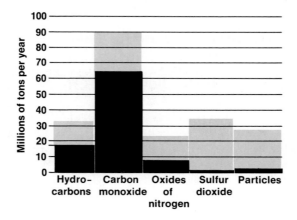

unit 2 chemistry in our world

tants in the Los Angeles area, even though neither smoke nor fog was the real problem there. Scientists now refer to this type of pollution as *photochemical smog*.

Photochemical smog is formed when unburned hydrocarbons react with oxides of nitrogen in the presence of *ozone* (a form of oxygen, O_3) and sunlight. Since the chief source of these pollutants is the exhaust from motor vehicles, modern cars are now equipped with devices that greatly decrease these harmful gases. One such device, called a *catalytic converter*, changes the hydrocarbons and carbon monoxide to harmless water vapor and carbon dioxide before the exhaust gases leave the vehicle. It also changes the oxides of nitrogen back into oxygen and nitrogen gases.

Another pollutant from auto exhaust comes from the use of a compound called *tetraethyl lead*, which was often added to gasoline. This lead in the gasoline reduced "knocking" in engines and increased gas mileage. Unfortunately, tetraethyl lead, even in small amounts, is also an air pollutant. New car engines are designed to use lead-free gasoline, but many older cars still use leaded gasoline. Car makers are now trying to build a pollution-free engine that will also provide better mileage. See Fig. 9-6.

Pollution comes from factories, power plants, and homes. One major source of air pollution is the burning of fuels. This burning gives off pollutants that you can see (smoke), and pollutants that you cannot see (harmful gases). The smoke comes from the partial burning of fossil fuels, especially soft coal. Smoke contains free carbon particles that can darken the air and cover objects with soot.

The harmful gases include sulfur dioxide, carbon monoxide, and other gases. Sulfur dioxide comes mostly from burning soft coal and to a lesser extent from some fuel oils. Sulfur dioxide harms the nose and throat, and can cause diseases of the lungs. Sulfur dioxide reacts with oxygen in the air to form sulfur trioxide, which in turn reacts with moisture in the air to form weak levels of sulfuric acid. (Write the chemical equations for these reactions.)

Some industries give off hydrogen sulfide (H_2S) gas, which has an odor like rotten eggs. Ammonia, chlorine, and other chemicals with bad odors that are irritating

9-6 Heavy traffic in large cities adds to the air pollution problem. In what way can this type of pollution be controlled?

or poisonous are also produced. Liquids used in fermenting processes and from the decay of plant and animal matter often have foul odors. Yet people may become so used to some odors that they hardly notice them.

Major industries that release dust into the air include grain and feed, lumber, cement, mineral and rock, as well as metalworking and mining. Certain urban areas have a high level of asbestos particles. These areas are mostly near construction sites. Asbestos particles are long, narrow fibers of a mineral that can easily be broken into tiny bits of dust. These particles have been known to cause lung cancer. They are caused by grinding, cutting, sanding, mixing, crushing, milling, and blasting. Whenever a solid is broken down into finer particles, dust is released into the air. If the particles are small enough, they may drift in the air for a long time.

Sodium and calcium fluorides are also factory wastes. These fluorides come mostly from the making of steel and ceramic products, and to a lesser extent from the making of aluminum and superphosphate fertilizer. When fluoride dust settles on food eaten by livestock, it may injure them or make them ill.

Other industrial pollutants include organic refuse (from dairies, canneries, and meat-packing plants), arsenic compounds (produced by tanneries), and sulfur compounds and wood fibers (from pulp mills).

activity

Solid particles in the air. Place a piece of filter paper over the end of the hose of a tank-type vacuum cleaner. Turn on the switch and allow the motor to run for several minutes while holding the hose in midair. Stop the motor and examine the filter with a lens under a bright light. Compare the filter with an unused piece of filter paper. Describe any differences you see. Repeat this test indoors and outdoors several times at different hours and on different days.

Noise can also be harmful. People who live near airports, factories, and busy traffic areas are exposed to loud noises. City noise, and even loud music, can cause damage to the ear after a long exposure. Loud noises may cause adrenalin to be released into the bloodstream, thereby increasing blood pressure and putting an added strain on the heart. These noises can even disturb bodily functions. Millions of people are partially or totally deaf because of noise.

The sound level in front of a rock band has been measured at 120 **decibels** (db). *A decibel is a unit of sound loudness (energy).* A noise level of 100 db can cause temporary deafness, and one of 140 db can kill small animals. Compare these values with the level of very soft music (about 30 db) and normal talking (about 65 db). You can see that noise is just as much of a pollutant as dust and smoke because it is harmful to people. (See Table 22-2, in Chapter 22, for a list of some common noise levels in decibels, and Table 22-3 for some limits to noise exposure.)

Air pollution is affected by weather. Wind speed, wind direction, and air turbulence are factors that either spread or concentrate pollutants in the air.

If there is no wind, pollutants can concentrate in one area. However, a strong breeze will blow pollutants away. The amount of air pollution around you depends on whether the wind is blowing the pollutants away from or toward you.

The wind also tends to mix and scatter pollutants. Winds that blow over level ground usually move smoothly. However, winds that blow over tall buildings and over hilly ground tend to swirl around. See Fig. 9-7.

Another kind of mixing of the air is caused by differences in temperature between layers of the air. On a clear day, the sun heats the ground and warms the layer of air near the surface. The warm air is less dense than the cold air and so rises while cleaner, cooler air comes from above. This mixing action, called an *unstable* conditon, helps to scatter a polluted air mass.

However, on a clear night, the earth's surface cools quickly by radiating its heat into space. This radiating process cools the air near the ground while the air above this layer remains warm. This air pattern tends to be very *stable*. The cool air stays near the ground while the warm air stays above it. As a result, pollutants are not scattered and concentrate in the air near the ground.

9-7 Mechanical mixing of the air caused by land features (terrain) and high buildings. How does this mixing compare with wind over level land areas?

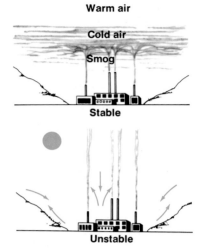

Warm air

Cold air

Smog

Stable

Unstable

9-8 Where the atmosphere remains stable, as in the diagram on the top, a "pollution trap" is likely to develop. Where unstable conditions exist, in the diagram on the bottom, a cleaning of the polluted area can be expected. What is a stable condition of the atmosphere called?

9-9 A layer of early-morning air hangs over Santiago, Chile. What is this condition called?

Such a condition, called a *temperature inversion*, is fairly common. See Fig. 9-8.

How long will a temperature inversion last? That depends on how quickly the earth cools during the night and warms up during the morning. Clouds, land forms such as mountains or valleys, and seasons of the year all affect the inversion. Temperature inversions tend to occur more often and last longer during the fall and winter months than during the spring and summer months. An inversion over a large city may remain for three or four hours after sunrise, and collect the pollutants from the morning rush-hour traffic. The results can be dirty air and smog, causing a health hazard. See Fig. 9-9.

When air is unstable, a mass of cool air forces warm air upward. These updrafts sweep away the dirty air and replace it with cleaner air at ground level.

Rain and snow can also help remove pollutants from the air. Can you explain why this is so?

Acid rain may cause problems. Rain is not pure water. As rain falls to earth, it picks up all kinds of gases and particles. The particles include dust from roads, farms, and surface mining operations, as well as from volcanoes, plants, and animals. In addition, rain picks up carbon dioxide from the air and becomes a very weak form of carbonic acid.

Pure water has a pH of 7, which is neutral. However, impurities almost always cause the value to be below 7, which is on the acidic side of the pH scale. **Acid rain** *is defined as rain with a pH of less than 5.60.* In extreme cases, rain has reached a pH as low as 3.0.

It is widely believed that acid rain harms some forms of plant and animal life. Some lakes, for instance, in the Adirondack Mountains of New York State, are so acidic that fish life is endangered. Acid rain has even been suspected of damaging the paint on cars and buildings.

Many people believe that some of the acid in rain comes from the burning of coal. However, scientific evidence is not clear on this point, because the acid level of rain often changes sharply from one rainstorm to the next, and from one place to another, even in the same storm. Because of these sharp changes, it is difficult to pinpoint causes. Perhaps the evidence will

unit 2   chemistry in our world

become clearer as further studies are made. Be alert to news stories concerning this problem.

Air pollution affects health. Of course, we cannot stop all industrial production in order to stop pollution. If we did, we would not have the food, clothing, and shelter to meet our needs. We must strive to maintain a balance between the risks of pollution and the benefits we obtain from industry.

When you inhale, your lungs take in some of the oxygen in the air. Your body needs oxygen to work properly. If you breathe too much polluted air for too many years, your lungs become less able to take in oxygen. Your chances of getting certain diseases, including cancer, will increase. See Fig. 9-10.

Even tobacco smoke adds to air pollution in enclosed areas. Carbon monoxide levels in smoke-filled rooms can reach levels much higher than those found along busy highways and traffic tunnels. Heavily polluted air will also harm or kill plants and animals as shown in Fig. 9-11.

Doctors have found that polluted air causes watery eyes, stuffy noses, coughs, sneezes, headaches, and sore throats. Carbon monoxide presents a special danger to people with lung and heart problems. The list of ailments that may be caused by polluted air includes asthma, emphysema, tuberculosis, bronchitis, and certain types of lung cancer.

Nitrogen compounds are plentiful in the air and they play a key role in the life processes of all living things. Most nitrogen compounds come from the decay of organic matter caused by bacteria. However, nitric oxide and nitrogen dioxide are pollutants. These pollutants are formed by manufacturing processes. Nitrogen dixoide, the chief irritant in smog, can cause damage to the lungs.

High levels of lead from leaded gasoline have been found in people living near areas of high auto traffic. Lead poisoning affects the central nervous system.

Most pollutants are produced by chemical reactions and can be controlled only by chemical means. Federal law requires car makers to meet exhaust standards. The catalytic converter is one common way of meeting these air-quality standards. Scientists are helping to develop

9-10 This device measures the capacity of the lungs. Breathing polluted air can injure the lungs and reduce their capacity. What are some diseases that can be caused by polluted air?

9-11 Research has shown that when plants are exposed to polluted air, they do not grow as well as plants grown in clean air. This grape leaf clearly shows the harmful effects of polluted air on growing plants.

safe pesticides and clean sources of energy. A constant search is under way for new materials and resources that will not harm the environment.

Air pollution can be controlled. Of course, the best way to control air pollution is to keep impurities from getting into the air in the first place. Most cities and towns forbid open burning of trash and garbage. You have seen how auto makers have reduced impurities. Industry is aware of the pollution problem also, and is spending billions of dollars yearly on research and on pollution-control equipment. Of course, all of these costs are paid for by the consumer.

In many industries, the *Cottrell device* is used to remove smoke and dust particles from waste gases that pass through chimneys into the air. In this device, wires are charged with a high voltage. Dust particles become charged with electricity when they pass near the wires and are repelled toward the opposite terminal. See Fig. 9-12.

Smaller filters of the same type have been developed for homes and offices. One such device gives the dust particles passing through it a positive charge. These positively charged dust particles are then attracted to negatively charged collector plates. The dust is held firmly to the plates by a sticky coating that removes about 90% of the dust from the air as it passes through.

Another method of cleaning waste gases from a smokestack is by the use of scrubbers. A fine spray of water is directed into a chamber to catch and wash away the soluble gases from the exhaust. The gases can later be reused. However, the resultant liquid residue, along with lime which is used in the scrubbing process, forms an enormous amount of a toothpaste-like mass. This substance is itself a disposal problem. Other methods of removing gases include adsorption by charcoal and filtering. Special machines are also used to separate materials of different densities.

Can air pollution be eliminated entirely? In an industrial country such as the United States, this goal is probably unrealistic. However, pollution of the atmosphere can be controlled. Since 1970, the levels of air pollutants, such as sulfur dioxide, carbon monoxide, and dust particles, have been greatly reduced.

solid wastes

Vast amounts of solid waste need disposing. As consumers, we are used to the idea of throwing things away. In one year alone in this country, people discard about 8 million cars, 20 million metric tons of glass, 180 million tires, 40 million metric tons of paper, and 80 billion cans. Paper makes up about 60% of the litter scattered along roads and highways in the United States. Consider that with only 6% of the world's population, the United States consumes about 40% of all the world's resources.

Junk cars present a major disposal problem for many cities and towns. It can cost more to recycle materials from junk cars than the materials are worth. Rubber tires and plastics cause pollution problems.

Worn out rubber tires are hard to dispose of. Burning the tires can produce air pollutants. Buried tires cannot be compacted and slowly tend to rise to the surface. Old tires are used less and less in making new tires. Therefore, there are more old tires to dispose of than ever before.

Many plastics cannot be broken down by natural causes. Nor can they be burned because they melt and tend to clog the burners. Some plastics may also give off poisonous gases when burned. In the future, new kinds of plastics may be used that will decompose in direct sunlight or in topsoil.

The pesticide DDT has been found in the fat tissues of many animals and humans. DDT moves up the food chain to larger fish and fowl, and finally to humans. Although DDT has been used by millions of people for over 30 years, it has not been found to be poisonous to human tissues. However, there is evidence that its use affects the reproductive cycle of many animals. Several countries that banned the use of DDT have started using this chemical again in the belief that the benefits obtained exceed the harm.

Preparing the soil for crops is becoming more and more of a problem as the soil becomes loaded with fertilizers and pesticides. Although these chemicals are needed to provide healthy crops, some of them are washed into rivers and streams, causing water pollution. The problem is not a simple one.

Most solid wastes are burned or buried. Two common methods of getting rid of solid waste are by *burning* or *burying*. Some solid waste, such as garbage, paper, and wood products, can be burned. However, burning these materials can cause air pollution.

One advantage of the burning method is that the heat released during the burning process can be used to produce steam to make electricity. This heat can also be used to heat homes and other buildings. Another advantage to burning is that it is easier to sort out materials that do not burn, such as pieces of glass, iron, and other metals. These materials can be used again to make new products.

Currently, the most popular way of disposing of waste is by burying it in a *sanitary landfill*. Garbage and trash are spread on the ground and a thin layer of clay is added as a cover. Then the soil containing the waste is compacted by a bulldozer. Other layers are added until the area is completely filled. Landfills are needed to get rid of solid wastes, but they often have a messy appearance and offensive odors. Unless properly located, water seeping through the garbage can pollute underground water supplies. See Fig. 9-13.

Some waste materials can be recycled. *Recycling* is a process in which wastes are reused to make new products. For instance, used aluminum cans are collected, melted down, and used to make more cans or other aluminum products. Old newspapers can be made into usable paper products. Copper and glass, as well as other materials, can also be recycled. By recycling, some substances in solid waste can be made into usable items again, thus saving raw resources and reducing the amount of litter.

Recycling is vital since mineral resources are being used up and cannot be replaced. These resources include many common metals such as iron, copper, tin, lead, and zinc.

Not all materials can be recycled. Organic matter, for example, is attacked by bacteria and broken down into compounds that enrich the soil. Fossil fuels cannot be recycled once they are burned. Rubber and plastics are not easily recycled either. However, to the extent that products can be reused, raw resources can be preserved.

9-13 An open garbage dump. Why is this form of disposal not satisfactory? What improvements would you suggest?

unit 2 chemistry in our world

radiation exposure

Radiation is found around the world. Radioactive substances are scattered throughout the entire crust of the earth. They are found as solids, liquids, and gases. Any hole that is drilled or carved into the earth is likely to hit some radioactive matter. Water, oil, and natural gas from wells are likely to contain some radioactivity. Likewise, mines for coal or other minerals will also contain some radioactive matter.

Another major source of radiation is outer space. Cosmic rays constantly bombard the earth. Taken together, radiation from the earth and space produce the natural *background radiation* that all people receive. There is not much that one can do to reduce such natural radiation, which is not part of the problem of radioactive exposure that is discussed in this chapter.

Radioactive wastes come from a number of human activities. By far the greatest source of radiation caused by humans is from the medical field. In fact, radiation exposure from all sources other than natural background and medical uses amounts to less than 2% of the total. See Table 9-1.

Radiation doses are measured in REMs. Radiation doses can be measured in several different ways. *The most accurate way of measuring the effect of radiation on the human body is by a unit called a* **REM.** A small fraction, only one-thousandth of a REM, is called a **millirem.** The average radiation dose received by people in the United States is 250 millirem (0.25 REM) per year. The annual dose varies greatly from place to place and depends on the activities of each person. Average doses and sources as well as doses based on special activities are listed in Table 9-2.

As can be seen from Tables 9-1 and 9-2, medical uses produce by far the greatest amounts of radiation exposure from nonbackground sources. It should also be noted that the medical uses listed in Table 9-2 deal with diagnosis, that is, the use of X rays to find out what may be wrong with a person. Radiation doses in *treatment* of diseases, such as cancer, are many times higher than the doses used in diagnosis.

Table 9-1:

Sources of radiation	
Natural background	67.6%
Medical	30.7%
Fallout from bomb testing	0.6%
Miscellaneous uses	0.5%
Occupational exposure	0.45%
Nuclear industry	0.15%

Table 9-2:

Radiation Dose Levels in Millirem (annual, unless stated otherwise)	
Cosmic rays at sea level*	35
Cosmic rays at 1000-meter altitude*	50
Air*	5
Building materials*	34
Food*	25
Ground*	11
Coast-to-coast jet flight (6 hr)*	5
One chest X ray (bone-marrow dose)†	44
One chest X ray (skin dose)†	1,500
One fluoroscopic X ray†	20,000
From nuclear power plant (within 80 km)*	0.01
From coal-fired power plant (within 80 km)‡	180 to 380

Sources: *Atomic Industrial Forum, Inc.
†United Nations: Ionizing Radiation: Levels and Effects, 1972.
‡Environmental Protection Agency Report: EPA 520-1-77009.

Nuclear power plants and safety. Unlike nuclear weapons, nuclear power reactors cannot explode. The uranium in a weapon must be enriched to more than 90% for the uranium to explode. The fuel rods of a power plant are enriched to only about 3%–5% and cannot explode.

An atomic power plant has many layers of protection built into it so that unforeseen events, human error, equipment failure, or accidents will not easily lead to problems. Even in the event of the "worst possible" accident, a core melt-down, some experts believe that there would be little, if any, danger to the general public.

unit 2 chemistry in our world

Since the fuel rods are located below ground and are surrounded by layers of protection, less radiation is released to the public from the normal operation of a nuclear power plant than from most building materials. In fact, the people living near a nuclear power plant receive less radiation than from a coal-fired plant. Coal, like most substances, contains small amounts of radioactive matter. This matter is released into the air along with smoke and gases when the coal is burned in a power plant.

Radioactive wastes can be disposed of safely. In a typical nuclear power plant, about one-third of the fuel is removed every year and replaced by fresh fuel. The used fuel rods still contain some useful fuel but are no longer good enough to be used in the reactor. These spent rods are temporarily stored in on-site pools to await reprocessing. Reprocessing allows much of the fuel in the rods to be recycled for further use. Only a very small amount of final waste material is not suitable for use and must be disposed of. The total volume of final waste from a large nuclear power plant amounts to about 2 m^3 per year. That amount could fit under a cafeteria table. By comparison, the waste from a modern coal-fired power plant fills an area of 2 km^2 to a depth of more than 1 m per year.

The high-level nuclear waste is very radioactive and toxic. However, unlike many other substances that can remain toxic forever, nuclear wastes decay to lower levels with the passage of time. See Fig. 9-14 and notice the curved line representing the toxic level of nuclear wastes. Compare nuclear waste with some natural ores found in the earth's crust as well as the wastes from a coal-fired power plant. How many years does it take for the toxic level of nuclear waste to drop to the level of uranium ore from which it was mined?

The final nuclear waste is turned into a non-soluble glass-like mass and then encased in concrete and lead. Some scientists recommend burying radioactive wastes in deep salt mines. Salt is a very stable compound and has a high melting point. Also, salt domes are found in regions that are less likely to be disturbed by earthquakes.

Some years ago small amounts of nuclear wastes from weapons production were placed into special containers and dumped into the ocean. However, this practice has stopped because of possible leakage from the containers.

The waste products from weapons production, medical uses, and nuclear power are mostly low-level wastes and may take up a large volume. Once recycled in a reprocessing plant, however, the wastes are high-level wastes, and take up very little space. Until the government decides where to build a reprocessing plant, all storage of radioactive wastes is temporary.

9-14 The curved line shows how the toxic level of nuclear waste declines with the passage of time. The straight dashed lines show that the toxic levels of some other substances do not decline. How long will it take for nuclear wastes to reach a level no more toxic than pitchblende? than coal wastes? than natural uranium ore?

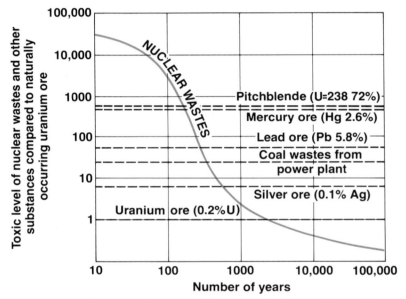

Sources: Lawrence Livermore Laboratories and United Kingdom Atomic Energy Authority

Population: Too Many or Too Few

The earth's population has increased greatly since the early part of this century. The increase is the result of an increase in the length of the human life span. People who formerly would have died of diseases now survive much longer.

Population growth has not been caused by an in-

         unit 2 chemistry in our world

crease in the birth rate. In fact, the birth rate seems to go down when a country's standard of living goes up. Many countries with high standards of living, including the United States, have a birth rate that is actually below the replacement level. This means that the population of a country will not be fully replaced and will begin to decline.

The decline in population does not occur as soon as the birth rate dips below replacement. Instead, the population continues to increase for several decades because a very large portion of the population is at a reproductive age. Once this large segment of our population ages and is no longer able to reproduce, the actual decline in the population will begin.

Already, the number of young people in this country has declined greatly in the last ten years. Check your school. Is your school enrollment increasing or decreasing? Find out if there are more or fewer first grade pupils than high school seniors. Do you know of any school buildings that have been closed because of a lack of students?

summary

1. Water pollutants include improperly treated sewage, chemicals, and oil.
2. Air pollution comes from motor vehicles, factories, power plants and industrial plants, and homes.
3. Air pollutants include carbon monoxide, sulfur dioxide, oxides of nitrogen, unburned hydrocarbons, and dust particles.
4. Catalytic converters, Cottrell devices, scrubbers, and special filters are used to prevent pollutants from getting into the air.
5. Wind, temperature inversion, snow, and rain affect air pollution levels.
6. Prolonged exposure to air pollutants may cause emphysema, tuberculosis, asthma, and certain types of lung cancer.
7. Solid wastes can be controlled by proper burning or by burying in a sanitary landfill.
8. Many waste materials can be collected and recycled into new products.
9. Most radiation exposure is caused by background radiation from the earth or from space.

review

Match each item in the left column with the best response in the right column. *Do not write in this book.*

1. acid rain
2. Cottrell device
3. cosmic rays
4. temperature inversion
5. photochemical smog
6. landfill
7. nitrate ions
8. sodium and calcium fluorides

a. radiation from outer space
b. used for solid waste disposal
c. water pollutant
d. waste that comes largely from steel and ceramic industries
e. air pollution control device
f. thermal pollution
g. reaction of unburned hydrocarbons
h. lead poisoning
i. concentrates polluted air
j. caused by dissolved carbon dioxide

questions

Group A
1. What is a pollutant?
2. How can water in streams and rivers become polluted? In what way do oil spills contribute to water pollution?
3. How do mercury compounds, sometimes found in polluted waters, get into the human body?
4. What is the cause of thermal pollution? Is it always harmful? Justify your answer.
5. What are the major sources of air pollution? List five air pollutants.
6. Why do new cars use lead-free gasoline?
7. What pollutants are given off by the incomplete burning of coal?
8. Is noise a pollutant? Why?
9. What is meant by temperature inversion? What effect does it have on air pollution?
10. What effect does polluted air have on people? What diseases can be caused by long exposure to polluted air?
11. What are three devices used to control air pollution?
12. Why is it hard to dispose of most plastic materials? What is being done about this problem?
13. What is a sanitary landfill? Describe how it works.
14. What is meant by recycling? Why is it important?
15. What is background radiation? How does it compare to radiation from human sources?

unit 2   chemistry in our world

16. What precautions are taken for the safe disposal of radioactive waste from nuclear reactors?

Group B
17. Describe how phosphate and nitrate ions can cause water pollution.
18. How can water pollution be controlled?
19. What is photochemical smog? How is it formed?
20. What effect does sulfur dioxide in the air have on people?
21. Describe how acid rain is formed.
22. Which condition of the air (stable or unstable) causes the scattering of pollutants in the air? Justify your answer.
23. What effect do lead particles in the air have on people?
24. What are "scrubbers"? What problem is created by their use?
25. Describe several methods of solid waste disposal.
26. Compare radiation exposure from medical uses to all other sources of radiation produced by humans.
27. Compare the volumes and kinds of wastes from a nuclear power plant with wastes from a coal-fired power plant.
28. Why do salt mines show great promise for disposal of radioactive wastes?
29. Why is pollution a greater problem today than in past years? What are some solutions?
30. Describe how the population can increase even though the birth rate is below the replacement level.

further reading

Air Pollution and Your Health, American Council on Science and Health (booklet), 1981.

Bassow, H., *Air Pollution Chemistry: An Experimenter's Sourcebook.* Rochelle Park, N.J.: Hayden, 1976.

Beckmann, P., *The Health Hazards of Not Going Nuclear.* Boulder, Colo.: Golem Press, 1976.

Kabbe, F., et al., *Chemistry and Ecology of Man.* Boston: Houghton Mifflin, 1976.

Lapp, Ralph, *The Radiation Controversy.* Greenwich, Conn.: Reddy Communications, 1979.

Murphy, A. W., ed., *Nuclear Power Controversy.* Englewood Cliffs, N.J.: Prentice-Hall, 1976.

Storin, Diane, *Investigating Air, Land, and Water Pollution.* Boulder, Colo.: Pawnee, 1975.

Turner, Stephen C., *Our Noisy World.* New York: Messner, 1979.

10

chemistry and the home

All people have certain basic needs. You have studied about two of these needs, air and water. You have seen how clean air and water are essential for life. Other basic needs include shelter, food, good health, and clothing. In this chapter, you will see how chemistry is involved in building and maintaining a modern home, in keeping people healthy, and in growing good food.

objectives:

☐ To identify the uses and properties of several kinds of building materials.

☐ To identify the materials used in making glass, and to describe the glass-making process.

☐ To identify the materials used in the making of paints and varnishes, and to describe the uses of each.

☐ To describe the chemical reactions in several types of fire extinguishers.

☐ To explain how soaps and cleaning agents work, and to describe the soap-making process.

☐ To describe the chemicals used for fertilizers, insecticides, and fungicides.

☐ To relate the uses of some drugs to personal health.

natural building materials

Lumber is widely used for construction. When the first colonists came to North America, their first need was for homes that would provide shelter and safety. The huge forests in their new country provided wood and wood products to build these homes. There were so many trees that the early settlers had to clear them away to raise food crops. They were even able to send wood back to England.

Trees are among the most useful of our natural resources. Throughout history, people have depended on wood for shelter, fiber, chemicals, furniture, and many other items in common use today.

Lumber must be dried before it can be used. Lumber *is wood that has been sawed from large tree logs.* Unlike most wood products, lumber is ready for use as it comes from the sawmill. As these boards are sawed from logs, they are called *rough lumber.* See Fig. 10-1. When the rough lumber is trimmed to various lengths, widths, and thicknesses, and properly smoothed and dried, it is called *finished lumber.*

10-1 A giant saw cuts a log into rough lumber. What is the next step in processing the wood into lumber?

10-2 Lumber is stacked for storage and air drying. Why is it necessary to dry lumber?

10-3 When trees are properly harvested, new forests usually grow through reseeding. In this nursery, seedlings are grown to restock a forest. Describe how these forests can be renewed.

Freshly cut logs produce *green* lumber. "Green" refers to the high sap moisture content in the lumber, and not to its color.

Since green lumber will shrink and may warp as it dries, it could easily ruin a building project. Therefore, lumber is dried before it is used.

Some lumber is stacked in lumber yards for *air drying*, as shown in Fig. 10-2. Other lumber is placed in a drying kiln (*kil*), a special kind of oven. The method used in drying depends on the kind of wood, its thickness, and specific requirements for the finished product.

In another part of the sawmill, boards are cut into flooring, siding, molding, and other forms of building materials. This finished lumber is also graded for quality before it is shipped.

There are two major classes of lumber. Of more than 1000 species of trees in the United States, only about 100 can be cut into lumber. These trees can be divided into two groups: *softwoods* and *hardwoods*.

Most softwoods are *conifers* (cone-bearing), such as pine, fir, and hemlock. Conifers normally keep their leaves or needles all year and are also called evergreen trees. Trees in the second group, the hardwoods, are mostly *deciduous* (di-*sid*-you-us). Deciduous trees are those that shed their leaves in the fall of the year. Examples of hardwoods are oak, gum, ash, birch, walnut, and maple.

In building, softwood lumber is used mostly as framing lumber and plywood. Hardwood lumber is used mostly as flooring and furniture.

Despite the large volume of trees cut during the past 200 years, the forests continue to produce a plentiful supply of wood products. The trees that supply lumber are a crop, which can be grown over and over again on the same land. See Fig. 10-3. Trees must be cut when they are mature. Harvesting makes way for the next crop, just as corn and wheat must be cut to make the land available for another crop. Since they can be renewed, forests can last for a long time if proper cutting practices are used.

The earth provides many materials that are used in the building industry. Among these materials used for building are limestone, marble, granite, slate, and sandstone.

unit 2 chemistry in our world

Limestone, a natural *sedimentary* (sed-ih-*men*-tuh-ree) *rock*, consists mostly of calcium carbonate ($CaCO_3$). This rock is formed when layers of materials are deposited by water, ice, wind, and other agents over long periods of time. Limestone is found in almost every state in the country. In some places, entire mountains are made of limestone. Most deposits of limestone are gray because of impurities. Limestone is used in the making of Portland cement, glass, iron, and steel, and in building roads.

Marble is a crystal formation of either limestone or dolomite ($CaCO_3 \cdot MgCO_3$). Marble was first formed as sedimentary rock. By heat and pressure over a long period of time, this rock changed to *metamorphic* (met-uh-*mor*-fik) *rock*. The word "metamorphic" means "changed in form." Marble is usually white but may also be brown, red, green, or black because of impurities in the rock. Marble can take a high polish because of its extreme hardness. This hardness also makes marble a good building material. See Fig. 10-4.

Granite and slate are two other natural building materials. Granite is an *igneous* (*ig*-nee-us) *rock*. This kind of rock was formed when hot molten material cooled and hardened. Granite is a mixture of the minerals feldspar, quartz, and mica. It is found in a range of colors including white, brown, green, and gray. Granite lasts a long time, has great strength, and can take a polish.

The building industry at one time used slate, a metamorphic rock, to cover roofs because it splits smoothly into pieces with flat surfaces and is very durable. Today, slate is seldom used for that purpose because it is too costly and heavy. Slate is a natural rock derived from clay. Pennsylvania is one of the leading slate-producing states.

Sandstone is a sedimentary rock that was formed when grains of sand were cemented together by pressure. Because it breaks easily, sandstone does not stand up well as a building material.

10-4 Marble, a form of crystallized limestone, is an excellent material for buildings and monuments. Here it is shown being quarried in Vermont. What property of marble makes it a good building material?

other building materials

Portland cement is used to make concrete. Cement is made mostly from two abundant raw materials, limestone and clay. Both of these materials are low in cost

10-5 Some rotating cement kilns are 150 m long and 4 m in diameter. Can you describe the operation of this cement kiln?

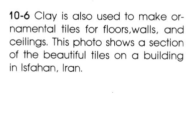

10-6 Clay is also used to make ornamental tiles for floors, walls, and ceilings. This photo shows a section of the beautiful tiles on a building in Isfahan, Iran.

and are widely distributed. However, large amounts of fuel, power, and heavy equipment are needed to produce cement.

"Portland" is the name given to the cement-making process. (The name comes from the Isle of Portland, England; the cement resembles the limestone found there.) A mixture of crushed limestone and clay is poured into the upper end of a slow-turning cement kiln as shown in Fig. 10-5. The mixture slowly melts into lumps about the size of peas, which are called *clinker*. After the clinker cools, it is ground into a fine powder. The cement is then placed in bags and is ready for use.

The most important use of cement is in the making of *concrete*. Concrete is made by mixing cement, sand, gravel, and water. The concrete is poured into forms that hold it in place until it hardens. The concrete sets, or becomes firm, in an hour or two but continues to gain strength for days and weeks. This hardening is caused by the formation of crystals. The crystals lock together and make a very hard, artificial stone. For many needs, steel rods or steel mesh are placed in the form before the concrete is poured. A much stronger product, called reinforced concrete, is the result.

Bricks, tiles, and pottery are made from clay. Bricks have been used since ancient times. In those days, bricks were blocks of clay, baked in the sun. Today, much harder bricks are made by heating them in a very hot furnace. See Fig. 10-6.

In the brick-making process, dried clay and sand are sifted together. Then these materials are mixed with

water to form a stiff paste. This paste is molded into bricks and then dried for a few days. Finally, the bricks are fired in a kiln. The temperature and length of firing depend upon the kind of clay used and the hardness wanted.

Common bricks and tiles are baked at low heats, while glassy bricks are baked at higher temperatures. Different materials in the clay produce bricks of various colors. Iron compounds, for instance, cause a red color.

Clay is also used in making pottery. A paste of clay is first made into the desired shape, then fired until it becomes hard and porous. Chinaware and porcelain contain a white clay, called kaolin, that is mixed with powdered quartz and a mineral called feldspar.

Mortar holds together blocks, bricks, and stones. Mortar is a material used to bond brick, concrete blocks, and other similar building materials. It is a blend of Portland cement, sand, slaked lime [$Ca(OH)_2$], and water. Masons spread mortar on bricks or concrete blocks to build a wall. The mortar sets and becomes hard. See Fig. 10-7.

The process by which mortar sets to a rock-like mass is not fully understood. Scientists believe that some of the slaked lime reacts slowly with the sand (SiO_2) to form calcium silicate ($CaSiO_3$) as follows:

$$Ca(OH)_2 + SiO_2 \rightarrow CaSiO_3 + H_2O$$

In addition to the above reaction, carbon dioxide from the air slowly unites with some of the slaked lime to form limestone ($CaCO_3$) as shown below:

$$CO_2 + Ca(OH)_2 \rightarrow CaCO_3 + H_2O$$

Plaster and dry wall are used on ceilings and walls. Putting plaster on the walls or ceiling of a building is a two-step process. First, an undercoat consisting of mortar containing hair or shredded fiber is applied. Then a second, or finish, coat is applied over the undercoat. This coat is a mixture of slaked lime, water, and a powder called *plaster of paris*. The finish coat dries quickly with a smooth, hard surface.

Gypsum (*jip*-sum), a mineral found in large deposits in some western states, is calcium sulfate ($CaSO_2 \cdot 2\ H_2O$). The dot (·) in this formula indicates that water molecules are loosely connected to the rest of the compound.

10-7 A worker using mortar to help build a brick wall. What materials are used to make mortar?

When gypsum is heated, it loses part of this water. A fine, white powder of plaster of paris [$(CaSO_2)_2 \cdot H_2O$] remains.

When water is added, plaster of paris forms a paste. This paste hardens, or sets, in a few minutes. As it sets, water molecules combine with the plaster of paris to form gypsum. When you use plaster of paris, you must work fast or the paste will set before you are finished. Obviously, plaster of paris must be stored in moisture-proof containers. Plaster of paris was used for mortar in building the pyramids.

Plaster is no longer used in most buildings. Instead, large panels of gypsum, called *dry wall*, are used. Dry wall is very smooth and is much easier to use, faster to install, and much less costly than plaster. See Fig. 10-8.

activity

Making a plaster of paris cast. Make a paste of plaster of paris and water, and spread it in a thin layer on a flat surface. Now place an oiled key or other small object on the paste before it hardens. After about 30 min, remove the object and note the impression that is left in the hardened plaster. Describe the hardening process that takes place.

10-8 This worker is installing sheets of dry wall on the wall and ceiling of this room. What are some advantages of using dry wall instead of plaster?

Glass has been made for hundreds of years. The Egyptians made glass centuries before the Christian era. One kind of natural glass is *obsidian*, produced by volcanic action. Obsidian was used by early cave people, who shaped it into tools and spearheads. In the past 80 years, scientists have developed many new and useful kinds of glass.

Glass is a mixture, having no fixed boiling and freezing points. It is sometimes called an "undercooled" liquid. Glass can be made by melting together limestone ($CaCO_3$), soda ash (Na_2CO_3), and white sand (SiO_2). These materials go into making about 90% of all glass used in the home, such as drinking glasses, bottles, and window panes. Broken scrap glass, called *cullett*, is often added to the glass mixture. The addition of scrap glass lowers the melting point of the raw materials. It also provides a low-cost way of recycling glass that otherwise would litter the land.

The raw materials for making glass are mixed and poured into a tank furnace. The furnace looks like a small swimming pool, except that it is very hot and filled with melted glass. Some furnaces hold as much as 1200 metric tons of glass.

The hottest part of the furnace is about 1500°C. At this point, glass is a pasty liquid, like molasses. It takes about a week for the raw materials to move from the shallow end of the furnace to the deep end. During this time, bubbles of carbon dioxide gas escape from the glass and some of the raw materials slowly change into a mixture of silicates. These reactions are shown in the following equations:

$$CaCO_3 + SiO_2 \rightarrow CaSiO_3 + CO_2 \uparrow$$

$$Na_2CO_3 + SiO_2 \rightarrow Na_2SiO_3 + CO_2 \uparrow$$

Glass is used in many ways. The molten glass from the furnace goes into various glass-forming machines. From the machines, the glass objects are carried to *annealing* ovens where they are slowly cooled. The slow cooling reduces strains in the glass that are caused by uneven cooling.

The bubbles you see in some glass bottles are caused by escaping carbon dioxide gas. The temperature in the furnace was not high enough to drive off all of the gas. Flawed glass cannot be used for making lenses or other optical instruments. Can you see why?

To make plate glass, a machine dips a horizontal iron rod into the furnace. The rod is lifted straight up, and the melted glass clings to it, forming a large sheet.

The thickness of the sheet is determined by controlling the speed of the rising rod and by keeping the glass in the furnace at the proper temperature. After the glass cools, the edges are trimmed.

A newer method of making plate glass is the *float* process. In this process, molten glass is floated on the surface of a bath of molten tin. Gravity keeps the liquid tin very flat so that the glass layer also takes on this flat shape. See Fig. 10-9.

Most kinds of glass cannot withstand sudden changes in temperature without breaking or cracking. *Pyrex* brand glassware is an exception; it is often used in the lab or as baking dishes in the home. Pyrex glass expands much less than ordinary glass when heated. Boric oxide,

10-9 A ribbon of newly cast plate glass is shown moving into a polishing and grinding machine.

10-10 This method of glassblowing is still used for specialized glassware. Can you suggest several items that may be made in this way?

added to the glass in small amounts, gives Pyrex this property.

Glass bricks are useful building materials. Since they are hollow, the air space inside acts as a good heat insulator. Because of their design, glass bricks transmit "soft" light but do not permit clear visibility.

If a glassmaker wants to produce a clear, colorless glass, pure ingredients must be used. Traces of iron oxide in the sand will give glass a pale green color. This type of glass is used in making some kinds of bottles.

Nearly all glass products are machine-made. Bottle-making machines imitate the process used by the glass-blower, whose method is shown in Fig. 10-10. These machines have a great many pipes that dip into the melted glass. Compressed air is used to blow the glass. One machine can turn out thousands of bottles an hour.

In many cities, glass windows in buildings are being replaced by a tough plastic material developed for the face shields of astronauts' helmets. This material is 250 times stronger than safety glass and can stop a 0.45-caliber gun bullet.

Paints and varnishes protect surfaces. Paints were used by ancient people to decorate their caves, to coat tools, and even to decorate their faces. Today, people use many kinds of paint, not only to protect surfaces, but also to make their surroundings more beautiful.

Oil paint is mostly a mixture of the following types of substances: (1) a *drying oil*, such as linseed oil, to act as a binder and to absorb oxygen of the air as it dries; (2) a *pigment*, such as zinc oxide, to give the paint body, covering power, and the desired color; (3) a *thinner*, such as turpentine, to make the paint spread easily; and (4) a *drying agent*, such as manganese dioxide, to act as a catalyst to speed up the drying process.

Drying oils in paints react with oxygen of the air and form elastic solids. The rubber-like coating that soon appears on the surface of an open can of paint is the result of this process. Common drying oils include linseed oil, which is pressed from ripe flaxseeds; tung oil, from the nut of the China tung trees; and soybean oil. Other oils for this purpose come from seeds of castor, safflower, sunflower, and hemp.

When you spread paint on a surface, you expect to cover the color underneath. The thinner you can spread paint and still cover the color underneath, the better the paint. Paint pigments are a mixture of several substances. Three of the best white pigments are white lead, zinc oxide, and titanium oxide.

For colored paints, small amounts of colored matter are added to the white pigments. Red lead oxide is used to protect steel structures from rusting. Metal pigments, such as powdered aluminum, are also added to paint.

Emulsion paint products have largely replaced oil-based paints for most uses. These are water-thinned paints that dry rapidly. When a resin emulsion is used as a binder in water-thinned paints, the paints are called *latex* paints. These paints have good covering properties and form easily cleaned, satin-like surfaces. For most uses, latex paints are better than oil paints.

Varnish is made by boiling certain gum-like materials, called resins, in oils. No pigment is added so that the varnish will not hide the grain of the wood. However, pigments can be added to varnish to make *enamel*. Enamel gives a higher shine and often a hard surface.

Lacquers are solutions of resins that dry to form a hard, shiny surface. Although oil paint hardens by taking in oxygen of the air, lacquers dry by evaporation of the solvent. The drying rate of lacquers depends upon the type of solvent used. Some lacquers dry so fast that you cannot use a brush to apply them. These lacquers must be sprayed on.

activity

Treating metals for rust protection. Using a pair of tongs, carefully dip four nails into dilute sulfuric acid for a few minutes. CAUTION: *Sulfuric acid is corrosive. Use a glass container. Do not touch the acid or get it on your clothes.* Remove the nails with a pair of tongs. Rinse and dry the nails. Coat one of them with oil paint and one with oil. Dip the third nail into copper(II) sulfate solution until it turns a copper color. Leave the last nail untreated as a control. Hang all the nails on a support outdoors for at least a week. Record your observations. Were all the nails protected from rust? Explain.

Three things are needed to permit a fire to burn. Fires have three basic needs: (1) *heat,* which brings materials to kindling temperature; (2) *fuel,* or something that will burn; and (3) *oxygen* to support the burning. To put out a fire, remove any one of these ingredients.

Flooding a fire with enough water cuts off the oxygen supply and also cools the burning material below its kindling temperature. Sand is another common material that can be used to cut off oxygen to a fire.

Water should not be used on oil fires because it may spread the oil, and thus spread the fire. It is also dangerous to use water near electrical wiring because water conducts an electric current. Therefore, water is not an ideal substance to put out all fires.

Chemicals help to put out fires. The oldest chemical fire extinguisher still in common use today is the *soda-acid* type. This extinguisher consists of a strong copper tank nearly full of a solution of sodium bicarbonate ($NaHCO_3$). Hanging from the inside top of the tank is a bottle of sulfuric acid (H_2SO_4) with a loose stopper. See Fig. 10-11.

When the tank is turned upside down, the stopper falls from the acid bottle. The acid spills into the baking soda solution. The reaction of the acid and baking soda forms sodium sulfate (Na_2SO_4), water, and carbon dioxide (CO_2) as shown in the following equation:

$$H_2SO_4 + 2\,NaHCO_3 \rightarrow Na_2SO_4 + 2\,H_2O + 2\,CO_2 \uparrow$$

The carbon dioxide is released so fast that a high pressure builds up inside the tank. This pressure forces a stream of solution out of the hose. Some of the carbon dioxide also dissolves in the liquid. The solution helps to put out the fire by cooling the burning material. In addition, the carbon dioxide coming out of solution forms a heavy blanket of gas. This blanket shuts off oxygen from the fire. Can you see why this stream of solution is more effective than plain water?

However, this type of extinguisher also has several faults: (1) The solution always contains some acid that has not reacted with the soda. This acid could cause damage to clothing and furniture. (2) This type of ex-

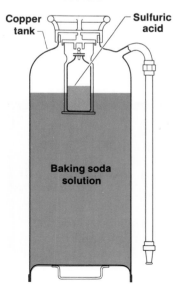

10-11 When a soda-acid fire extinguisher is turned upside down, the acid reacts with the baking soda solution to give off carbon dioxide. Write the chemical equation for this reaction.

Copper tank
Sulfuric acid
Baking soda solution

tinguisher cannot be used around electric wiring because of the danger of an electric shock. (3) The soda-acid type of extinguisher is not effective for oil or gasoline fires. The stream of liquid tends to sink below the blazing oil, thus causing the fire to spread. (4) The tank is large and heavy, and once started, there is no way of stopping it until the tank is empty.

A much better type of extinguisher is the *foam* type. This type produces a mixture of aluminum hydroxide and carbon dioxide. Foam extinguishers are not often used around the house because the foam will damage rugs and furniture. They are, however, effective in fighting oil fires.

Carbon dioxide extinguishers have a strong, steel cylinder that holds liquid carbon dioxide under great pressure. When you open the valve, carbon dioxide rushes out through a cone-shaped nozzle. The cooling effect of the sudden expansion changes most of the escaping gas into carbon dioxide "snow." See Fig. 10-12.

The "snow" is about $-80°C$. This is so cold that it lowers the temperature of the burning material below its kindling point. The heat from the fire causes the "snow" to change to heavy CO_2 gas, which prevents oxygen from reaching the fire. The carbon dioxide does not harm home furnishings. The liquid CO_2 extinguisher works well against oil and electrical fires.

Underwriters' Laboratories (UL) is a national non-profit testing group. The UL labels on fire extinguishers mean that they have been carefully tested for effectiveness and safety. Many low-cost, dry-chemical fire extinguishers are available for home use.

10-12 What type of fire extinguisher is effective against electrical fires?

chemistry and cleaning agents

Many cleaning agents are used in the home. Cleanliness is essential to good health. Of the several cleaning agents in the home, soap and other **detergents** are the most common. *A detergent is a substance that removes dirt.* Household chemicals used for cleaning also include scouring powders, solvents, and special purpose cleaners for metals, glass, and textiles.

Scouring powders loosen dirt by grinding or wearing

it away. Powdered sand and pumice (a kind of volcanic rock) are often used. Cleaning solvents dissolve grease and are often used for dry cleaning.

Household ammonia is used for cleaning purposes. Ammonia has the advantage of not leaving a film when it dries. This weak base is a good solvent for grease, although like all other bases, it is harmful to most painted surfaces.

Soap is a common detergent. Soap is made by boiling a fat, such as lard, with a solution of sodium hydroxide (lye). Fats are compounds of carbon, hydrogen, and oxygen. The most common form of fat for making laundry soap is tallow, a by-product of the meat-packing industry.

The word equation for the soap-making process is

$$\text{fat} + \text{lye} \rightarrow \text{glycerin} + \text{soap}$$

Glycerin remains in some soaps. In others, however, the glycerin is recovered as a by-product. Glycerin is used in making cellophane, printer's ink, medicines, cosmetic lotions, and some explosives. See Fig. 10-13.

Soapmakers add oils to improve the quality of soap. Oils used to make toilet soap include coconut, palm, and olive. Soybean, corn, castor, and linseed oils are also used. Perfume is added to give the soap a refreshing smell. Water softeners, dyes, and germicides are also added to the finished product.

Soaps clean by chemical action. In the cleaning process, soap molecules surround greasy dirt particles. A typical soap molecule has a rather long structure. The "sodium end" of the molecule is water soluble and therefore dissolves in water. The other end of the molecule is oil soluble and tends to dissolve in the oily covering of dirt particles. Figure 10-14 shows how the soapy layer acts on the oily covering of the dirt. After the oil-soluble end weakens the greasy film, the dirt particles are easily washed away by the water.

Synthetic (soapless) detergents have largely replaced soap for laundry purposes. These detergents are derived from petroleum and other sources. Like soap, all detergents on the market today are biodegradable.

The role of a soapless detergent is that of a wetting agent. Soapless detergents make water "wetter" by

10-13 Fats and oils are boiled with sodium hydroxide to make soap. What important by-product is produced in the soap-making process?

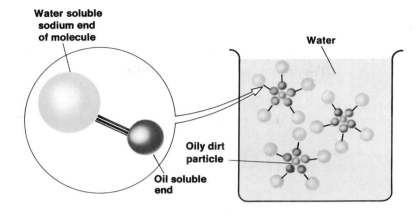

Water soluble
sodium end
of molecule

Water

Oily dirt
particle

Oil soluble
end

10-14 The oil-soluble end of the soap molecule prepares the way for the water-soluble sodium end to work away the dirt particles. Why is soap called a detergent?

reducing the force of attraction between water molecules. That is, they allow the water to spread through cloth fibers more readily, loosening the dirt.

When a garment is dry-cleaned, it is rinsed in a solvent other than water. The use of special dry-cleaning solvents avoids shrinking and keeps the colors from running together.

chemistry and food

Fertilizers increase crop yields. Fertile soil contains organic matter, plant and animal life, and minerals. Decayed plant and animal life accounts for the presence of organic compounds in the soil. Bacteria and other forms of simple plant life, however, make up the greatest part of the "living" portion of fertile soil. The soil should also contain soluble minerals. If these minerals are missing or scarce, soils must be fertilized in order to insure healthy crops. See Fig. 10-15.

About 16 elements are needed for the growth of healthy plants. Nitrogen, phosphorus, and potassium are the major elements needed in large amounts. Nitrogen is necessary for the making of proteins in plants. Phosphorus is needed for the formation of healthy roots. Potassium is used as a catalyst in the making of starches and sugars. These and other elements must be in an ionic form before they can be used for the growth

10-15 In addition to proper nutrients, many factors affect plant growth. As shown in this photo, experiments in plant genetics could lead to improvements in the efficiency of food production.

chapter 10   chemistry and the home   177

and development of plants. Most fertilizers contain soluble compounds of the three major elements.

Many states have laws that require the components of commercial fertilizers to be stated on the label. For example, the label "4-6-8" means that the fertilizer contains 4% nitrogen, 6% phosphorus, and 8% potassium.

Some people believe that manure and compost (organic matter) produce more nourishing foods than do chemical fertilizers. According to many studies, however, it makes no difference whether the plant nutrients come from organic sources or are produced by chemicals.

Chemicals protect crops from insects and diseases. To grow healthy plants, harmful insects must be controlled. Two common types of insect pests are the sucking insects and the chewing insects. These unwanted insects can be destroyed by chemicals called *insecticides.*

For many years, DDT (Dichloro-Diphenyl-Trichloroethane) was used to kill insect pests. This compound kills insects by destroying their nerve tissues. Even though DDT is very effective, it is no longer used in the United States. A potential hazard may exist because DDT remains active for years. It moves up the food chain to concentrate in the fatty tissues of animals, including humans.

Arsenic compounds of lead and calcium are often used in fruit sprays. Since these compounds, as well as many other insecticides, are poisonous, fruits and vegetables must always be thoroughly washed before they are eaten. See Fig. 10-16.

A fungus is a microscopic plant related to bacteria. The fungi include rusts, molds, rot, mildew, and smut. All of these fungi can injure plants. Chemicals that destroy these fungi are called *fungicides.* Sulfur, applied as a fine dust, is used as a fungicide. Lime-sulfur spray is also used. Bordeaux mixture, containing copper(II) sulfate, lime, and water, is a standard spray for many fungus diseases.

Leavening agents are used in baking. Years ago, bakers would let bread dough stand in the warm air for several hours before baking. During this period, the dough would be kneaded several times to assure a more

10-16 This photo shows the spraying of crops with an insecticide. What is the purpose of this spraying?

thorough mixing of the ingredients. This practice was needed for the bread to "rise." A lump of the dough was mixed with the next batch of bread. Although they knew that these practices were needed, people did not understand what caused the bread to rise.

Louis Pasteur, a French scientist, solved the mystery of this process. Pasteur found that while the dough was in warm air, tiny plant cells, called yeast, became embedded in it. The yeast cells produced chemicals called *enzymes* (*en*-zimes), or organic catalysts. The enzymes set off a chemical reaction in the dough. This reaction, known as *fermentation*, produces bubbles of carbon dioxide gas. In baking, the heated bubbles increase in size, causing the dough to expand or rise. *A substance that causes dough to rise is called a* **leavening agent.**

The enzymes released by yeast cells start the fermentation in the dough. When a batch of dough is made, a little sugar is added to the flour. The yeast cells then provide the enzyme that changes the sugar to alcohol and carbon dioxide. The equation for this reaction using glucose (sugar) is

$$C_6H_{12}O_6 \rightarrow \quad 2\,C_2H_5OH \quad + \quad 2\,CO_2 \uparrow$$

sugar   grain alcohol   carbon dioxide

The heat of baking drives off the alcohol from the finished bread. Carbon dioxide bubbles cause air holes in the bread. These air holes make the bread lighter and more easily digested. One disadvantage of this method of baking is that it takes several hours for the

10-17 In baking bread, yeast first causes the bread to rise. Then the heat of baking produces the "oven rise." What is one disadvantage of using yeast in baking?

dough to rise when yeast is used as a leavening agent. See Fig. 10-17 on the preceding page.

To avoid the delay caused by using yeast, *baking soda* and *baking powders* are now commonly used. Baking soda reacts quickly with acid to produce carbon dioxide. Therefore, baking soda is sometimes mixed with flour and sour milk. The lactic acid in the sour milk causes the release of carbon dioxide gas, and the dough rises.

A disadvantage of this process is that it is hard to control the reaction, since sour milk varies in lactic acid content. Most people now use specially prepared baking powders for home baking.

Baking powders are leavening agents that contain a mixture of the following ingredients: (1) baking soda as a source of carbon dioxide gas; (2) a substance, such as cream of tartar, that produces an acid when water is added; and (3) corn starch to keep the mixture dry.

When water is added to the baking powder, it reacts with the acid substance to produce hydrogen ions. The hydrogen ions react with baking soda to produce carbon dioxide gas. Can you explain why baking powders are packed in moisture-proof containers?

The advantage of baking powder is that the reacting substances are present in just the right amounts. On the market today, you find "ready mixes" for baking breads and cakes. These products contain all the needed ingredients, including baking powder, and the directions for baking.

activity

The fermentation process. Prepare about 0.5 L of 10% unsulfured molasses solution. Add a few green peas to this solution. Then add about 1/4 package of dry yeast. Cover the container with a piece of cloth and keep it for several days in a warm, dark place. Describe the odor you detect after that time. Devise a way to collect the gas given off, and bubble it through lime-water solution. Observe and explain the results.

chemicals that protect your health

Disinfectants stop the growth of germs. Disinfectants *are chemicals that kill or stop the growth of bacteria.* When these chemicals are too strong to be used on the skin or other body tissues, they are called *germicides.* When these chemicals are safe to use on the skin but still strong enough to stop the growth of bacteria, they are called *antiseptics.* In many cases, antiseptics are dilute solutions of germicides.

One of the most traditional antiseptics is tincture of iodine, which is used for minor cuts. This solution is made by dissolving iodine crystals in alcohol, to which some potassium iodide(KI) is added. Although there are many home remedies for minor cuts and bruises, soap and water is one of the best first-aid treatments.

A 3% solution of hydrogen peroxide (H_2O_2) is also an effective household antiseptic. Since strong light will break down this compound, hydrogen peroxide is stored in brown bottles. Hydrogen peroxide slowly loses its strength by breaking down into water and oxygen as follows:

$$2 H_2O_2 \rightarrow 2 H_2O + O_2 \uparrow$$

Because of the readiness with which hydrogen peroxide gives up oxygen, it is used in bleaching organic materials, such as silk, wool, hair, and ivory. Concentrated hydrogen peroxide is used as a rocket-fuel oxidizer.

Germicides are too harsh to be used on living tissue. Sodium hypochlorite (NaOCl) is a germicide that is used as a household disinfectant and as a bleach. Chloride of lime ($CaOCl_2$) is used as a swimming-pool disinfectant. These compounds are poisonous and should be handled with great care.

Drugs, when used properly, can ease pain and fight disease. Primitive people found out by chance that some of the plants growing about them seemed to relieve pain, heal sores, and even cure diseases. These plants were the first **drugs.** *Drugs are chemicals that cure or prevent diseases, or ease pain.*

Drugs such as digitalis, morphine, belladonna, cascara, and quinine are still derived from plants. Other drugs are composed of such minerals as Epsom salts ($MgSO_4$) and milk of magnesia [$Mg(OH)_2$]. Still other drugs—vaccines and gland extracts, for example—come from animal sources. However, most drugs today are prepared in large chemical laboratories.

Drugs must be used wisely, under a doctor's care. Some drugs are habit-forming while others are poisonous. In fact, any drug can be poisonous if it is misused. Even aspirin causes about 100 deaths every year, mostly among small children.

Drugs used to relieve pain are called *analgesics* (an-al-*gee*-zicks). Perhaps the most common analgesic drug is aspirin. Aspirin, which was first used in Germany over 80 years ago, is the most common pain-killer available without a prescription. The idea that one type of aspirin works better than another is not supported by research. Americans consume millions of aspirin tablets every year.

Some pain-killing drugs called *narcotics*, although helpful, affect the nervous system and are habit-forming. Narcotics are obtained from the opium poppy and other plants or are made from chemicals. Physicians use these drugs to relieve pain after operations or during a painful illness, or to cause sleep. However, continued and unsupervised use of narcotics can lead to drug addiction. Some well-known narcotics are codeine, morphine, and the much more potent drug heroin.

Other drugs besides narcotics can be habit-forming. Among these drugs are *barbiturates* and *amphetamines*. Barbiturates act as a sedative, or depressant, to help you sleep or relax. Doctors prescribe these drugs to treat high blood pressure, insomnia, and mental disorders.

Doctors sometimes prescribe amphetamines in the form of inhalant sprays to relieve head colds and hay fever. Amphetamines are also used as stimulants to relieve fatigue. They can cause an increase in heart rate and an increase in blood pressure as side effects.

Since 1935, physicians have known that certain organic sulfur compounds, called *sulfa drugs*, produce outstanding results in controlling certain diseases. It was found that sulfa drugs interfered with the ability of germs to digest their own food. This gave the white blood cells in the body a chance to attack and destroy the bacteria.

The first sulfa drug, *sulfanilamide* (sul-fa-*nil*-uh-mide), was introduced in time to save countless wounded soldiers from serious infections during World War II. There are now many other sulfa drugs available for treatment of various infections. Sulfa drugs are highly effective for treating certain types of pneumonia, strep throat, dysentery, venereal diseases, and blood poisoning.

Many diseases are controlled by antibiotics. Penicillin, produced in 1929 by Dr. Alexander Fleming, was the first successful **antibiotic.** *Antibiotics* (an-tih-by-*ott*-icks) *are chemicals produced by living organisms that prevent the growth of bacteria.* Penicillin is a crystalline substance taken from molds similar to those that sometimes grow on bread. Penicillin is most effective against bacteria that cause local infections, certain types of pneumonia, and other bacterial growths. See Fig. 10-18.

A serious problem in the use of sulfa drugs and antibiotics is that bacteria tend to build up a resistance to them. After long use, the drug may become useless. Sometimes, however, an infection that has become resistant to one drug can be treated by switching to another drug.

Cosmetics have been used throughout history. The Egyptians of several thousand years ago were the earliest known users of cosmetics. See Fig. 10-19. Cosmetics are materials applied to make the body more attractive. Among the many cosmetics found on the market are face and bath powders, rouge, facial creams, nail polish, and many kinds of lotions.

Talc (magnesium silicate) is used in making face and bath powders. It is mixed with calcium carbonate to absorb moisture. Kaolin (a basic material in clay) is then added to help the powder stick to the skin. Titanium oxide and zinc oxide, finely ground and thoroughly mixed, are used as white pigments. Coloring matter and perfumes are then added.

Rouge is a face powder to which some red iron(III) oxide has been added. Lipsticks are mixtures of fat or wax with dyes, to which an oil is added as a softener.

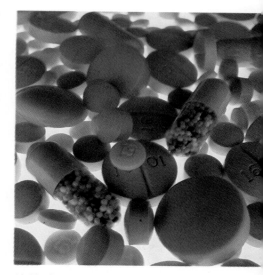

10-18 Pure, crystalline antibiotics are packaged in capsule form ready for the commercial market. Where do antibiotics come from?

10-19 Cosmetics, such as the elaborate eye makeup shown in this photo, were used by the Egyptians thousands of years ago.

About 50% of lipsticks are made using castor oil, perfume, and coloring matter.

Facial creams are emulsions of fats, waxes, and oils in water. Borax and other substances are used to make the oils and water blend together. The oils used to make creams include olive, almond, and mineral. Beeswax, mineral oil, and borax are contained in most cold creams. Lanolin, a grease from the wool of sheep, is an excellent skin softener and is often used in creams, lotions, and hair preparations. Some creams contain potassium soap for cleaning action.

Nail polish is made from lacquers, coloring matter, and a solvent. Nail polish remover contains organic solvents such as acetone.

The proper use of toothpaste or tooth powder helps prevent tooth decay by keeping the teeth clean and by slowing bacterial growth. You can make a good low-cost tooth cleaner by using one part of baking soda, two parts of table salt, and a little chalk ($CaCO_3$) as a mild abrasive. Some flavoring and a small amount of detergent may be added. Commercial toothpastes and powders, however, contain a small amount of chalk, calcium phosphate or calcium sulfate, flavoring, and a detergent. Fluoride toothpastes include very small amounts of tin(II) fluoride to harden tooth enamel.

Louis Pasteur
Discoverer of the Germ Theory

Louis Pasteur (1822–1895), a French scientist, was noted for his work on fermentation and decay. He developed the germ theory of disease and found that food could be made sterile by heat treatment (pasteurization).

Pasteur also showed that milk could be soured by adding organisms from buttermilk or beer. The milk would remain unchanged if these organisms were absent.

One of the puzzling problems of Pasteur's day was the theory of "spontaneous generation," which stated that all living things originate from nonliving matter. Pasteur observed that fermentation of certain liquids was hastened by their exposure to air. This caused him to wonder whether invisible organisms were always present in the atmosphere, or whether they were spontaneously generated. Pasteur performed a series of experiments. In one experiment, he filtered air and exposed unfermented liquids to the clean air. Pasteur was able to show that the organisms causing fermentation were not spontaneously generated. Instead, they came from similar organisms present in ordinary air.

The wine makers of France found that their wine often became sour. The wine turned to vinegar after the fermentation process went too far. Pasteur discovered that certain kinds of bacteria caused the spoilage. When the bacteria were not present, no acids were formed and good wine was produced.

The harmful organisms spoiling the wine could have been killed by boiling, but this would have also destroyed the flavor of the wine. Pasteur found a solution by heating the fermented liquid high enough to kill the bacteria while keeping the temperature low enough to preserve the flavor of the wine. This discovery saved the French wine industry and introduced the process of pasteurization to the world.

Pasteur also reasoned that if bacteria could cause wine to become "sick," perhaps bacteria could also cause sickness in animals and people. His idea later became known as the germ theory of disease.

summary

1. Softwood lumber, such as pine, fir, and hemlock, is used mostly in construction.
2. Limestone is a sedimentary rock used as a building material.
3. A mixture of clay and limestone is heated in a kiln to make Portland cement.
4. Concrete is made by mixing cement, sand, gravel, and water.
5. Common glass is made by mixing sand, soda ash, and limestone.
6. Paints, varnishes, enamels, and lacquers protect wood and metal surfaces.
7. Fires need three things to keep burning: (1) heat to raise the material to its kindling temperature, (2) something to burn (fuel), and (3) oxygen of the air.
8. Soapless detergents lessen the force of attraction between water molecules, permitting the molecules to spread through the cloth and remove the dirt.
9. Leavening agents in dough cause it to rise by forming bubbles of carbon dioxide gas.
10. Nitrogen, phosphorus, and potassium are major elements needed for the proper growth of plants.
11. Drugs are substances used to fight disease or to ease pain.
12. Cosmetics are used to alter one's appearance.

review

Match each item in the left column with the best response in the right column. *Do not write in this book.*

1. granite
2. concrete
3. glass
4. Portland cement
5. plaster of paris
6. glycerin
7. sulfur
8. germicide
9. analgesic
10. lanolin

a. obtained by heating gypsum
b. made from limestone and clay
c. igneous rock
d. metamorphic rock
e. made from sand, soda ash, and limestone
f. mixture of sand, cement, water, and gravel
g. found in some fungicides
h. skin softener
i. sodium hypochlorite
j. by-product of soap-making process
k. anesthetic
l. used to relieve pain

questions

Group A

1. What is "green" lumber? How is it treated?
2. What is the main difference between conifers and deciduous trees? Give several examples of each.
3. Name the three groups of rock used as building materials.
4. How is mortar made? plaster of paris?
5. Why is hair or shredded fiber added to plaster?
6. How are bricks made?
7. What are the raw materials used in making glass?
8. What causes the bubble-like flaws in some glass?
9. Name four items commonly found in oil paint.
10. What is latex paint? What are the advantages of using it?
11. Describe three ways of putting out a fire.
12. List some advantages and some disadvantages of the soda-acid type fire extinguisher. of the foam-type extinguisher.
13. What is a detergent? List several.
14. Write the word equation for making soap. Describe how soap removes dirt.
15. What is the value of using household ammonia as a cleaner?
16. What major elements are needed for proper plant growth? A fertilizer is labeled "5-10-5." What does this mean?
17. If you cut your finger, should you use an antiseptic or a germicide? Why?

Group B

18. In what important ways does the wise use of forests differ from the use of other natural resources, such as oil and coal?
19. Describe the making of Portland cement.
20. What is the difference between cement and concrete?
21. Describe how glass is made. Why are glass objects cooled slowly?
22. Describe the difference between the drying action of ordinary oil paint and that of lacquer.
23. Describe how the soda-acid fire extinguisher works.
24. How does a liquid CO_2 fire extinguisher put out a fire?
25. Describe how synthetic detergents clean clothes.
26. Name and describe the action of three important ingredients of baking powder.
27. Why is baking powder useless after standing in an open container for a long time?
28. Describe the action of hydrogen peroxide as an antiseptic. Why must it be stored in dark bottles?
29. Describe the reaction that takes place when yeast acts on sugar.
30. What effects do each of the following classes of drugs have upon the body: (a) narcotics, (b) barbiturates, (c) amphetamines?
31. Describe the difference between the source of sulfa drugs and that of most antibiotics.

further reading

Akroyd, T. N., *Concrete—Its Properties and Manufacture.* Elmsford, N. Y.: Pergamon Press.

Angeloglou, M., *A History of Make-up.* New York: Macmillian, 1970.

Asimov, Isaac, *The Chemicals of Life.* New York: Abelard-Schuman.

Berger, G., *Home Economic Careers.* New York: Franklin Watts, 1977.

Deman, J. M., *Principles of Food Chemistry.* Westport, Conn.: Avi Publishing, 1976.

Gregor, T., *Manufacturing Processes: Ceramics.* Englewood Cliffs, N. J.: Prentice-Hall, 1976.

Hammersfahr, J. E., *Creative Glass Blowing.* San Francisco: W. H. Freeman.

Mitchell, L. *Ceramics: The Stone Age to Space Age.* New York: McGraw-Hill.

Platt, R., *Discover American Trees.* New York: Dodd, Mead.

Prescott, F., *The Control of Pain.* New York: Thomas Y. Crowell.

Shreve, R. N., *Chemical Process Industries.* New York: McGraw-Hill, 1967.

White, Paul D., *My Life and Medicine.* Boston: Gambit, 1971.

Weiner, J., et al., *Bread.* Philadelphia: J. B. Lippincott, 1973.

metallurgy

Think of the many things you see and use every day that are made of metals such as iron, aluminum, or copper. To make a list of all the things made of metal would be an almost endless task. Can you imagine a world without metals? It would be hard to find substitutes for them. In this chapter, you will study the **metallurgy** of some common metals. *Metallurgy is the science of taking useful metals from their ores, refining them, and preparing them for use.* Not every piece of rock that contains a useful metal is an **ore.** *An ore is a rock or mineral from which a metal can be obtained profitably.*

objectives:

☐ To describe the process of making pig iron in the blast furnace and of making steel in the basic-oxygen furnace.

☐ To gain an understanding of the major processes in the heat treatment of steel.

☐ To explain how iron rusts and how to keep it from corroding.

☐ To develop an understanding of the electrolytic process in obtaining active metals from their ores.

☐ To apply the principles of chemistry to the processes of roasting and reduction.

☐ To examine some general properties of metallic alloys, and to identify the metals that make up several common alloys.

the metallurgy of iron

There are several common ores of iron. Iron, a well-known metal, is the fourth most abundant element in the earth's crust. See Table 2-1 (p. 20). The iron used by people thousands of years ago did not come from the earth but from *meteorites,* rocky materials flying through space striking the earth. At first, people believed that iron was a gift of the gods and used it in their religious services. Objects made of iron taken from meteorites date back to about 3000 B.C. However, it was not until about 1500 B.C. that people discovered how to obtain iron from its compounds in the earth. This discovery marked the beginning of the Iron Age.

The major ore of iron today is a reddish-brown mineral called hematite (*hem*-uh-tite) (Fe_2O_3). This iron ore contains about 52% iron. For many years, the main supply of hematite was the Lake Superior region. See Fig. 11-1. Today, the ore is imported from huge deposits in Quebec, Labrador, and South America.

In addition to the hematite deposits in the Lake Superior region, there are also large supplies of low-grade ore called taconite (*tack*-uh-nite), a hard rock containing about 25–50% iron. Taconite is crushed, refined, and concentrated before it is put to use. This ore is becoming more and more important as higher-grade ores are used up.

Limonite ($Fe_2O_3 \cdot x\ H_2O$), magnetite (Fe_3O_4), and siderite ($FeCO_3$) are less vital sources of iron. Magnetite is so named because it is magnetic. That is, magnetite attracts iron like a common magnet. Large supplies of iron ore are important to a country because steel, which is made from iron, is extremely important to modern industry.

11-1 An open-pit mine near Lake Superior. Discuss some environmental problems of open-pit mining and their solutions.

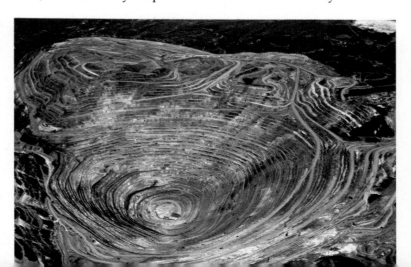

Iron is obtained by reducing its ore in a blast furnace.
Hematite is not at all like iron. The formula, Fe_2O_3, shows that oxygen is combined with the iron in the ore. To get rid of this oxygen, the ore must be heated with a **reducing agent.** *A reducing agent is a substance that takes away the oxygen from a compound.* The blast furnace is used to bring about the chemical reactions needed for making iron from its ore. See Fig. 11-2.

Coke, which is made from soft coal, is used in the reduction process. The coke, when burned with oxygen of the air in a blast furnace, is changed to carbon dioxide (CO_2) and then to carbon monoxide (CO). Carbon monoxide acts as the reducing agent in taking away the oxygen from hematite. This is shown by the following reactions:

Step 1. Burning of coke, which is mostly carbon (C):

$$C + O_2 \rightarrow CO_2 \uparrow$$

Step 2. Formation of carbon monoxide from carbon dioxide and coke:

$$CO_2 + C \rightarrow 2\ CO \uparrow$$

Step 3. Carbon monoxide acts as the reducing agent and takes away the oxygen from the hematite ore:

$$Fe_2O_3 + 3\ CO \rightarrow 2\ Fe + 3\ CO_2 \uparrow$$

The carbon dioxide formed as one of the products in Step 3 reacts with coke to form more carbon monoxide.

Hematite is not pure iron. The ore always contains varying amounts of impurities. The most common impurity is sand. Sand is mostly silicon dioxide (SiO_2). When heated, silicon dioxide reacts with limestone ($CaCO_3$) to form calcium silicate ($CaSiO_3$) and carbon dioxide. The easily melted calcium silicate is a glassy waste product called *slag*. The reaction is

$$SiO_2 + CaCO_3 \rightarrow CaSiO_3 + CO_2 \uparrow$$

Large amounts of limestone are added to the ore and coke in the blast furnace to bring about the above chemical reactions. The mixture of iron ore, limestone, and coke that is placed in the blast furnace is called a *charge*. When the materials are hot enough, the iron becomes liquid and sinks toward the bottom of the furnace. Because it is less dense, the slag floats on top of the liquid iron. Both of these liquids are drawn off at the bottom of the blast furnace, as shown in Fig. 11-2.

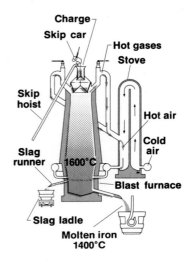

11-2 Section of a blast furnace. A blast furnace is about ten stories high. Can you tell what the "skip car" in the diagram is used for?

The blast furnace is a very efficient device. Almost everything that comes from the furnace is useful in some way. Some of the slag is made into a fiber-like material, called mineral wool, which is used as an insulating material. Slag is also used for fire-proofing, for making roofing materials, and for building roads. The "waste" heat from the furnace is used to warm the cool fresh air used to obtain the oxygen. The waste gases are burned as a fuel. Even the waste dust is made into fertilizer.

Iron from the blast furnace is used in making steel. The iron drained from the blast furnace is called *pig iron*. It contains about 92–94% iron. The rest is made of carbon, manganese, phosphorus, silicon, and sulfur as impurities. For many uses, a greater purity of iron is needed. See Fig. 11-3.

To change pig iron into steel, the impurities must be removed, or at least greatly reduced. An important difference between steel and iron is that steel contains a small but measured amount of carbon. Steel may also contain small but controlled amounts of other metals added to obtain certain properties. Impurities such as

11-3 Pig iron taken from a blast furnace is being poured into a basic-oxygen furnace. What is the purpose of this process?

unit 2 chemistry in our world

sulfur, silicon, and phosphorus are removed from the pig iron in the steel-making process. Scrap steel from car bodies and engines is often used in making new steel.

The basic-oxygen furnace is widely used in steel making because it turns out large amounts of high quality steel in a short time. As shown in Fig. 11-4, the egg-shaped furnace is mounted on pivots. The furnace is filled with a charge consisting of molten pig iron from the blast furnace, limestone, and scrap steel. A water-cooled oxygen pipe is lowered into place. Then oxygen is blown into the molten material to burn away the unwanted matter. No fuel is needed because the reaction between oxygen and impurities is so strong that it keeps the mixture in a molten state from the heat produced.

After about 40–60 min, the furnace is tilted and about 80 metric tons of finished steel are poured into large blocks. The steel can also be made directly into finished products.

Carbon and heat treatment give steel certain desired properties. Changing pig iron into steel involves getting rid of unwanted substances. Measured amounts of other substances are then added to produce steel with the required properties.

The amount of carbon in steel varies from about 0.05 to 2%. If the percentage of carbon is low, the product is very soft steel, such as that used in making paper clips. Steel with a high amount of carbon is hard and very brittle. The carbon crystals make the steel hard because they keep the atoms of iron from sliding past each other. The hardness of steel depends not only on the amount of carbon but also on how the carbon is joined with the iron. This is determined by the crystal structure after heat treatment.

Almost every type of steel must be given some form of heat treatment. Some steels need a high degree of hardness or toughness. Some steel objects must be highly elastic. This means they can resist strong forces. Other steel objects must be able to take on a new shape without breaking. One example is a rod drawn through a die to produce wire as shown in Fig. 11-5.

The three main methods used in the heat treatment of steel are quenching, tempering, and annealing.

Quenching is a heat treatment in which steel is heated to a given temperature and then plunged into water or oil

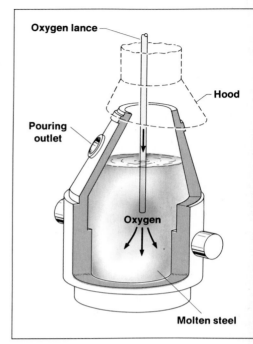

11-4 The basic-oxygen furnace turns molten pig iron into steel in a short time. What is the purpose of the oxygen lance shown in the diagram?

11-5 Cross-section of a steel rod being forced through an opening (die) to produce wire. What property of steel is needed to make wire?

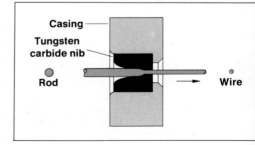

for sudden cooling. This process makes the steel strong, hard, and brittle.

When high-carbon steel is quickly cooled, it becomes very hard. Steel made this way would be too hard and brittle for many needs. To get the right degree of hardness, the quenched steel is reheated to a point below the quenching temperature and then recooled. The higher the reheating temperature, the softer the finished steel becomes. This process, called *tempering*, relieves the stresses in the steel and increases its toughness. Both of these steps control the size and shape of the crystals in the metal.

Annealing consists of slowly cooling heated steel, first in a furnace and then in air. This process softens the steel so that it can be made into wire.

activity

Heat treatment of iron. Heat the bent portion of two bobby pins to redness in a Bunsen flame. Dip one heated pin into cold water so that it cools quickly. Remove the second pin gradually from the tip of the flame to cool slowly in the air. Now test the springiness of each of the two bobby pins by pulling the points apart. Compare the two pins with one that has not been heated. Describe the results.

Active metals can prevent iron from rusting. Unprotected iron and steel objects rust in moist air. The corrosion, or rust, that forms is brittle so the rust flakes off. This exposes the metal below it to further rusting until the iron has rusted throughout.

The rusting of iron is an electrochemical process. Iron contains carbon and small amounts of other impurities. Carbon particles in contact with moist iron cause the iron to rust quickly. Rain water containing dissolved carbon dioxide reacts with the iron. The carbon acts as the positive part of a small dry cell; the iron becomes the negative part. The impure water becomes the electrolyte. See Fig. 11-6. When iron rusts, the iron atoms lose two electrons and become iron(II) ions (Fe^{++}):

$$Fe - 2 \text{ electrons} \rightarrow Fe^{++}$$

11-6 Iron and steel contain impurities that cause rusting unless covered by a protective coating. What acid is produced by water and carbon dioxide?

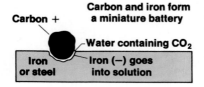

Carbon +
Carbon and iron form a miniature battery
Water containing CO$_2$
Iron or steel
Iron (−) goes into solution

unit 2 chemistry in our world

If conditions favor the loss of electrons, the rate of rusting is increased. If iron can be kept from losing its electrons, rusting is slowed or stopped. Also, if iron can be protected from oxygen, carbon dioxide, and moisture, the surface will not rust.

If a more active metal, such as magnesium, is connected to a less active metal, such as iron, the more active metal will rust. The iron is protected from going into solution as iron(II) ions. Thus, rusting of the iron is prevented. For example, blocks of magnesium bolted below the water line to the steel hull of a ship protect the steel from rusting. The more active magnesium metal goes into solution as Mg^{++} ions, leaving the iron unchanged and free of corrosion. Of course, the magnesium metal is used up in the process and must be replaced. Can you think of a way to prevent iron pipes from rusting in moist ground?

metallurgy of other common metals

Aluminum is obtained by the Hall process. The aluminum that was first produced was so costly that it could only be used for jewelry. In fact, Napoleon had aluminum knives, forks, and spoons, which he used in place of gold and silver for special occasions. As late as 1852, the cost of aluminum was more than $1100 per kilogram. Today, the cost is only a tiny fraction of that amount.

Aluminum is made by passing an electric current through a solution of melted aluminum oxide (Al_2O_3) and cryolite (Na_3AlF_6). Aluminum oxide is the main compound in an ore called bauxite (*bawks*-ite). Bauxite contains 45–60% aluminum oxide.

Bauxite ore is often found in shallow deposits close to the earth's surface. The richest deposits are found in a wide band along either side of the equator. At present, much bauxite comes from Jamaica, Surinam, and Guyana. In this country, bauxite is mined mostly in Arkansas, Georgia, and Alabama.

In 1886, Charles Martin Hall, a 23-year-old American, found an easy way of separating aluminum from its oxide ore. He found that when aluminum oxide and cryolite were heated to about 1000°C, the molten mixture con-

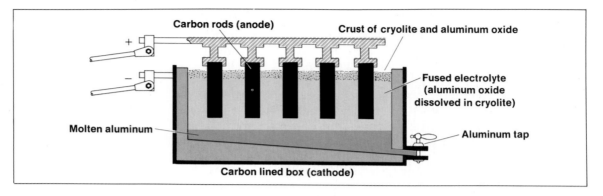

Carbon rods (anode)

Crust of cryolite and aluminum oxide

Fused electrolyte (aluminum oxide dissolved in cryolite)

Aluminum tap

Molten aluminum

Carbon lined box (cathode)

11-7 Aluminum is obtained by electrolysis of purified aluminum oxide and cryolite (sodium aluminum fluoride). What is given off at the anode?

ducted an electric current. Aluminum metal formed at the negative pole of the cell, or *cathode*. At the positive pole, or *anode*, oxygen was set free. As shown in Fig. 11-7, aluminum liquid sinks to the bottom of the cell, from which it is removed from time to time. This reaction is shown by the following equation:

$$2 \ Al_2O_3 \rightarrow 4 \ Al + 3 \ O_2 \uparrow$$

Part of the oxygen set free at the anode combines with the hot carbon rods. Carbon monoxide, which burns on contact with the air, is formed. As electrolysis continues, the carbon rods are used up and must be replaced. This electrolysis process uses very large amounts of electric energy.

Aluminum is a good conductor of heat and is widely used for cooking utensils. As a conductor of electricity, aluminum is not as good as copper. On the other hand, aluminum is much lighter than copper, so it is used to some extent in electric power lines. Aluminum foil is used in the walls and ceilings of buildings to keep heat out in the summer and hold heat in during the winter. Aluminum foil is used in this way because it reflects up to 90% of the heat energy that hits it. Since aluminum does not absorb heat, it also does not give off much heat. The greatest user of aluminum is the building industry. Aluminum shingles, siding, pipes, window frames, and roofs are only a few examples of aluminum products on the market. See Fig. 11-8.

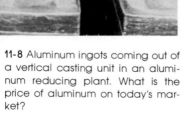

11-8 Aluminum ingots coming out of a vertical casting unit in an aluminum reducing plant. What is the price of aluminum on today's market?

Copper is sometimes found in a free state. Copper was one of the first metals to be used by primitive people. It can be found as a free metal, that is, not combined

196

chemically with other elements. Free copper is still found in Egypt, though it probably was more plentiful there thousands of years ago. Scientists have dug up copper dishes dating back to 4000 B.C.

In this country, about 95% of the copper deposits are found in Michigan, Arizona, Utah, Montana, Nevada, and New Mexico.

Free copper is found deep underground and is difficult to mine. It is mixed with various rocks that must first be crushed to a coarse powder. The lighter rock is then removed by a stream of water, and the heavier copper metal remains. This step concentrates the ore. Concentrating increases the amount of copper in the batch.

The second step mixes the concentrated ore with coke and limestone in a small blast furnace. In the furnace, limestone unites with silica impurities to form slag, in the same way slag is formed in making iron in a blast furnace. The burning coke melts the copper, which collects in a pool at the bottom of the furnace. The molten copper is then drawn off and cast into large plates. Further refining by electrolysis makes the copper ready for commercial use.

Copper is also found in compounds. In addition to free copper, several ores of copper are compounds. Examples are chalcopyrite (kal-koh-*pie*-rite) ($CuFeS_2$), chalcocite (*kal*-koh-site) (Cu_2S), and cuprite (*coo*-prite) (Cu_2O). Mineral experts believe that about half of the world's copper resources are in the form of chalcopyrite.

The following steps describe how copper is obtained from chalcocite, a sulfide of copper:

1. The *ore-flotation* process is used to separate copper(I) sulfide from the rock. In this process, crushed ore, water, and clean oil are mixed in huge tanks. When air is blown through the mixture, froth collects on top. The bits of ore cling to the air bubbles on top, while the wet rock settles to the bottom. Then the froth with the copper(I) sulfide is skimmed off into other tanks. See Fig. 11-9 on page 198.

2. The ore is then roasted. **Roasting** *is a process in which a sulfide ore is heated in oxygen-enriched air to change it into an oxide.* The Cu_2S is heated with limestone to change most of the ore into copper(I) oxide (Cu_2O). Limestone combines with silica, forming a slag that floats on top of the molten copper oxide-sulfide mixture.

3. The next step is the *reduction* of the mixture to the

11-9 In the ore-flotation process, the particles of Cu_2S stick to the air bubbles and are skimmed off. What settles to the bottom of the tank?

11-10 Blister copper being poured into 2750-kg slabs from a blast furnace. What is the next step in making refined copper?

copper metal. A special blast furnace is used to cause a reaction between the two copper compounds in the mixture.

$$Cu_2S + 2\,Cu_2O \rightarrow 6\,Cu + SO_2 \uparrow$$

As the molten copper cools into slabs, the escaping gases leave the metal with a blistered look. The slabs are therefore called *blister copper*. See Fig. 11-10.

4. The final step in preparing copper for the market is refining by *electrolysis*. Impurities in crude copper make it a very poor conductor of an electric current, so it must be purified even more.

All crude copper is refined by electrolysis, as shown in Fig. 11-11. The cell is a tank of copper(II) sulfate solution containing a small amount of sulfuric acid to help conduct an electric current. The anodes are slabs of crude copper. Thin sheets of pure copper serve as cathodes on which the pure copper is deposited.

An electric current attracts the copper ions (Cu^{++}) in the solution to the cathode sheets. There, they are plated out as pure copper (Cu). The impure copper anodes supply fresh copper ions for the solution. Chemists write these reactions as follows:

At the cathodes (−): $Cu^{++} + 2$ electrons $\rightarrow Cu$

At the anodes (+): $Cu\ \ - 2$ electrons $\rightarrow Cu^{++}$

unit 2 chemistry in our world

During the plating process, the anodes slowly get smaller as the atoms of copper go into solution as copper ions. Meanwhile, the cathodes increase in size as copper ions from the solution are deposited as pure copper. The copper obtained by this process is over 99.9% pure. The impurities from the anodes drop to the bottom of the tank as a muddy deposit. This mud often contains gold and silver as valuable by-products.

activity

Electroplating of copper. Using distilled water, prepare about 50 mL of a dilute solution of copper(II) sulfate in a beaker. Add several drops of sulfuric acid. Clean two carbon rods, like those in a flashlight cell. Hold them upright and dip the ends vertically into the solution about 4 cm apart. Across the two carbon poles, connect a 6-volt battery. After several minutes, remove the battery and the carbon rods from the solution. What do you observe on the anode? the cathode? Explain. Why was sulfuric acid added?

Copper is a reddish metal that is often covered with a brownish coat of tarnish. If you rub off the tarnish with sandpaper, you will see the red color of pure copper underneath. When copper is exposed to the weather, it becomes covered with a greenish coating of several copper compounds. Unlike the rusting of iron, these coatings protect the copper underneath from further corrosion.

Copper is most widely used for making electric wire. It is also used for water pipes and roofing. Next to silver, copper is the best conductor of electric currents.

11-11 Crude copper is refined by electrolysis. Can you write the equations for the reactions at the anode? cathode?

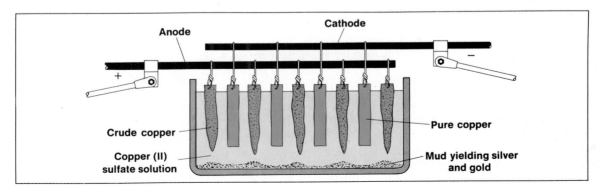

11-12 Silver bullion cast in this factory will be stored in huge vaults for safekeeping. Can you name several other precious metals?

Silver is a precious metal. Precious metals are those that are useful but are scarce and costly. You know about silver, gold, and platinum. Other precious metals that are not so well known are palladium, osmium, and iridium.

Silver is found as a free metal and also as silver sulfide (Ag_2S), called argentite. Most commercial silver, however, is produced in processing copper, lead, and zinc ores. The best source of silver comes from the muddy deposits that sink down to the bottom of the tank in the refining of copper.

When this mud is treated with sulfuric acid, silver sulfate (Ag_2SO_4) is formed. During this reaction, the sulfuric acid separates the silver from the gold. Strips of copper are dipped into the solution and pure silver collects on the strips. The single replacement reaction is

$$Ag_2SO_4 + Cu \rightarrow CuSO_4 + 2\ Ag \downarrow$$

Carefully weighed bars of silver, called bullion, are obtained by melting down the silver. See Fig. 11-12.

Silver is a soft metal that shines with a beautiful luster. Although silver is the best conductor of electric current, it is too costly for most electric wiring. You have probably seen how silver tarnishes and turns brown or even black. Sulfur compounds present in such foods as mustard and eggs, and the fumes of sulfur in polluted air, will cause silverware to tarnish. This tarnish is a coating of silver sulfide, produced by the reaction of silver with sulfur compounds.

Large amounts of silver are used in jewelry and other ornaments. The best table silver is sterling, which is 92.5% silver and 7.5% copper. Photography is a billion-dollar industry that makes use of most of the silver in this country.

Mexico is the largest silver-producing country, followed by the United States and Canada.

activity

Cleaning tarnished silver. Find a badly tarnished silver object such as a spoon, and boil it for several minutes in an aluminum pan containing a dilute solution of salt and baking soda. Remove the object from the solution and look at it closely. What do you see?

Other metals are obtained from their ores. You have studied the methods of obtaining iron, aluminum, copper, and silver from their ores. Other common metals are also obtained in a similar way. See Tables 11-1 and 11-2.

Table 11-1:

The Metallurgy of Other Common Metals					
Metal	Ore or Source	Formula	Process Used	Country Found	Uses
Lead	galena	PbS	ore-flotation, roasting, reduction	United States, Russia	making car batteries, pipes
Magnesium	sea water	$MgCl_2$	electrolysis	worldwide	car and plane parts
Tin	cassiterite	SnO_2	ore-flotation, reduction	Malaysia, Bolivia	making tin plate for cans
Zinc	sphalerite	ZnS	roasting, reduction	United States, Canada, Russia	making galvanized iron; brass articles

Table 11-2:

Other Common Metals and Their Uses	
Metal (symbol)	Uses
Cadmium (Cd)	control rods for nuclear reactors; plating
Chromium (Cr)	plating; making stainless steel
Cobalt (Co)	mixed with other metals; treatment of cancer
Gold (Au)	jewelry; ornaments; standard of wealth
Mercury (Hg)	lamps; switches; thermometers and barometers
Nickel (Ni)	hardens steel; plating; catalyst
Platinum (Pt)	catalyst; electronics; lab-ware; jewelry
Tantalum (Ta)	surgery; making chemical equipment
Titanium (Ti)	combustion chambers for rockets and jet aircraft
Tungsten (W)	filaments for light bulbs; mixed with other metals

metal alloys

Most metal objects are mixtures, not pure metals. To get a metal that is totally pure is a costly process and one that is not usually needed. Two major exceptions are the pure aluminum used in pots and pans, and the pure copper used for electric wires. Even a 0.5% impurity in copper will cut down its ability to carry current by 50%. Few other metal objects in common use are made of a single, pure metal. Instead, people use tailor-made mixtures of metals that have properties designed for special uses.

A substance made up of two or more metals melted together is called an **alloy.** Over 35 metal elements go into making thousands of alloys. Different alloys are obtained from the same elements by mixing them in different proportions. See Fig. 11-13.

There are several ways in which metals can be blended in an alloy. One way is by making a *solid solution.* If one metal dissolves in another when they are melted together, and stays dissolved when cooled, the two metals form a solid solution. For instance, copper and zinc mix in all proportions, as do alcohol and water. On the other hand, only a limited amount of zinc will mix with lead to form an alloy. If more than the needed amount of zinc is added, the excess will simply form a layer on top. Gold and copper will also form a solid solution. In fact, most common alloys are solid solutions.

11-13 Alloys of metals such as titanium, tungsten, and chromium enable missles, jets, and the space shuttle to resist very high temperatures.

unit 2   chemistry in our world

Sometimes metals form *metallic compounds* that do not follow the pattern of valence numbers discussed earlier in the text. These alloys have crystal structures with unusual formulas, such as Cu_5Zn_8, Na_4Pb, or Al_2Cu.

Still another class of alloys is called *metallic mixtures*. In this class, crystals of one substance are scattered throughout the mass, somewhat like raisins in a cake. Alloys in this class have a wide range of properties. Steels are an example of the metallic mixture class of alloys. The way in which the substances are mixed in alloys is shown in the crystal pattern. This pattern can be learned by looking at the polished surface of an alloy with a microscope and by X-ray studies.

Some metals will not form alloys with certain other metals. For example, copper does not mix easily with iron. However, copper forms some of the most widely used alloys when mixed with tin and zinc.

Alloys have desirable properties. The properties of alloys are usually quite different from those of the basic metals that they contain. For one thing, an alloy is often harder than the basic metals. Brass, which is composed of zinc and copper, is harder than either of these metals. The ability of brass to resist stretching is more than twice that of copper and more than four times that of zinc.

Alloys are poorer conductors of heat and electricity than the pure metals. It has already been mentioned that impurities in copper greatly reduce the amount of electric current it can carry.

Further, the melting point of an alloy is often lower than that of any of the metals of which it is made. Many other properties of metals are changed in alloys. These properties include color, elasticity, expansion, and magnetism.

Alloys have many uses. The use of an alloy of copper and tin, called *bronze,* began thousands of years ago. People were making objects of copper without knowing that tin was present in the copper. Later, they learned to mix tin with copper. The bronze alloy that resulted was harder and more durable than either copper or tin. More than 3000 years ago, people looked for tin to make bronze utensils. You may have seen bronze memorial tablets and statues in parks and other public places. Bronze bearings are used in the rotating parts of electric motors.

11-14 Bearings shown here are made of an alloy that helps reduce friction from a rotating shaft. Can you suggest a reason why a bearing may "burn out"?

Have you ever heard someone talk about "burning out a bearing" in a car? A bearing is a hollow metal tube that allows turning parts to rub against each other with minimum friction. See Fig. 11-14. An alloy, called *babbitt*, which includes tin, copper, and antimony, is often used for such bearings. The friction produced between the alloy and the steel is much less than between steel against steel. Babbitt bearings are used on the crankshaft of car engines.

Another widely used alloy is *brass*, an alloy of copper and zinc. Brass hardware is used on ships and also around the home. Since foods affect it, brass cannot be used for cooking utensils. However, brass does make long-lasting water pipes.

As you have learned, a measured amount of carbon is usually added to iron to make steel. Most steel contains a small amount of a number of other metals.

The kind of metal added to the steel depends on what is needed in the finished product. For instance, nickel is added to increase the heat and acid resistance of steel. Chromium can be added to prevent rust. Some steel used today may contain as many as 15 different kinds of metals, depending on the properties needed. See Fig. 11-15.

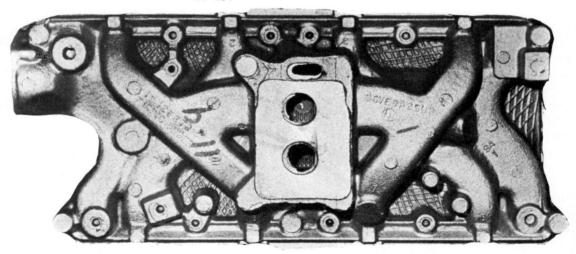

11-15 This automobile manifold is made of a light-weight aluminum alloy. What would be some advantages in using a lighter block in a car engine?

Table 11-3 lists some additional alloys along with their properties and uses. The table also shows the usual makeup of each alloy.

Table 11-3:

Alloys and Their Uses			
Alloy	Composition	Property	Uses
Alnico	aluminum nickel cobalt iron copper	permanent magnet	magnets for telephones, loudspeakers, hearing aids
Duralumin (doo-rahl-uh-min)	aluminum magnesium copper manganese	strong, lightweight	tools, ladders, building materials, frames
Monel	copper nickel	does not tarnish easily	sinks, drainboards, ice cream cabinets
Nichrome	nickel chromium iron manganese	high electrical resistance	heating elements in electric irons, toasters, and ranges
Wood's metal	bismuth cadmium tin lead	low melting point	plugs for sprinklers of automatic fire extinguishers
Solder	lead tin	low melting point	to join two pieces of metal together
Stainless steel	chromium steel	does not rust	building materials, cabinets, sinks, kitchen utensils.

activity

Preparation of an alloy. In a crucible, heat 20 g of bismuth, 10 g of lead, 5 g of cadmium, and 5 g of tin. Stir with an iron wire until the mixture has melted. Using forceps, pour the molten mixture carefully into a mold to harden. What is the name of the alloy you made? Observe what happens when this alloy is placed in boiling water. From the masses you used, find the percentage of each element in your sample. Suggest some uses for this alloy.

The Flame Test for Metals

There are certain metallic elements that can be recognized by the color they give to a flame.

When electrons in the outer shells of atoms are "excited" enough by energy received from a suitable source, they are raised to higher energy levels. When these electrons fall back to their normal energy levels in the atom, this energy is given off in the form of light. The light produced has a special spectrum, or color pattern, for each given element. No two elements have the same spectrum.

The outer electrons of the Group I family of metals are easily excited to higher energy levels by the use of a Bunsen flame. These elements give a distinct color to the flame that can be easily seen. The colors can be used to identify the metals. For instance, lithium compounds give off a bright red color in a Bunsen flame; sodium compounds, yellow; and potassium compounds, violet.

Several metals in the Group II family also give color to the flame. Among these are calcium, which produces an orange-red color; barium, yellowish-green; and copper, green.

To perform the flame test, hold a platinum wire attached to a glass rod in a Bunsen flame until no color is observed beyond the usual blue color of the flame. This procedure cleans the wire. Next, touch the tip of the wire to a small sample of a compound containing one of the metals mentioned. Then hold the tip in the outer edge of a Bunsen flame and observe the color produced.

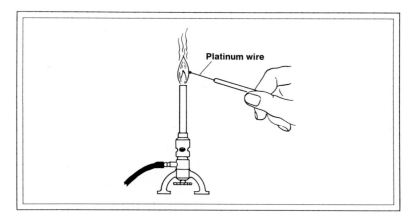

Platinum wire

summary

1. Metallurgy is the science of refining useful metals from their ores.
2. An ore is a rock or mineral from which a metal can be obtained in worthwhile amounts.
3. In making iron, a charge of iron ore, coke, and limestone is heated in a blast furnace.
4. The basic-oxygen furnace makes iron into steel.
5. Steel contains regulated amounts of carbon for different properties.
6. The hardness of steel is controlled by the amount of carbon it contains and by special heat-treatment processes.
7. Various ways used to obtain metals from their ores include ore-flotation, roasting, reduction, and electrolysis.
8. Roasting is a process in which sulfide minerals are heated in air, or oxygen-enriched air, to change them into oxides.
9. Reduction is a process in which an oxide ore gives up its oxygen to produce a free metal.
10. An alloy is made up of two or more metals melted together.
11. Some metals can be identified by their color in the flame test.

review

Match each item in the left column with the best response in the right column. *Do not write in this book.*

1. calcium silicate	a. tin ore
2. carbon monoxide	b. changing sulfide ores to oxides
3. hematite	c. a magnetic material used in telephones
4. roasting	d. lead ore
5. electrolysis	e. negative terminal of an electrolytic cell
6. galena	f. excellent conductor of electricity
7. cathode	g. bronze
8. silver	h. solvent for Al_2O_3
9. alloy	i. slag obtained from a blast furnace
10. cryolite	j. iron ore
	k. process used in refining of copper
	l. reducing agent used in a blast furnace

questions

Group A

1. What is meant by metallurgy?
2. Are all metal-bearing rocks ores? Justify your answer.
3. Name three important ores of iron.
4. What substances are contained in a blast furnace charge?
5. What is the difference between iron and steel?
6. What charge is used in the basic-oxygen furnace?
7. What compound is used to remove silica in steel-making?
8. What function does coke serve in the steel-making process?
9. What effect does sudden cooling have on hot steel?
10. When iron rusts, what change takes place in the iron atoms?
11. What is meant by roasting an ore? Why is this done?
12. List five important uses for aluminum. Why is aluminum used in these ways?
13. What are the major ores of copper? Where are deposits of copper ores found?
14. How is blister copper refined? What is the purpose of the refining process?
15. What is a precious metal? List five precious metals.
16. List three sources of silver. Where is silver mined?
17. What was probably the first alloy known to exist? How was this alloy probably discovered?
18. How is monel metal used? Wood's metal?
19. What metals are used to make solder?

Group B

20. Explain by means of chemical equations how pig iron is produced in a blast furnace.
21. Liquid-air plants and coke ovens are often located near steel mills. Why?
22. Describe the steel-making process in a basic-oxygen furnace.
23. What agents cause iron to rust?
24. It is not a good practice to join copper tubing with iron fittings. Suggest a reason why.
25. How would you show that aluminum and copper tarnish?
26. What did Charles Martin Hall discover? Describe his process.
27. Describe the chemical action that takes place at the anode during the electrolysis of aluminum.
28. Describe how free copper is separated from a rock mixture.
29. Describe how copper is obtained from its sulfide ore. Give reasons for each step.
30. Describe how ore-flotation is used for concentrating the Cu_2S ore.
31. What causes silver to tarnish? How can you remove the tarnish easily?
32. List three ways in which alloys differ from the pure metals of which they are made.
33. Why is bronze or babbitt metal used for bearings?
34. Why are alloys often used in place of pure metals?

problems

1. What is the minimum amount of iron you would expect to get from a metric ton of taconite? (A metric ton is 1000 kg.)

2. If hematite were pure Fe_2O_3, what percentage of iron would it contain? (Atomic masses: Fe = 56; O = 16.)

3. Why would the percentage of iron obtained in problem 2 be different from the value given in this chapter?

4. On a separate sheet of paper, write the equations below and fill in the blanks. *Do not write in this book.*

   a. Zn^{++} + 2 electrons → _____

   b. Pb _____ → Pb^{++}

   c. $2Cl^-$ _____ → Cl_2

   d. _____ −1 electron → Na^+

   e. K^+ + 1 electron → _____

   f. Mn^{++++} _____ electrons → Mn^{++}

5. Pure gold is marked 24K (carat). If a bracelet having a mass of 50 g is marked 14K, what percentage of gold does it contain? Find the number of grams of gold in the bracelet.

further reading

Burt, O. W., *The First Book of Copper.* New York: Franklin Watts.

Fisher, D. A., *Steel: From the Iron Age to the Space Age.* New York: Harper & Row.

Knauth, P., *The Metalsmiths.* New York: Time-Life Books, 1974.

Massey, A. G., et al., *Chemistry of Copper, Silver and Gold.* Elmsford, N.Y.: Pergamon Press, 1975.

McCabe, C. L. , et al., *Metals, Atoms and Alloys.* Washington, D.C.: National Science Teachers Association, 1963.

Sinkankas, J., *Mineralogy for Amateurs.* New York: Van Nostrand, 1964.

Sorrell, Charles, *Minerals of the World.* Racine, Wis.: Western Publishing, 1974.

Tracy, E. B., *The World of Iron and Steel.* New York: Dodd, Mead, 1971.

12

fossil fuels

Anything that burns gives off heat. However, this trait alone does not make a good **fuel.** *A fuel is a substance that can be burned to produce heat at a reasonable cost.* Fuels contain potential energy that is locked within their chemical bonds. Good fuels should be easy to store, leave little unburned ash, and burn cleanly. Further, the burning should produce no unwanted by-products. Coal, petroleum, and natural gas meet nearly all these conditions. All common fuels contain carbon either as a free element or in compounds. In the last chapter, you learned about the use of coke (carbon) to reduce oxide ores to useful metals. In this chapter, you will learn about fossil fuels. All of these fuels come from organic matter.

Today, the world is facing a shortage of fossil fuels. We are using more of these fuels than at any time in history. The supply is running out.

objectives:

☐ To describe the various stages in the formation of coal.

☐ To describe the making of charcoal and coke, and the formation of water gas and natural gas.

☐ To compare the products formed by complete and incomplete combustion of a hydrocarbon fuel.

☐ To describe the destructive distillation of soft coal and the fractional distillation of petroleum.

☐ To demonstrate understanding of the causes of developing energy shortages.

unit 2   chemistry in our world

coal

Coal is a product of the distant past. The origin of coal dates back some 300 million years. Scientists believe that coal is mostly the buried remains of tropical plants. In the past, swampy sections of the earth were covered by very dense plant life. These huge tree ferns and mosses built up many layers as they died. Finally, these layers of dead plants were covered with soil and rock. See Fig. 12-1. Heat from the earth's interior and pressure from the weight of the earth turned these remains into coal. This process of coal-making is probably going on even today. Unfortunately, we are burning coal faster than new coal is forming.

In the United States, coal mines average less than 100 m in depth. However, many mines in other countries go down from 1000 to 1300 m. Much coal in this country is strip-mined, where it is close to ground level and can be exposed by removal of a layer of soil.

Peat is the first stage in the making of coal. Peat is a soft, brown, spongy material composed of plant life that has been changed through the action of heat and pressure. The fiber structure of peat clearly shows its plant origin. Peat bogs are found in Michigan, Wisconsin, and Pennsylvania. Since peat contains a large amount of moisture, it produces a great deal of smoke as it burns. Thus, it is smoky and is not a good fuel. Despite this, peat has been used as a fuel in Europe for hundreds of years because of its abundance and its low cost.

As peat continues to decompose over a long period of time, it loses most of its fiber. The peat becomes *lignite,* or "brown coal." Lignite is harder than peat, contains less moisture, and has a higher percent of carbon. The amount of moisture in lignite is about 50%.

Heat and pressure slowly change some lignite into *bituminous* (by-*too*-min-us), or soft coal. Bituminous coal contains even less moisture than lignite. However, it contains much less *volatile* matter; that is, matter that, when heated, forms gases that burn easily.

In some places, such as eastern Pennsylvania, the folding of the earth's crust has produced still greater pressure. This increase in pressure changes some bituminous coal into *anthracite* (*an*-thruh-syte), or hard coal. Anthracite is often deeper in the ground than bituminous

12-1 A fossil imprint of an ancient fern plant is clearly recorded in this sample of coal. What is a fossil?

coal. In addition, it contains a higher percent of carbon and a lower percent of volatile matter.

It may be hard to picture in your mind the chemical changes that changed plants into black, glossy coal. Remember that these changes take place over millions of years.

The moisture and carbon content of different kinds of coal vary greatly. Even different samples of the same kind of coal can have different amounts of moisture and carbon. A high-moisture content in fuel lowers its heat value. Peat and lignite contain more moisture than coal, so their heat values are lower.

Because of its value to industry, coal is sometimes called "black gold" or "black diamond." All solid fuels, like coal, contain matter that does not burn but remains as *ash*. In general, low-ash fuels have more heating value.

Most of the coal used in this country contains 6–8% sulfur. Unfortunately, the burning of high-sulfur coal is a major source of air pollution.

activity

Amount of ash in coal. Crush and find the mass of a sample of coal. Heat the sample in an open porcelain crucible until it is completely burned. Then find the mass of the ash residue and of the empty crucible. Determine from these data the percent of ash in the coal sample.

Anthracite is more costly then bituminous coal, but it is a cleaner-burning fuel. It also gives off less smoke and gas with a strong odor. Anthracite has a high heat value, although high-grade bituminous coal gives off more heat per kilogram than any other coal.

Bituminous coal, in addition to being the most important solid fuel, is also the most plentiful. Fortunately, large desposits of this resource are found in the United States. The United States is the largest coal-producing country in the world, followed by Russia and China. Bituminous coal is mined in some 37 states in this country and is the chief fuel used in steam-electric generating plants. Coal and its products supply about 20% of the country's energy needs. See Figs. 12-2 and 12-3.

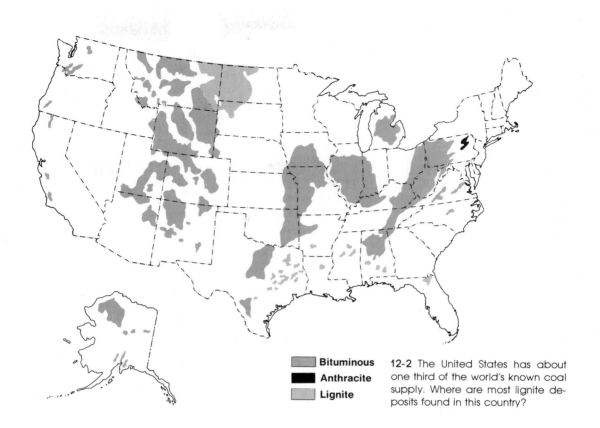

Bituminous
Anthracite
Lignite

12-2 The United States has about one third of the world's known coal supply. Where are most lignite deposits found in this country?

Many by-products come from soft coal. Unless soft coal is properly burned, a loss of fuel and heat will result. Black smoke and soot are caused by the incomplete burning of volatile matter in the coal. If the coal is properly fed into the fire chamber, it results in less smoke and more heat. Today, in many factories and electric power plants, soft coal is ground into powder and blown into the fire chamber with a blast of air. In this manner, the coal burns like a gas, with great efficiency.

If soft coal is heated to between 700°C and 1000°C in the absence of air, it breaks down into (1) coal gas, (2) a mixture of liquids, and (3) a solid called *coke*. Breaking down coal in the absence of air is a process called *destructive distillation.*

Coke is a hard, porous, solid fuel that contains more carbon than coal because most of the volatile matter is driven off. Coke varies in color from gray to black.

12-3 This power shovel can scoop out 23 metric tons of coal. The air-conditioned cab prevents the operator from inhaling coal dust. Why is this safeguard important?

12-4 Red-hot coke made from soft coal is being discharged into a railroad car. Can you describe the coke-making process?

12-5 Destructive distillation of soft coal or wood can easily be demonstrated in the lab. In each case, what is the solid residue called?

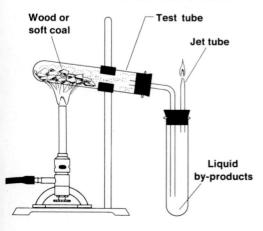

Wood or soft coal

Test tube

Jet tube

Liquid by-products

Thousands of metric tons of this fuel are prepared daily in huge coke ovens. See Fig. 12-4.

Most of the coke produced today is used by the iron and steel industry as a fuel and reducing agent. Coke ovens also produce many useful by-products. These products include ammonia, which is used in making fertilizers; coal gas, which is used as fuel; and coke-oven tar, used in making dyes, perfumes, flavors, and medicines. Synthetic rubber, antiseptics, nylon, and many thousands of other products are made from materials obtained from soft coal.

Coke and charcoal can be made in the laboratory. Coke can be made in the lab by using the setup shown in Fig. 12-5. Fill the test tube almost full with small pieces of soft coal. Heat the tube with a Bunsen burner. Light the coal gas that comes from the jet tube. The tar-like material that condenses in the vertical tube contains ammonia, oils, and coal tar. Over a hundred consumer products are obtained from the coal tar alone. After heating the coal for about 10 min, break open the test tube and examine the coke. Ignite the coke and observe how it burns.

You have probably seen *charcoal* left over from a wood fire or have used charcoal in outdoor grills. To produce charcoal in large amounts, furnaces such as the one shown in Fig. 12-6 are used. Wood is loaded on steel cars and pushed into a huge oven. Around the chamber are flues where gas is burned. The heat drives off gases from the wood. The solid that remains is charcoal. Beside charcoal, this process yields wood alcohol, acetic acid, wood tar, pitch, and creosote. These by-products are as useful as the charcoal itself.

Like coke, charcoal burns with almost no flame because it contains no volatile matter. Charcoal, though it gives a hot, smokeless flame, is too costly to use as a fuel for heating homes. However, there are many industrial uses for charcoal, such as the making of special steels and as a filter to clean dirty water and air.

To prepare charcoal in the lab, place some small pieces of wood in a Pyrex test tube that is arranged as shown in Fig. 12-5. Heat the wood for 10 to 15 min. Burn the gas that comes out of the jet tube. After most of the volatile matter has been driven off, look closely at the charcoal.

The heat values of several solid fuels are compared with alcohol and gasoline in Table 12-1.

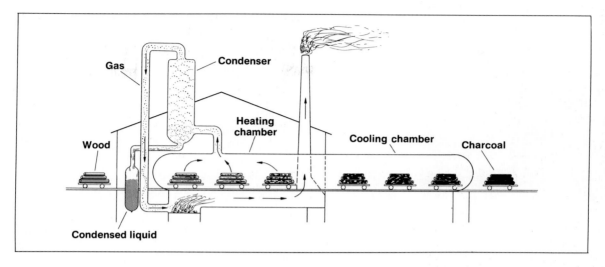

12-6 Diagram of the commercial preparation of charcoal from wood. Can you describe this process based on the diagram?

Table 12-1:

Heat Values of Several Fuels Are Compared	
Fuel	Calories/Gram*
Wood	5,000
Alcohol	6,400
Coal	7,300
Coke	7,800
Charcoal	8,000
Gasoline	10,800

*The calorie is a unit of heat measurement.

petroleum

Petroleum may come from marine life. The word *petroleum* is derived from the Latin "petra," meaning rock, and "oleum," meaning oil. Petroleum was used very early in history. Certain primitive people had "sacred" fires that burned oil and natural gas that seeped from the ground. In fact, the Chinese used petroleum as a fuel as early as 200 B.C.

chapter 12   fossil fuels   215

12-7 Workers setting up an advertising road sign for one of the first motor oils made of petroleum. Can you name several other products that come from petroleum?

12-8 The first oil well in this country. How much oil did this well produce?

One theory about the origin of petroleum, or *crude oil*, is that it comes from marine life buried in sediments millions of years ago. Many scientists believe that bacterial action helped change this marine life into oil and gas. Heat and pressure under layers of soil, rock, and water helped to bring about chemical changes.

Crude oil is a slippery mixture with a strong odor made up of thousands of compounds. It consists mostly of hydrocarbons (compounds of hydrogen and carbon). Compounds of oxygen, nitrogen, and sulfur are also included. Crude oil is seldom used as it comes from the ground. Rather, crude oil is separated into a large number of products by *refining*.

Many people heat their homes with fuel oil that comes from petroleum. Lubricating oils, greases, gasoline, asphalt, kerosene, and tar are only a few of the thousands of products that come from this resource. See Fig. 12-7.

Obtaining petroleum from the ground is a costly process. The first oil well, only 21 m deep, was dug by Col. Edwin L. Drake in 1859 near Titusville, Pennsylvania. The well produced only a few barrels of oil a day for many years. See Fig. 12-8.

Today, oil and natural gas together provide about 75% of this country's energy needs. Russia, the United States, Saudi Arabia, and Iran are the leading oil-producing countries. Texas, Louisiana, California, and Alaska produce most of the oil in this country.

Not all drilling operations produce oil. About one drilling in nine hits oil or gas. The "dry" holes, or "dusters," are a total financial loss.

Petroleum, natural gas, and salt water fill the oil-bearing sands under domes of "cap" rock (a layer of shale or clay), as shown in Fig. 12-9. When a drill reaches the oil-bearing sands, the pressure of the natural gas may force the oil upwards to the surface. As the pressure of the natural gas decreases, pumps obtain the oil.

Oil wells are usually drilled with rotary bits. A heavy bit equipped with a diamond or hard steel tip is fastened to a length of steel pipe. See Fig. 12-10. The pipe is then attached to a round drill table. The power to rotate the table, pipe, and bit is supplied by an engine. New lengths of pipe are attached to the main shaft as the bit bores its way down. Each time that the worn-out bit must be replaced, the entire shaft has to be raised and taken from

unit 2 chemistry in our world

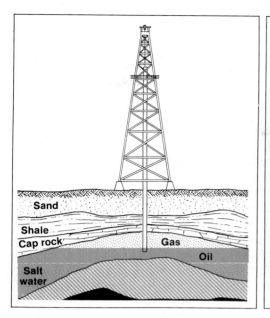

the hole. Rock cores removed from the hollow steel pipe tell what kind of rock is in each layer.

During the drilling, a soupy mud is pumped through the pipes to flush out the cuttings. This muddy liquid also serves to cool the drill. When the drill reaches an oil dome, the oil may be forced out as a "gusher" by pressure from trapped gases. To keep from losing too much oil from a gusher, an arrangement of valves is used to control the flow of crude oil and natural gas. After the drilling rig has been removed, a "Christmas tree" of valves, as shown in Fig. 12-11, is finally placed in position.

Crude oil from the well is taken to the refinery where it is separated into many products. A network of pipe lines carries the oil to the refinery or to a port where it is loaded on ocean-going tankers. Pumping stations on land, located every 130–145 km, force the crude oil to refineries that are often thousands of kilometers away. Pipelines also carry gasoline, fuel oil, and natural gas. Branch lines connect different oil fields with main lines, forming a huge underground network.

Petroleum is refined by distillation. A pure liquid compound has a definite boiling point at a given pressure. Ethyl alcohol, for example, boils at 78°C at sea level. But the boiling point of a mixture is not fixed. For instance, a

12-9 (left) Crude oil, natural gas, and salt water saturate the oil-bearing sands. Only two out of 100 wells produce enough oil to be profitable. What is a duster?

12-10 (right) Workers prepare to lower a bit into an oil well. What is the bit made of?

12-11 The "Christmas tree" is a series of valves designed to control the flow of oil from the well. Why is this device used?

mixture of alcohol and water starts to boil at 78°C, but the temperature slowly rises to 100°C as boiling continues.

When the petroleum mixture is heated, the liquids with the lowest boiling points boil off first. *The process of separating a mixture of liquids having different boiling points is called* **fractional distillation.** The initial refining of petroleum into various products is done by this process.

The crude oil from the wells is separated by an apparatus divided into three main parts: (1) a pipe still, (2) a fractionating tower, and (3) a condenser. See Fig. 12-12.

The crude oil is heated under pressure in a furnace, called a pipe still, to between 370° and 430°C. Pressure in the tubes keeps the crude oil from becoming a vapor. The hot oil flows from the pipe still to the bottom of a tall fractionating tower. Here the pressure is reduced and the liquid becomes a vapor.

The vapors rise up the tower, which may be more than 30 m high. Those hydrocarbons with a low boiling point range, such as those in gasoline, continue to move up the tower as a vapor. The condenser cools the vapors and they become liquids. Shelves collect the condensed liquids within the tower. These shelves hold the liquids until they are drained off into separate storage tanks as shown in Fig. 12-12. Table 12-2 lists the boiling point ranges of various crude oil products.

12-12 The basic parts of a fractional distillation plant for petroleum refining. Can you describe the function of each part?

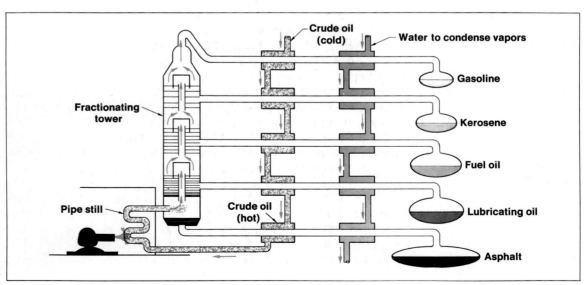

unit 2   chemistry in our world

For better separation, each of the liquids obtained in the table is distilled again. In addition, each liquid is chemically treated to remove unwanted materials.

Table 12-2:

Petroleum Products*			
Number of Carbon Atoms	Boiling Temperature Range	Name of Product	Principal Use
1 to 5	below 40°C	gas	fuel
6 to 10	40° to 180°C	gasoline	fuel
11 to 12	180° to 230°C	kerosene	fuel, cracking stock
13 to 17	230° to 300°C	light gas oil	diesel fuel, cracking stock
18 to 25	300° to 405°C	heavy gas oil	lubricant stock
26 to 60	405° to 515°C	residue	wax, residual oil, asphalt

*(*Chemistry and Petroleum,* American Petroleum Institute)

gaseous fuels

Natural gas is a useful fuel. Many homes and factories in the United States use natural gas as a fuel for cooking and heating. Natural gas is almost entirely made up of the hydrocarbon called methane (CH_4). Methane is formed when plant life breaks down underwater. Natural gas is sometimes called *marsh gas* because it is found bubbling from the water in warm, marshy areas.

A region that has coal and petroleum may also have deposits of natural gas. Wells are drilled through rock to porous sandstone layers where the gas is found. These gas wells range in depth from 100 to 1000 m. Natural gas is an excellent fuel, giving more heat per cubic meter than any other fuel gas. Natural gas is convenient to use, leaves no ash, and burns with almost no smoke.

The use of natural gas in this country has increased sharply over the past 25 years. Because of its wide use, apparent natural gas shortages have developed. Natural gas is supplied to most cities and towns by a network of pipe lines over 650,000 km in length. See Fig. 12-13 (p. 220).

12-13 Gas pipelines have to be constructed over hills and mountains. Can you give some reasons why these pipes are often buried?

12-14 These tanks are efficient for storing butane, propane, and other fuel gases under pressure. Can you give a reason why these tanks are spherical?

Water gas is an industrial fuel. It may sound strange to talk about *water gas,* since water will not burn. Yet water in the form of steam combines with hot coke (carbon) to make a useful fuel.

Burning coke is first heated white-hot by forcing air through it. This process is called the "air run." The air is then cut off and steam is allowed to pass through the hot coke ("steam run"). After a short time, the steam is shut off and the coke is reheated by another blast of air. This back-and-forth process is continued until all the coke is used up. Water gas is produced only during the "steam run." When the coke and steam (water) react, carbon monoxide and hydrogen are given off:

$$C + H_2O \rightarrow CO \uparrow + H_2 \uparrow$$

The mixture of carbon monoxide and hydrogen is called *water gas.* Since both carbon monoxide and hydrogen can burn, this mixture can be used as a fuel. Carbon monoxide and hydrogen can also be sold separately for consumer use.

The fuel value of water gas may be enriched with propane (C_3H_8) or butane (C_4H_{10}). Propane and butane are gases obtained from the refining of petroleum, or from natural gas. See Fig. 12-14. Since both hydrogen and carbon monoxide are odorless, a dangerous leak in the gas line might otherwise go undetected. Therefore, materials are added to water gas to give it an odor.

Propane, butane, or a mixture of the two gases can be compressed into a liquid and stored in steel cylinders. Gas stored in this way is known as "bottled gas," or LPG

(Liquefied Petroleum Gas). LPG provides a fuel of high-heat value to places that are remote from regular gas-line service. The liquid fuel changes to a gas when the pressure is reduced at the outlet valve. The bottles shown in Fig. 12-15 may be refilled from a delivery truck or simply replaced by filled ones.

A burning hydrocarbon provides energy. As you have seen, liquid and gaseous fuels are made up mostly of hydrocarbons. Even soft coal, which contains carbon in a free state, contains hydrocarbons. Coke and charcoal are largely composed of noncrystalline carbon.

When a hydrocarbon burns completely, the products formed are carbon dioxide and water. Heat and light energy are released. *Complete combustion* takes place when a fuel burns with enough oxygen to support the burning process. For instance, when methane burns under this condition, carbon dioxide, water, and energy are the products formed:

$$CH_4 + 2\,O_2 \rightarrow CO_2 \uparrow + 2\,H_2O + energy$$

On the other hand, when there is a shortage of oxygen, *incomplete combustion* occurs. The resulting products are carbon monoxide, water, and energy:

$$2\,CH_4 + 3\,O_2 \rightarrow 2\,CO \uparrow + 4\,H_2O + energy$$

In any normal burning of a hydrocarbon, both carbon dioxide and carbon monoxide are produced. Carbon dioxide is nonpoisonous. Breathing carbon monoxide, however, can cause death or permanent body damage. The gas combines with the hemoglobin of the red blood cells more readily than oxygen. Thus, these cells are useless for carrying the oxygen that the body needs.

12-15 With bottled gas, homes located away from city gas lines can have the same gas service. What fuel gases might be in these cylinders?

Products formed in the burning of a hydrocarbon. Light a candle and hold a dry, cold beaker (open end down) about 5 cm above the flame for a few seconds. What do you observe inside the beaker? How did it get there? Why must the beaker be cold? Now hold the bottom of the cold beaker in the yellow part of the flame for several seconds. What do you observe on the bottom of the beaker? Explain.

12-16 Synthetic diamonds are made from noncrystalline carbon. These industrial diamonds are used to coat a wheel for shaping lenses. What form of carbon is contained in pencil lead?

other forms of carbon

Diamonds are crystallized carbon. Carbon occurs in two crystal forms, diamond and graphite. These crystal forms are in addition to the noncrystalline forms of carbon in coke, charcoal, and other fossil fuels you have just studied. Soot is another example of free carbon, formed by incomplete burning of fossil fuels. Although diamond and graphite are not used as fuels, they are discussed here because they are well-known forms of carbon.

When carbon is placed under great heat and pressure, it is crystallized to form diamond. Diamond is the purest form of carbon and the hardest natural substance known.

After many years of research, scientists have learned to make small diamonds for industrial uses. These synthetic diamonds have the same properties as natural diamonds. They are used widely in industry. See Fig. 12-16. Industrial diamonds are used for cutting and grinding metals, and for bits in drilling oil wells and tunnels. Recently, the hardness of diamond has been equaled by a synthetic substance called borazon, or boron nitride (BN).

The major source of diamond is a region in South Africa. Diamonds, as they are mined, do not have the shape or luster of gems. Their brilliance and beauty depend upon the skill with which they are cut and polished.

Graphite is also crystallized carbon. The word graphite comes from a Greek word meaning "to write." The "lead" in your writing pencil is really graphite mixed with clay. Bits of graphite suspended in oil are often used as a lubricant. Graphite is also used as a moderator in nuclear reactors, to slow down fast-moving neutrons.

Graphite is a soft, grayish-black, "greasy" substance. The carbon atoms of graphite form a crystal pattern that differs from that of the carbon atoms of diamond. See Fig. 12-17. This unusual arrangement of carbon atoms accounts for the very different properties of graphite. Graphite is very soft; diamond is very hard. Graphite is a fairly good conductor of an electric current; diamond is a nonconductor.

Natural graphite is mined in Sri Lanka (Ceylon), an island off the coast of India, and in Russia. In the United States, graphite is mined in New York and Pennsylvania.

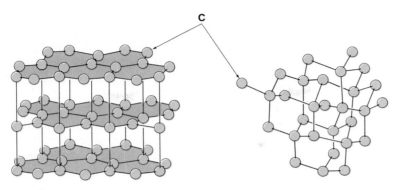

12-17 The planes of carbon (C) atoms in graphite (left) move freely over each other. Can you describe why the arrangement of carbon atoms in diamond (right) makes diamond very hard?

Synthetic graphite is made from hard coal or coke in an electric furnace.

The Quest for More Energy

This country is rapidly running out of low-cost fuels. For many years, the United States has been almost self-sufficient in the mining of fossil fuels, the stored energy of the sun. These fuels, as you have seen, are petroleum, natural gas, and coal.

Oil supplies most of the energy that powers our machines and labor-saving devices. These machines have helped bring about the highest standard of living in history. Domestic oil is in short supply today. This shortage seems less obvious because of the ease with which oil is being imported. At present, about one half of the oil used in the United States is imported. The use of oil as a source of energy keeps growing rapidly.

You have seen that petroleum and natural gas provide about three fourths of all energy needs. But as the demand for natural gas has increased, the discovery of new gas sources has decreased. The low-cost and easy-to-find gas has been used up. The deeper and harder-to-find gas will be more costly.

Coal provides about 20% of all present energy needs. Some experts say that there is enough coal to last for 300 years. However, most coal has a high-sulfur content, which would add to air pollution. It will be necessary, therefore, to find efficient and profitable ways of removing the sulfur from coal before it is burned. Unlike oil and gas, coal has a high labor cost connected with its mining.

Miners must enter and work the mines to obtain the coal. It is dangerous work because of potential mine disasters, and a health hazard because of black-lung disease. Coal has the potential of supplying all energy needs, but at a much higher cost than is now being paid.

Other sources of energy await future development. These sources include: (1) solar radiations (sunshine); (2) use of heat from the earth's interior; (3) energy of the wind through the use of windmills; (4) the use of ocean tides; (5) energy obtained from garbage and other waste products; and (6) nuclear energy (fission and fusion). Nuclear and solar energy appear to be the only long-term solutions to the huge growth in the demand for energy.

A model of Shiva, an experimental laser fusion device, is shown below. Experiments performed with Shiva will help scientists determine if laser fusion could be an energy source for the future. The fusion fuel in the blue target chamber will be hit by 25 trillion watts of laser power. Temperatures of 100,000,000°C will cause the fuel to ignite in a fusion reaction.

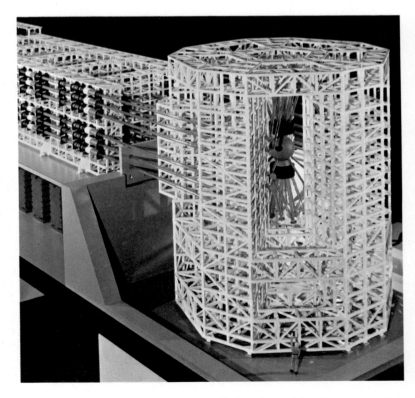

summary

1. Fuels give off heat at a reasonable cost.
2. Coal is formed from the buried remains of plants.
3. Coal is composed mostly of free carbon.
4. Coke is made by the destructive distillation of soft coal; charcoal is made by the destructive distillation of wood.
5. Fractional distillation is the process of separating a mixture of liquids having different boiling points.
6. Petroleum was probably formed from marine life acted upon by bacteria, heat, and pressure through the ages.
7. Liquid and gaseous fuels consist mostly of hydrocarbons.
8. When a hydrocarbon burns with enough oxygen, the products formed are water, carbon dioxide, and energy (heat and light).

review

Match each item in the left column with the best response in the right column. *Do not write in this book.*

1. bituminous
2. coke
3. water gas
4. graphite
5. petroleum
6. natural gas
7. peat
8. H_2O and CO
9. fractional distillation
10. charcoal

a. composed of H_2 and CO
b. liquid composed mostly of hydrocarbons
c. composed mostly of methane (CH_4)
d. first stage in coal formation
e. process of separating a mixture of liquids having different boiling points
f. products of complete combustion of a hydrocarbon
g. hard coal
h. formed by heating wood in absence of air
i. form of crystallized carbon
j. formed by destructive distillation of soft coal
k. soft coal
l. second stage in coal formation

questions

Group A
1. What are some traits of a good fuel? List three advantages of gas as a fuel.
2. Trace briefly the stages in the formation of coal.
3. List two advantages of anthracite over bituminous coal.
4. How is charcoal made? List three by-products in making charcoal.
5. Name two widely used household fuels in your community.
6. List several advantages and disadvantages of soft coal as a fuel.

7. Where is anthracite mined in the United States? bituminous coal?
8. What group of compounds is found in petroleum?
9. Name the three main parts of the distillation apparatus in a refinery and describe their functions.
10. List five major products obtained directly from the distillation of petroleum in order of increasing boiling-point.
11. Which states lead in the production of petroleum? Which countries?
12. How are oil wells drilled? In what way is crude oil transported to the refineries?
13. What is the composition of natural gas? Discuss some of its advantages as a fuel.
14. How is water gas prepared? How is it enriched?
15. What products are formed by the complete burning of a hydrocarbon fuel? incomplete burning of a hydrocarbon fuel?
16. List two examples of noncrystalline carbon. crystallized carbon.

Group B
17. Why did the plant matter of the coal-forming period not decay completely?
18. Compare the carbon and moisture content of anthracite with that of bituminous.
19. Discuss one theory of how petroleum was formed.
20. Describe what happens in a fractionating tower.
21. Describe why the oil sometimes "gushes" from newly drilled wells. What is done to prevent this from happening?
22. Why is the "air run" needed in the making of water gas?
23. What is the composition of bottled gas? Why is it useful?
24. Explain why breathing carbon monoxide gas is dangerous.
25. Why do diamond and graphite have such different properties?
26. If diamond is strongly heated in enough oxygen, what compound would you expect to be formed?

further reading

Asimov, Isaac, *The World of Carbon.* New York: Abelard-Schuman.

Fowler, John, *Energy-Environment Source Book.* Washington, D.C.: National Science Teachers Association, 1975.

Gesner, A., *Practical Treatise on Coal, Petroleum and Other Distilled Oils.* Clifton, N.J.: Augustus M. Kelley, Publishers.

O'Connor, R., *The Oil Baron: Men of Greed and Grandeur.* Boston: Little, Brown, 1971.

Rosen, S., *Future Facts.* New York: Simon & Schuster, 1976.

Sell, G., *The Petroleum Industry.* New York: Oxford University Press, 1963.

organic compounds

13

Earlier in the text, you saw how elements could be grouped into metals and nonmetals, and you studied the common gases of the air. You saw that compounds could be acids, bases, and salts. All of these substances are studied in the branch of physical science called *inorganic chemistry*. In the last chapter, you studied fossil fuels. In that chapter, you learned that fossil fuels are composed of carbon and its compounds. *The study of carbon compounds is the subject of* **organic chemistry.** Carbon compounds make up the structure of all living things. Chemists estimate that there are over 3 million known carbon compounds. Thousands more are being discovered each year. In this chapter, you will see how carbon atoms can combine with other carbon atoms as well as with atoms of other elements to form a large number of long-chain compounds.

objectives:

☐ To distinguish between the properties of organic and inorganic compounds.

☐ To develop skill in writing and interpreting structural formulas.

☐ To explain the reason for the existence of a very large number of carbon compounds.

☐ To develop understanding of the formation of compounds from unsaturated hydrocarbons.

☐ To define the following terms: straight-chain compound, branched-chain compound, ring compound, isomers, and octane rating.

organic chemistry

There are differences between organic and inorganic compounds. The basic laws of chemistry hold both for organic and inorganic compounds. There are, however, a few differences in the reactions of organic compounds compared with inorganic compounds. In general, these differences are as follows:

1. Most organic compounds are insoluble in water. They will, however, dissolve in such organic liquids as alcohol, ether, or chloroform. Many inorganic compounds dissolve, more or less readily, in water.

2. Organic componds will decompose by heat more readily than inorganic compounds. An inorganic compound, such as sodium chloride, will vaporize without breaking down into sodium and chlorine when heated to a high temperature.

3. Reactions involving organic compounds proceed at a much slower rate than do reactions between inorganic compounds. Organic reactions often require hours or even days for completion. If conditions are right for the reaction, most inorganic reactions will take place at once.

4. Organic compounds, existing as molecules, are formed from the elements by covalent bonding, that is, sharing of electrons. Most inorganic compounds are formed as a result of ionic bonding, that is, transfer of electrons.

Organic compounds can be placed into groups. The compounds of carbon can be divided into a number of characteristic groups to make them easier to study. You are already familiar with *hydrocarbons*, found in fossil fuels. Hydrocarbons are compounds that are composed only of hydrogen and carbon. The hydrocarbon group is the simplest group of organic compounds.

Other important organic groups include *alcohols, aldehydes, organic acids, ketones, ethers,* and *esters.* Compounds in each of these basic groups have similar properties and molecular structure. Other organic compounds found in foods are proteins, carbohydrates, fats, and vitamins. However, in this chapter, you will study the hydrocarbon and alcohol groups.

hydrocarbons

The hydrocarbons include many important compounds. The number of known hydrocarbon compounds runs into the tens of thousands. New hydrocarbons are being discovered and prepared all the time. Fortunately, chemists have made the study of hydrocarbons easier by separating them into various *series*.

The simplest and most abundant series of hydrocarbons is called the *alkane series*. Physically, alkanes resemble long straight chains. Alkanes differ from one another chiefly in the number of links in the chain. The first compound in this series is the gas *methane* (CH_4), or marsh gas, the main component of natural gas.

The next compound in the alkane series is *ethane* (C_2H_6). Ethane is followed by *propane* (C_3H_8) and *butane* (C_4H_{10}).

Table 13-1 lists some of the common members of the alkane series. Notice in the table that the name of each compound in the series ends in *-ane*. Also notice that each compound differs from the one before it by the addition of CH_2. To find the order of the compounds in this series, the general formula C_nH_{2n+2} is used. The first compound, methane, starts with $n = 1$.

Table 13-1

The Alkane Series of Hydrocarbons		
Name	Formula	Phase at Room Temperature
Methane	CH_4	gas
Ethane	C_2H_6	gas
Propane	C_3H_8	gas
Butane	C_4H_{10}	gas
Pentane	C_5H_{12}	liquid
Hexane	C_6H_{14}	liquid
Heptane	C_7H_{16}	liquid
Octane	C_8H_{18}	liquid
Nonane	C_9H_{20}	liquid
Decane	$C_{10}H_{22}$	liquid
Eicosane	$C_{20}H_{42}$	solid
Hexacontane	$C_{60}H_{122}$	solid

The maximum value of n is not known. However, some known members of the alkane series have values of n over 1000. Many members of the alkane series, together with other hydrocarbons, are found in petroleum.

Structural formulas are used to represent organic compounds. In many organic compounds, several arrangements of atoms in a molecule are possible. For this reason, a number of carbon compounds have the same simple chemical formula. To avoid confusion, chemists often show the arrangement of atoms in a molecule by means of a *structural formula*. The structural formulas for each of the first four members of the alkane series are as follows:

$$
\begin{array}{cc}
\quad\;\; \text{H} & \quad\;\; \text{H}\quad\text{H} \\
\quad\;\; | & \quad\;\; |\quad\; | \\
\text{H}-\text{C}-\text{H} & \text{H}-\text{C}-\text{C}-\text{H} \\
\quad\;\; | & \quad\;\; |\quad\; | \\
\quad\;\; \text{H} & \quad\;\; \text{H}\quad\text{H} \\
\text{methane, CH}_4 & \text{ethane, C}_2\text{H}_6
\end{array}
$$

$$
\begin{array}{cc}
\;\; \text{H}\quad\text{H}\quad\text{H} & \;\; \text{H}\quad\text{H}\quad\text{H}\quad\text{H} \\
\;\; |\quad\; |\quad\; | & \;\; |\quad\; |\quad\; |\quad\; | \\
\text{H}-\text{C}-\text{C}-\text{C}-\text{H} & \text{H}-\text{C}-\text{C}-\text{C}-\text{C}-\text{H} \\
\;\; |\quad\; |\quad\; | & \;\; |\quad\; |\quad\; |\quad\; | \\
\;\; \text{H}\quad\text{H}\quad\text{H} & \;\; \text{H}\quad\text{H}\quad\text{H}\quad\text{H} \\
\text{propane, C}_3\text{H}_8 & \text{butane, C}_4\text{H}_{10}
\end{array}
$$

The dash (—) in the above formulas shows that the outer electron of one carbon atom is paired with the electron of a hydrogen atom or another carbon atom. The pairing of electrons between atoms, as indicated by the dash, is called a *chemical bond*. Hydrogen and carbon atoms form molecules by sharing electrons in a covalent bond. In the structural formulas above, notice that each carbon atom is surrounded by four bonds. Carbon forms four bonds because each carbon atom has four outer electrons to be shared, either with hydrogen or with other carbon atoms. A lab model of the methane molecule, as it is often represented, is shown in Fig. 13-1.

Each member compound of the alkane series is said to be *saturated*. In other words, every carbon atom in the compound, as the structural formula shows, shares a *single bond* with another carbon atom or hydrogen atom. A single bond is shown by a dash in the structural formula. Therefore, no more hydrogen atoms can be added to the compound.

13-1 In the methane molecule, CH_4, a carbon atom forms covalent bonds with four hydrogen atoms. In the diagram, how are these covalent bonds represented?

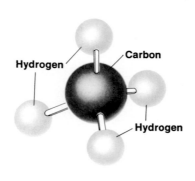

Hydrogen

Carbon

Hydrogen

230

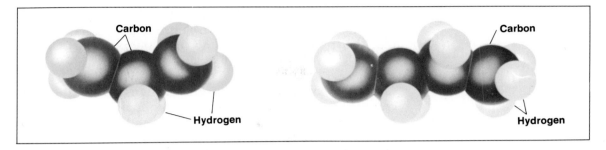

Carbon

Hydrogen

Carbon

Hydrogen

Compounds can be straight-chained or branched-chained. In certain hydrocarbons, the carbon atoms are linked or bonded together in long, *straight chains,* as shown in Fig. 13-2. The structural formulas for propane and butane already shown above indicate that these compounds form straight chains. However, many hydrocarbon compounds, starting with butane (C_4H_{10}), also form side chains, or *branched chains.* The carbon atoms can be attached in either a straight chain or a branched chain. Carbon atoms form a branched chain by bonding on to the middle carbon atom. The formulas for straight- and branched-chain butane are compared below:

13-2 Laboratory models of propane (left) and butane (right), both straight-chain compounds. How many hydrogen atoms are attached to the carbon atoms on each end of the molecules?

straight chain
normal butane (C_4H_{10})

branched chain
isobutane (C_4H_{10})

Although both of the above compounds have the same chemical formula, they are, in fact, different compounds. The branched-chain compound, isobutane, has properties that differ from those of normal, straight-chain butane. This difference is due entirely to the different arrangement of the atoms in the molecule. Therefore, butane has two **isomers.** *Isomers are compounds whose molecules have the same number and kind of atoms but with a different arrangement.*

The more carbon atoms in a hydrocarbon molecule, the more isomers it is likely to have. Pentane, the next

chapter 13 organic compounds

231

compound after butane in the alkane series, has three isomers. These isomers are called normal pentane (written as *n*-pentane), isopentane, and neopentane. Decane ($C_{10}H_{22}$) can have as many as 75 possible isomers.

Gasoline contains many hydrocarbons. Gasoline is a complex mixture of compounds of the alkane series from pentanes through nonanes, as well as other hydrocarbon compounds. Octane is one of the important hydrocarbons in gasoline. The amount and kind of isomers in gasoline affect its quality. For instance, gasoline with branched-chain hydrocarbons burns more slowly than gasoline containing mostly straight-chain compounds.

Sometimes the mixture of gasoline and air burns too quickly in a car engine. Instead of a smooth power stroke, the pistons in the cylinder of the engine receive sharp, hammer-like blows, resulting in lost power and possible engine damage. These blows are called "knocking."

Modern car engines compress the gasoline-air mixture to about 1/8 or 1/10 the original volume of the cylinder. This squeezing brings the air and gasoline particles very close together. The more the engine compresses the mixture before firing, the greater the amount of energy produced from the same amount of mixture. However, too much compression will cause the gasoline vapor to burn too quickly and cause knocking.

Straight-chain hydrocarbons burn quickly and produce knocking in gasoline. Branched-chain compounds burn more slowly under high compression. The reason for this difference is not hard to understand. Since burning means combining with oxygen, a long, straight-chain molecule is easily reached by the oxygen molecules in the gasoline-air mixture. The branched-chain molecule is more compact and less available to the oxygen.

Knocking can be prevented by (1) using a slower-burning fuel and by (2) adding catalysts to the gasoline. The catalyst that was most widely used for this purpose is *tetra-ethyl* (tet-ruh-*eth*-il) *lead*. This compound slows down the rate at which the fuel burns, thus improving its antiknock qualities. Since this lead catalyst causes air pollution, later model cars are designed with engines that have lower compression ratios and use no-lead gasoline.

Gasoline is rated by octane numbers. Engineers have established a system of comparing the knocking property

unit 2   chemistry in our world

of gasoline. The ability of a gasoline to resist knocking is expressed by a rating called the *octane number*. The higher the octane number, the more knock-resistant the gasoline.

An isomer of octane with three branched chains has excellent antiknock properties. This iso-octane hydrocarbon is used as a standard, with an assigned octane number of 100. Normal straight-chain heptane knocks very badly in high-compression engines. Therefore, heptane is given a rating of zero. Test mixtures of these two compounds are used as a standard by which gasoline can be compared.

For example, a mixture of 93% iso-octane and 7% normal pentane solution is assigned an octane number of 93. Any gasoline that has an antiknock quality as high as this standard test mixture is given the same octane number. Such a gasoline makes an excellent car engine fuel. Airplanes and some cars use 100-octane gasoline. This fuel performs the same as for pure iso-octane. Because of environmental concerns, lower-compression car engines are now being built, and therefore, gasoline with lower octane values can be used. Some low-lead and nonlead gasolines are in the 89–92 octane range. The best buy in gasoline is the one at the lowest price that will not "knock" or "ping" in your car.

Hydrocarbons can be unsaturated. You have seen that the alkane series of compounds is said to be saturated. That is, carbon atoms are connected by single bonds. There is another series of hydrocarbon compounds called the *alkene series*. The alkene series is composed of compounds in which two carbon atoms in the molecule are connected by two bonds. These bonds are shown in the structural formulas by two dashes ($=$). In this type of bonding, two outer electrons of one carbon atom are paired with two outer electrons of another carbon atom, forming a *double covalent bond*. The first compound of the alkene series of hydrocarbons is *ethene* (C_2H_4). The structural formula for ethene is

$$
\begin{array}{ccc}
H & & H \\
\diagdown & & \diagup \\
& C = C & \\
\diagup & & \diagdown \\
H & & H
\end{array}
$$

The names of the members of the alkene series are derived from the names of members of the alkane series having the same number of carbon atoms. The suffix -*ene* is substituted for the suffix -*ane*. For instance, eth*ane* is the alk*ane* with two carbon atoms. Thus, the alk*ene* with two carbon atoms is named eth*ene*.

The second member of the alkene series has three carbon atoms and is called *propene* (C_3H_6). The structural formula for propene is

$$\begin{array}{c} \quad\ \text{H}\ \ \ \text{H}\ \ \ \text{H} \\ \quad\ |\ \ \ \ \ |\ \ \ \ \ | \\ \text{H--C--C=C--H} \\ \quad\ | \\ \quad\ \text{H} \end{array}$$

The third member of the alkene series has four carbon atoms and is called *butene* (C_4H_8). The fourth member is called *pentene* (C_5H_{10}). The general formula for compounds in the alkene series is C_nH_{2n}, starting with $n = 2$. Notice that there are four bonds around each carbon atom.

Sometimes the symbols for two carbon atoms in a structural formula are connected by three dashes ($\equiv$), representing a *triple covalent bond*. Can you suggest what this triple bonding means? Triple bonding gives rise to a third series of hydrocarbons, called the *alkynes* (al-*kynes*). *Ethyne* (C_2H_2), commonly called *acetylene*, is the first compound of the alkyne series. Its structural formula is

$$\text{H--C} \equiv \text{C--H}$$

Acetylene burns with a very hot flame. It is used as a fuel in welding and cutting metals. See Fig. 13-3.

The second compound in the alkyne series is *propyne* (C_3H_4). The structural formula for propyne is

$$\begin{array}{c} \quad\ \ \text{H} \\ \quad\ \ | \\ \text{H--C} \equiv \text{C--C--H} \\ \quad\ \ | \\ \quad\ \ \text{H} \end{array}$$

The general formula for compounds in the alkyne series of hydrocarbons is $C_nH_{2n\text{-}2}$, starting with $n = 2$.

Organic compounds can form rings. Another important group of hydrocarbon compounds is the *benzene series*,

13-3 Acetylene is used in welding metals together. What is the formula for this compound?

unit 2 chemistry in our world

named after *benzene* (C_6H_6), the first member. The ends of the chain of carbon atoms in molecules of this series are linked together in a closed ring. These compounds are called *ring* compounds. Benzene is a typical ring compound in this group and is shown by the following structural formula:

$$
\begin{array}{c}
\text{H} \\
| \\
\text{C} \\
\diagup \quad \diagdown \\
\text{H---C} \qquad \text{C---H} \\
| \qquad\qquad | \\
\text{H---C} \qquad \text{C---H} \\
\diagdown \quad \diagup \\
\text{C} \\
| \\
\text{H}
\end{array}
$$

The general formula for member compounds in the benzene series is C_nH_{2n-6}. The first member, benzene, starts with $n = 6$. See Fig. 13-4. The fact that the ring compounds may also have branched chains makes great numbers of different hydrocarbons possible.

Benzene is a colorless, flammable liquid obtained from coal tar and petroleum. It is the basic raw material for the making of thousands of organic compounds used in plastics, dyes, drugs, perfumes, and explosives. Benzene is an excellent solvent for many organic compounds, including rubber.

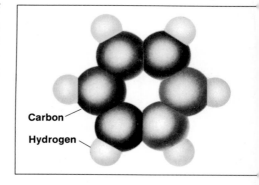

13-4 Laboratory model of the benzene (C_6H_6) molecule, the simplest of the ring compounds. Can you write the formula for the second member of the benzene series?

addition and substitution reactions

Unsaturated compounds can be added to. You have likely seen references to "polyunsaturated fats" in food commercials. An organic compound with double or triple bonds between carbon atoms is said to be *unsaturated*. All members of the alkene and alkyne series are unsaturated compounds. In certain reactions, the covalent bonds of an unsaturated hydrocarbon are broken. Other atoms can then be *added* to the molecule, making it saturated. The following equations show reactions that take place with the addition of hydrogen to unsaturated hydrocarbons:

$$\text{H}-\overset{\overset{\displaystyle H}{|}}{\text{C}}-\overset{\overset{\displaystyle H}{|}}{\underset{\underset{\displaystyle H}{|}}{\text{C}}}-\text{H} \quad + \text{H}_2 \rightarrow \text{no reaction}$$

saturated
compound
(alkane)

$$\underset{\text{H}}{\overset{\text{H}}{}}\!\!\diagdown\!\text{C}=\text{C}\!\diagup\!\underset{\text{H}}{\overset{\text{H}}{}} \quad + \text{H}_2 \rightarrow \quad \text{H}-\overset{\overset{\displaystyle H}{|}}{\text{C}}-\overset{\overset{\displaystyle H}{|}}{\text{C}}-\text{H}$$

unsaturated   saturated
compound   compound
(alkene)   (alkane)

$$\text{H}-\text{C}\equiv\text{C}-\text{H} \quad + \quad 2\text{H}_2 \rightarrow \quad \text{H}-\overset{\overset{\displaystyle H}{|}}{\text{C}}-\overset{\overset{\displaystyle H}{|}}{\text{C}}-\text{H}$$

unsaturated   saturated
compound   compound
(alkyne)   (alkane)

There is some evidence that saturated fats in the diet may be the cause of excess deposits of *cholesterol* (kol-*les*-ter-ol) on the walls of arteries. These deposits roughen and harden the walls of the arteries, may cause blood clots to form, and bring on heart disease. Much medical research is now going on in this area.

Atoms in organic molecules can be replaced. Double and triple bonds in unsaturated compounds can be broken, and hydrogen atoms can be added to form saturated (single bond) compounds. Another type of reaction, called *substitution,* is also possible. Substitution describes a type of chemical reaction in which one atom, or group of atoms, is replaced by another atom or group of atoms.

Under certain conditions, hydrocarbons react with fluorine, chlorine, bromine, and iodine. For example, methane reacts with chlorine to form methyl chloride (CH_3Cl):

$$CH_4 + Cl_2 \rightarrow CH_3Cl + HCl$$

Notice that methyl chloride contains a chlorine atom that has replaced the hydrogen atom from methane,

$$H-\underset{\underset{H}{|}}{\overset{\overset{H}{|}}{C}}-\boxed{H} \qquad H-\underset{\underset{H}{|}}{\overset{\overset{H}{|}}{C}}-\boxed{Cl}$$

methane   methyl chloride

Addition and substitution reactions are among the most important reactions in organic chemistry. See Fig. 13-5.

In theory, many other organic compounds can be made from a hydrocarbon simply by replacing hydrogen atoms with other elements and radicals. By substituting chlorine, for example, for one, two, three, and four hydrogen atoms in methane, a series of different compounds can be made:

$$H-\underset{\underset{H}{|}}{\overset{\overset{H}{|}}{C}}-Cl \qquad H-\underset{\underset{H}{|}}{\overset{\overset{Cl}{|}}{C}}-Cl \qquad H-\underset{\underset{Cl}{|}}{\overset{\overset{Cl}{|}}{C}}-Cl \qquad Cl-\underset{\underset{Cl}{|}}{\overset{\overset{Cl}{|}}{C}}-Cl$$

| methyl chloride (CH_3Cl) | methylene chloride (CH_2Cl_2) | chloroform ($CHCl_3$) | carbon tetrachloride (CCl_4) |

The above four compounds are also named *mono-*, *di-*, *tri-*, and *tetra*-chloromethane. In addition to chlorine, fluorine can also be substituted for one or more hydrogen

13-5 Methane and chlorine react to undergo a substitution, or replacement, reaction. Can you name another type of organic reaction?

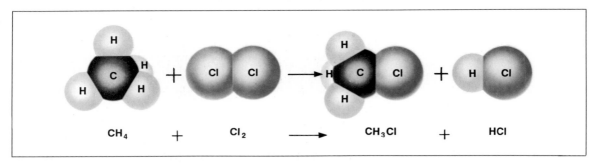

$$CH_4 \qquad + \qquad Cl_2 \qquad \longrightarrow \qquad CH_3Cl \qquad + \qquad HCl$$

atoms in methane. For instance, the chemical name for *Freon* (CCl_2F_2) is di-chloro-di-fluoro-methane. Freon is used as a refrigerant for home refrigerators and air-conditioning systems. Can you write the structural formula for this compound?

alcohols

Alcohols form a large group of organic compounds. All the straight-chain hydrocarbons can substitute the OH radical for hydrogen atoms. When one or more hydrogen atoms are replaced by OH radicals, an *alcohol* is formed. Alcohols make up one of the most important groups of compounds in organic chemistry. For example, methyl alcohol (CH_3OH), the simplest alcohol, can be derived by replacing one hydrogen atom in methane (CH_4) with one OH radical:

methane   methyl alcohol

Notice that although alcohol molecules contain OH groups, they are not bases. The reason is that alcohols do not ionize like inorganic bases you studied earlier in the text. Alcohols are not electrolytes; they do not conduct an electric current.

Methyl alcohol, or methanol, is sometimes called wood alcohol. This compound was formerly made by the destructive distillation of wood. Methanol is now prepared, using a catalyst, by the reaction of carbon monoxide and hydrogen:

$$CO + 2 H_2 \rightarrow CH_3OH$$

Methyl alcohol is very poisonous. If swallowed, it can cause blindness or death. Methyl alcohol is used as a solvent for lacquers and shellac, and in the making of many organic compounds.

Ethyl alcohol (C_2H_5OH), also called ethanol, can be made by replacing one hydrogen atom of ethane (C_2H_6) with one OH radical:

unit 2 chemistry in our world

$$\begin{array}{ccc} & H \quad H & \\ & | \quad\ | & \\ H-C-C- & \boxed{H} \\ & | \quad\ | & \\ & H \quad H & \end{array} \qquad \begin{array}{ccc} & H \quad H & \\ & | \quad\ | & \\ H-C-C- & \boxed{OH} \\ & | \quad\ | & \\ & H \quad H & \end{array}$$

ethane   ethyl alcohol

Ethyl alcohol is most often made from the fermentation of grain or fruit juices by yeast enzymes. Dextrose, or fruit sugar, is broken down by the yeast to yield alcohol and carbon dioxide:

$$C_6H_{12}O_6 \rightarrow \quad 2\ C_2H_5OH \quad + \ 2\ CO_2 \uparrow$$
dextrose   ethyl alcohol

The alcohol content of a fermented mixture may rise to about 15% maximum. At this point, the organisms producing alcohol are killed as a result of alcohol poisoning. An alcohol can be prepared in a more concentrated form by distilling the fermented mixture. Ethyl alcohol, commonly called grain alcohol, is the intoxicating agent in alcoholic drinks. Ethanol is also used in huge amounts in the chemical industry.

Ethyl alcohol is a fairly good antiseptic. Doctors use it to swab the skin before they give you an injection. For many purposes, ethyl alcohol is changed to *denatured* alcohol by adding foul-tasting or poisonous substances to make it unfit to drink. See Fig. 13-6.

13-6 Tanks of industrial alcohol are stored in the distillation plant for distribution to all parts of the country. What is denatured alcohol?

chapter 13   organic compounds   239

activity

Preparation of ethyl alcohol. In a large bottle, prepare a water solution of molasses and yeast. Allow the solution to stand at room temperature for several days. Identify the gas that is given off by passing it through limewater. Distill a portion of the fermented mixture. Dampen the inside of a beaker with the distilled liquid. Does the liquid have an odor? Does it burn? Identify the compound produced.

Some alcohols, such as ethylene glycol [$C_2H_4(OH)_2$], contain more than one OH radical. Ethylene glycol is widely used as an antifreeze in car radiators. It is marketed under such brand names as Prestone and Zerex. Ethylene glycol, added to water, will not corrode the radiator. Water freezes at 0°C. If four parts of water are mixed with six parts of ethylene glycol, the freezing temperature is lowered to −49°C.

Another important alcohol that contains more than one OH group is glycerol [$C_3H_5(OH)_3$], commonly called glycerin. Glycerin is a by-product of soap making. This compound is used in the making of printer's ink, lotions, and the explosive nitroglycerin. Glycerol is made by replacing three hydrogen atoms of propane (C_3H_8) with three OH radicals:

propane   glycerol

Table 13-2 lists some of the members of the alkane series of hydrocarbons with the names and formulas of their related alcohols. Notice that in each case one hydrogen atom has been replaced by one OH group.

Table 13-2:

Some Alcohols from Methane Series			
Hydrocarbon		Alcohol	
methane	CH_4	methyl alcohol	CH_3OH
ethane	C_2H_6	ethyl alcohol	C_2H_5OH
propane	C_3H_8	propyl alcohol	C_3H_7OH
butane	C_4H_{10}	butyl alcohol	C_4H_9OH
pentane	C_5H_{12}	amyl alcohol	$C_5H_{11}OH$
hexane	C_6H_{14}	hexyl alcohol	$C_6H_{13}OH$

Dorothy Hodgkin
Nobel Prize Winner

Dorothy Crowfoot Hodgkin, British chemist, was born in Cairo, Egypt, in 1910. She attended both Oxford and Cambridge universities in England. In 1947, she was selected as a Fellow of the Royal Society. Hodgkin was awarded the Nobel Prize in chemistry in 1964 for discovering the structure of several biochemical compounds. Her work included crystal structure analysis of complex substances, including penicillin and vitamin B_{12}. The B_{12} molecule has the most complex vitamin structure yet known, with the formula $C_{63}H_{90}O_{14}N_{14}PCo$.

Vitamin B_{12}, when injected into the human body in an amount as small as 0.000006 g per day, can cure pernicious anemia. Pernicious anemia is a disease caused by failure of the liver to supply vitamin B_{12} to the bone marrow. Vitamin B_{12} is needed for the bone marrow to produce the needed red blood cells in the body.

So far as is known, vitamin B_{12} is not present in higher plants. However, it is required in the diet of all higher animals. The exact ways in which vitamin B_{12} functions in living things are not yet known. Hodgkin's research will continue to provide solutions to many problems related to complex biochemical reactions.

summary

1. Organic compounds include hydrocarbons, alcohols, aldehydes, organic acids, ketones, ethers, and esters.
2. Hydrocarbons are grouped into the alkanes (single bond), the alkenes (double bond), and the alkynes (triple bond).
3. Carbon atoms are linked together in straight chains, branched chains, and ring compounds.
4. Single-bond hydrocarbon compounds are saturated. No other hydrogen atoms can be added.
5. Isomers are compounds with the same number and kind of atoms but with a different arrangement of atoms in the molecule.
6. The octane rating is a number that compares the antiknock properties of a gasoline with that of a standard test fuel.
7. Double- and triple-bond compounds are unsaturated. Bonds can be broken and other atoms added.
8. Two important types of organic reactions are addition and substitution (replacement).
9. Alcohols are obtained from a hydrocarbon by replacing one or more hydrogen atoms with one or more OH groups.

review

Match each item in the left column with the best response in the right column. *Do not write in this book.*

1. isomers
2. octane rating
3. alkane series
4. unsaturated
5. alkyne series
6. alcohol
7. benzene series

a. catalytic cracking
b. methane, ethane, propane
c. contains OH groups
d. hydrocarbons having double and triple bonds
e. process of producing aviation gasoline
f. first member is called acetylene
g. compounds whose molecules contain the same number and kind of atoms but with a different arrangement
h. ability of a gasoline to resist knocking
i. series of ring compounds

questions

Group A
1. What is organic chemistry? Name five different groups of organic compounds.
2. Draw the structural formula for pentane. What does this formula tell you about the molecule?
3. Draw the structural formulas for normal heptane and normal octane. What are the differences between them?
4. What is an isomer?
5. What is meant by a saturated compound? an unsaturated compound? Give an example of each.
6. Do straight-chain or branched-chain compounds make better gasoline for high-compression engines? Justify your answer.
7. Name three hydrocarbons from the alkane series in gasoline.
8. What causes a car engine to knock?
9. Write the molecular formula for the second member of the benzene series.
10. List several uses of benzene.
11. Describe two important types of organic reactions.
12. How would you recognize the formula for an alcohol?

Group B
13. Distinguish between the properties of organic and inorganic compounds.
14. Each compound in the alkane series is saturated. What does this mean?
15. Write the molecular formula of the 50th member of the alkane series.
16. Draw the structural formulas for two branched-chain isomers of pentane.
17. How is the octane rating of a gasoline determined? What value of octane do most new cars use?
18. Distinguish between the structure of the alkene and alkyne series.
19. Draw the structural formulas for four substitution compounds that methane forms with iodine.
20. How is methyl alcohol made commercially? ethyl alcohol?
21. How does an inorganic base differ from an alcohol?

further reading

Asimov, Isaac, *The World of Carbon.* New York: Abelard-Schuman.

Campaigne, E., *Elementary Organic Chemistry.* Englewood Cliffs, N. J.: Prentice-Hall.

Metcalfe, H. C., et al., *Modern Chemistry.* New York: Holt, Rinehart and Winston, 1982.

Seymour, R. B., *General Organic Chemistry.* New York: Barnes & Noble Books.

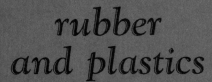

14

rubber
and plastics

Rubber and plastics, like fossil fuels, are organic compounds. Rubber is a long-chain hydrocarbon that comes from the rubber tree or is made from petroleum. Petroleum also provides the raw materials for making plastics and synthetic fibers.

objectives:

☐ To describe the process by which crude rubber is made from latex of the rubber tree.

☐ To distinguish between the properties of natural and synthetic rubber.

☐ To describe the differences between the making of thermoplastic and thermosetting plastics.

☐ To identify some properties of and uses for the various families of plastics on the commercial market.

rubber

Rubber is a natural product. Very ancient picture writings show Egyptians and Ethiopians using balls and other objects made from rubber. Later, when explorers came to the New World, they found Indians in Central and South America playing games with balls made of raw rubber. Today, rubber is anything but a plaything. In the United States alone, over 40,000 different products are made from rubber.

Rubber is easy to stretch. It is a nonconductor of electricity, making it a good cover for electric wiring.

Rubber is waterproof; it absorbs shocks; and it is air-tight, making it a suitable material for car tires. Special types of rubber resist heat and cold, and the corrosive action of many chemicals.

Natural rubber comes from the rubber tree. This tree grows wild in the Amazon Valley of South America but is also planted in tree farms in Southeast Asia. Over 90% of the world's supply of natural rubber comes from Southeast Asia. A rubber tree must grow for five to eight years before it is large enough to tap. A warm climate and from 170 to 250 cm of rainfall per year are needed for its proper growth. Rubber trees grow to a height of 12 to 16 m and will produce rubber for about 40 years.

A sap, called *latex*, drips from V-shaped cuts made in the trunk of the rubber tree, as shown in Fig. 14-1. Latex is a milky fluid that contains about 35% rubber. Trees are tapped each day during the early-morning hours. Have you ever broken or cut the stem of a goldenrod or a dandelion? Did you notice the sticky, white juice that came from the cut stem? This liquid is like latex, the raw material that comes from the rubber tree.

The rubber in latex appears as small particles scattered in the liquid, somewhat like butterfat in cow's milk. The latex is first strained, as shown in Fig. 14-2, to remove the suspended matter. Next, the liquid passes into special tanks, where acetic acid is added. The acid causes the particles of rubber to clump together. The solid mass of rubber is then washed and made into sheets. Finally, the sheets are smoked, dried, and baled.

14-1 To obtain latex from the rubber tree, a worker makes a V-shaped cut on the tree trunk. What climatic conditions are necessary for the proper growth of rubber trees?

14-2 Latex from rubber trees is strained to removed suspended material. What is the next step in making raw, natural rubber?

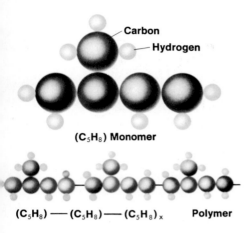

Carbon

Hydrogen

(C_5H_8) Monomer

(C_5H_8) —— (C_5H_8) —— (C_5H_8)$_x$ Polymer

14-3 Rubber monomers link together to form a polymer. What is a monomer? a polymer?

The simplest formula for rubber is $(C_5H_8)_x$. As you can see, rubber is a hydrocarbon, composed of carbon and hydrogen. The "x" in the formula is a number that is unknown at the present time. Some chemists believe that the value of x may be as high as 400,000. Rubber molecules are giants compared with simple molecules like H_2O and CH_4. Chemists call the C_5H_8 unit a *monomer* (*mon*-oh-mer), a single unit of a whole molecule. When many monomers link together, they form a giant molecule, called a *polymer* (*pol*-ih-mer), meaning many units. With the aid of an electron microscope, scientists have seen and studied some of these giant molecules.

Natural rubber, which is a polymer of C_5H_8, is shown in Fig. 14-3. The molecules of natural rubber form long zigzag chains. This property is probably one reason why rubber is so elastic.

Crude rubber can be improved through chemistry. Crude rubber is tough and sticky, but not strong and elastic. In 1839, Charles Goodyear found a way to make rubber less sticky, more elastic, and better suited for the many uses of rubber products.

By heating small amounts of sulfur with the crude rubber, Goodyear developed a product that could stand up to hard wear. The sulfur improved the properties of the raw rubber. This process, called *vulcanization*, launched the modern rubber industry that supplies rubber products for thousands of uses.

The properties of vulcanized rubber depend upon the amount of sulfur that is added. Auto tires and rubber bands usually contain about 4% sulfur. A harder rubber contains a much higher percentage of sulfur. During the vulcanizing process, sulfur joins with the polymer chains, forming rubber with better properties. Sunlight, wide temperature changes, and petroleum products are harmful to natural rubber.

Other substances may be added in making rubber articles. Used, or reclaimed, rubber from auto tires may be added to a batch of new rubber. As the name implies, reclaimed rubber is obtained from worn-out rubber products. Used rubber may be treated to get rid of the fiber content. Rubber goods made largely from reclaimed rubber are not as flexible as those made from pure natural rubber. However, reclaimed rubber goods are tougher and longer-lasting, as well as less costly.

Organic catalysts are added to rubber to speed up the vulcanizing process. Other organic compounds are added to rubber to keep it from getting stiff and brittle.

Most rubber is used in making auto tires. A great deal of rubber is used in making cars. In fact, there are over 300 rubber parts in a single car. About two thirds of all the rubber used in the United States goes into making car tires.

Strong cotton, rayon, and nylon cords are embedded in soft rubber and used to strengthen a tire. Fiber glass, polyester, and steel are also used to make tires. Steel wires, called beads, are placed in the rim of the tire to make sure that the tire holds firmly to the wheel.

A tire is built up by wrapping layers, or *plies,* of rubberized fabric around a hollow steel drum. The plies may be bias, belted-bias, or radial, as shown in Fig. 14-4. Each ply, or layer, is applied in sequence with the tread layer being applied last. Finally, the tire is placed in a mold and heated to complete the vulcanizing process. See Fig. 14-5.

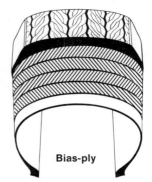

Bias-ply

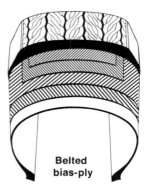

Belted bias-ply

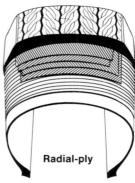

Radial-ply

14-4 Bias, belted-bias, and radial plies (layers) are used in automobile tires. Can you describe the difference in construction of each of these tires?

14-5 Newly made and vulcanized auto tires are sorted for shipment. What substance is used in the vulcanizing process?

chapter 14   rubber and plastics   247

14-6 These used tires were punched with holes to reduce buoyancy and weighted with cement. They were then sunk off the coast of Florida to provide new homes for fish. What other products can be obtained from scrap tires?

Car tires contain large amounts of carbon black for added strength and longer life. The carbon black also accounts for the color of tires. Carbon black also gives longer life to rubber footwear and other articles as well. In less costly rubber goods, clay is used as a filler. Zinc oxide is sometimes added to make rubber white.

In the past, worn-out tires could not be recycled on a large scale at a reasonable cost. However, recent advances have been made that show promise for recovering oil, carbon black, and steel from scrap tires. See Fig. 14-6.

activity

Removal of rubber from latex. Into a beaker half-full of latex, pour a small amount of acetic acid. Stir the mixture and observe the lumping of raw rubber. Wash several lumps to remove the acid and dry. Study the lumps of raw rubber. Squeeze and stretch them. Heat a portion of the sample in hot water. Examine its elastic properties. Cool another sample of the raw rubber in a refrigerator for several hours. Study its elastic properties. What conclusions can you draw from what you observed?

14-7 White foam rubber, after being whipped with air, is poured into a mattress mold. Can you describe one process by which foam rubber is made?

Foam rubber has many uses. Thousands of tiny air bubbles in *foam rubber* make it an excellent material for use in sofas and chairs. Foam rubber is used in making many other products as well. Blocks of foam rubber, covered with leather, cloth, or plastic sheet, make excellent cushions because they hold their shapes well even after long use. Some mattresses, for example, are thick blocks of foam rubber covered with cotton cloth.

One method of making foam rubber consists of whipping air into the latex, as is done when whipping cream. During vulcanization, the trapped air expands, inflating the mass into foam rubber. See Fig. 14-7.

Another method of making foam rubber is to add ammonium carbonate [$(NH_4)_2CO_3$] to a batch of raw rubber. When heated, ammonium carbonate changes into gaseous ammonia, carbon dioxide, and water vapor:

$$(NH_4)_2CO_3 \rightarrow 2\ NH_3 \uparrow\ +\ CO_2 \uparrow\ +\ H_2O \uparrow$$

248

The heat produced by the vulcanizing process breaks down the ammonium carbonate into gases, which inflate the dough-like mass of rubber.

Organic compounds are used to make synthetic rubber. The word *synthetic* means "put together" in Greek. For many years, scientists were unable to make a good substitute for natural, tree-grown rubber. Although each substitute was rubber-like, no single one had all the qualities of natural rubber. In 1931, *neoprene*, a suitable synthetic rubber, appeared on the market. The main raw material for making neoprene is the first member of the alkyne series of hydrocarbons, called acetylene. If you compare the structural formulas of natural rubber and neoprene, you can see that they are quite similar:

$$\left[\begin{array}{cccc} & H & & H \\ & | & & | \\ -C- & C & =C- & C- \\ & | & | & | & | \\ & H & CH_3 & H & H \end{array} \right]_x \qquad \left[\begin{array}{cccc} H & & H & H \\ | & & | & | \\ -C- & C & =C- & C- \\ | & | & & | \\ H & Cl & & H \end{array} \right]_x$$

natural rubber   neoprene rubber

Oil and greases, which destroy natural rubber, have little effect on neoprene. Therefore, neoprene is useful for making gasoline-pump hoses, gaskets, tank linings, and many other items in common use.

There are over 100 types of synthetic rubber made for special uses. One very vital type of rubber that is made in large amounts is *styrene-butadiene* (*sty*-reen-byoo-tuh-*dy*-een), better known as SBR rubber. Styrene and butadiene are two hydrocarbon compounds.

During World War II when the Japanese cut off the supply of natural rubber from Southeast Asia, a number of SBR plants were built. Since that time, the synthetic rubber industry has grown so that most of the rubber in use today is synthetic. The main rubber used today for making car tires is SBR rubber. See Fig. 14-8.

14-8 Inspecting "crumb" rubber in a synthetic rubber plant in Houston, Texas. What raw materials are used to make SBR rubber?

plastics

Plastics are synthetic products. The word *plastic* has two meanings. If used in a general way, it can mean any material that can be shaped or molded into a desired form. Glass, clay, wax, and concrete, therefore, are considered "plastic" materials under the general meaning. However, the term *plastic*, as used by chemists, refers to synthetic products made from coal, oil, or related raw materials and molded or shaped into final forms by heat and pressure. Rubber itself, either natural or synthetic, is a good example of a plastic. Most common plastics are formed by linking together monomers to form very large molecules, called polymers.

In 1868, while trying to find a substitute for ivory used in billiard balls, John W. Hyatt treated cellulose nitrate with camphor. This reaction formed a plastic material called *celluloid*. The discovery of celluoid marked the beginning of the plastics industry.

Some 40 years later, a second major step forward was taken. Dr. Leo Baekeland mixed phenol with formaldehyde. This reaction produced the first phenolic plastic, Bakelite.

Today, the plastic industry is one of the few billion-dollar industries in the United States. The growing use of plastic objects in almost every part of society is due to their wide range of colors, durability, weight, and ease of mass production.

Plastics are thermoplastic or thermosetting. Synthetic plastics can be divided into two main groups, *thermoplastic* and *thermosetting*. Those plastics that can be softened and shaped by gentle heating are thermoplastic. These plastics harden when cooled, no matter how often this process is repeated. Thermoplastics melt easily. Thermosetting plastics take their permanent shape when heat and pressure are applied during the forming process. Reheating will not soften these materials. Thermosetting plastics are difficult to melt.

Whether a plastic is thermoplastic or thermosetting depends on the kind of chemical reaction that takes place while the plastic is being made. When chains of molecules are formed by "head-to-tail" addition of monomer units, the result is usually a thermoplastic product. If, however, new bonds are formed between the chains and between the units, this cross-linking usually produces a thermosetting product.

Special molds are used in making plastic articles. Thermoplastic materials are shaped into rods, tubes, and

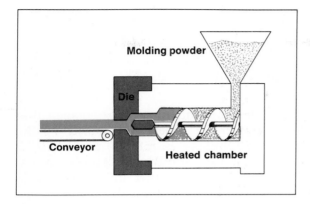

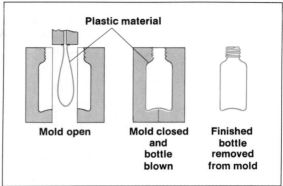

Plastic material

Mold open

Mold closed and bottle blown

Finished bottle removed from mold

Molding powder

Die

Conveyor

Heated chamber

sheets by a process called *extrusion*. First, a dry powdered plastic material is loaded into a hopper. Then a turning screw feeds the powder into a heated chamber, such as the one shown in Fig. 14-9. The powder is changed to a paste-like mass. This mass is forced through an opening onto a conveyer belt where it is cooled either by air blowers or by dipping the mass into water. This extrusion process is like squeezing toothpaste out of a tube.

Another method by which thermoplastic articles are made is by *injection molding*. In this process, a plunger forces the plastic powder through a heated chamber where it becomes a paste-like material. At the end of the heating chamber, there is a small opening through which the softened plastic enters a closed mold. As soon as the plastic cools, the finished product is taken from the mold.

Blow molding is also used in forming thermoplastic articles. This process consists of blowing a plastic material against a mold. As shown in Fig. 14-10, air is blown into the plastic, as into a balloon, to force the material onto the sides of the mold. How does this process compare with glass blowing?

There is still another kind of molding used. Thermosetting plastics are sold to factories in the form of molding powders. Finished products are then made from the powder by means of *compression molding*. Figure 14-11 shows how the powder is forced into a shaped metal mold. Using this process, many copies of the product can be made with ease. Wood that has been ground into a fine powder is often mixed with the molding powder to strengthen the finished product.

14-9 Thermoplastic molding powders can be made and shaped into various forms. What is this process called?

14-10 "Blow molding" is used to form plastic bottles. What kind of plastic is used in this mold?

14-11 Many identical plastic articles can be made from this compression mold. What is often added to the plastic molding powder to strengthen the finished product?

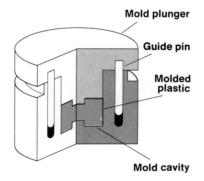

Mold plunger

Guide pin

Molded plastic

Mold cavity

Measured amounts of the molding powder are placed between the heated chambers in a giant press. When the press closes, the molding powder changes under heat and pressure to a plastic object in the shape of the mold. After a given time, the press opens and the finished product is ready.

Thermoplastic articles are plentiful. You have seen that cellulose nitrate and camphor were first used in the making of celluloid. Cellulose nitrate is prepared by treating cotton with a mixture of nitric and sulfuric acids. The soaked cotton is then dissolved in a mixture of alcohol and ether. Cellulose nitrate is left as a jelly-like mass when the solvent evaporates. Camphor is then added as a softening agent. The result is celluloid. Among the many things that are made from this plastic are shoe-heel covers and fabric coverings. The use of celluloid is quite limited, however, because this plastic burns readily, creating a fire hazard.

14-12 A bubble of Saran Wrap, a vinyl plastic. What use can be made of this plastic?

Another related plastic is *cellulose acetate*. The same reactions are involved in the making of cellulose acetate as in cellulose nitrate. The only difference is that acetic acid is used in place of nitric acid. Although similar to cellulose nitrate, cellulose acetate is used in vacuum cleaner parts, combs, toys, eyeglass frames, recording tapes, photo film, and lamp shades. Cellulose acetate does not burn readily.

All *vinyl* (*vye*-nil) plastics are tough and strong. The flexible types of vinyl can be bent back and forth without weakening or tearing. They resist chemicals, water, heat, and cold. Some well-known items that are made from vinyl plastics include raincoats, door frames, inflatable water toys, garden hoses, floor covering, shower curtains, and wall coverings. See Fig. 14-12.

Another group of widely used thermoplastics is *acrylic* (uh-*kril*-ick). This plastic is sold under such trade names as Lucite or Plexiglass. Acrylics are used as airplane and building windows, outdoor signs, car tail lights, see-through furniture, salad bowls, surgical instruments, skylights, and costume jewelry. Acrylics withstand exposure to weather. They are softer than glass, however, and are easily scratched. See Fig. 14-13.

Polyethylene plastics are flexible. The hydrocarbon ethene is used to make polyethylene. This thermoplastic is tough, waxy, moisture-proof, and is not acted on by most solvents. Polyethylene is used in making such items as ice-cube trays, moisture-proof bags for freezing foods, rigid and squeezable bottles, tumblers, bags for candy, and many other unbreakable articles. Paper milk cartons are coated with polyethylene to prevent leakage. Even when cooled to a very low temperature, polyethylene will not become stiff and brittle. See Fig. 14-14.

Nylon is another thermoplastic material available as a molding powder. Nylon can also be produced as sheets, rods, filaments, and tubes. The finished products can be shaped by injection, extrusion, or blow molding. Nylon is tough, flexible, absorbs very little water, and is not affected by many common chemicals, greases, and solvents. Nylon is a familiar material for gears, ropes, brush bristles, fishing lines, and faucet washers. The most common use of nylon, however, is in making clothing fabric, about which you will study in the next chapter.

A *fluorocarbon* plastic, called *Teflon,* is a stickless material that produces almost no friction with other surfaces. In fact, Teflon feels somewhat like a wet bar of soap. Teflon

14-13 Acrylic plastic was used to build the huge roof of the Astrodome in Houston, Texas. Can you list several properties of this type of plastic?

14-14 A single worker can easily handle this new lightweight plastic pipe. Can you name the type of plastic used in making these pipes?

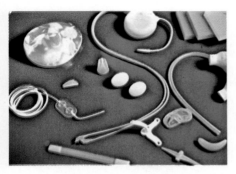

14-15 Because of its ability to resist reaction with body chemicals, this type of plastic has found new uses in surgery.

resists heat and cold, and can also withstand swift temperature changes from subzero to above boiling. Foods can be removed from Teflon-coated frying pans without sticking, even though no fat or cooking oil is used. Teflon is being used to make long-lasting, self-oiling piston rings, car bearings, and ball joints. Surgeons even use this remarkable plastic to replace parts of veins and arteries in the body. See Fig. 14-15.

activity

Making a plastic material. Pour about 100 mL of a saturated solution of aniline hydrochloride and 100 mL of 40% formaldehyde solution (formalin) into a large beaker. Heat gently and stir with a glass rod. Record what you observe. Can this material be poured into a mold? What effect does heat have on the material obtained?

Thermosetting plastics have many uses. The major families of thermosetting plastics include plenolics, melamines, polyesters, epoxies, and polyurethanes. Bakelite is a phenolic plastic and is considered by the industry to be the "workhorse" of the group. *Phenolics* are strong and hard. They do not readily absorb water and other chemicals. Phenolics easily stay rigid and keep their luster and strength. Car distributor heads and insulation, washing-machine agitators, cabinets, and telephone cases are some of the things that are made from this group of plastics. Some members of this family, such as Bakelite, are dark in color. Thus, the shades of color obtained by adding pigments to phenolics are limited.

A colorful, unbreakable dinnerware in common use is made from *melamine* plastics. Other products, including handles for kitchen utensils, buttons, and hearing-aid cases, are made out of melamine plastics. Melamines provide a glassy surface and are resistant to scratching and heat.

A plastic used to saturate cloth, paper, and mats of glass fibers belongs to a family of plastics called the *polyesters*. These plastics are highly resistant to most solvents, acids, bases, and salts. Polyesters absorb very little, if any, water and have an excellent resistance to weather. Polyesters tend to have a very hard surface.

Certain plastics are used to firmly bond metals, ceramics, glass, and other plastics. *Epoxy* resins are often used for this purpose. They are used for castings and coatings as well.

Interest has grown sharply in the *polyurethane* (poly-*your*-eh-thane) family of plastics in recent years. One type of polyurethane forms a foam, which is superior to foam rubber in many ways. This foam is also a good insulation for homes. Five centimeters of this foam have the same insulating properties as about 25 cm of fiber glass. Polyurethane plastic can also be molded and colored to produce wood graining in furniture.

Thermosetting plastics are used in making many *laminated* products. Laminated means that the article is built up in layers or plates, one on top of another. Sheets of cotton cloth or paper are dipped in Bakelite varnish and then piled one on top of another to any desired thickness. The layers are then put in a press, and heat and pressure are applied. This process turns out a very hard, tough sheet of material. Laminated plastics are used for making gears, table tops, electronic panels, and lightweight airplane propellers, as well as many other items.

Charles Goodyear 1800-1860

For many centuries, Indians in Central and South America obtained a milky liquid called latex from certain trees. They used this liquid to make elastic bouncing balls. The early explorers used the milky latex to waterproof their clothing. The latex worked until the sun melted some of the rubber on the fabric, causing it to become soft and sticky. During the cold weather, the material became hard and brittle. Latex was, as they learned, not suitable for making waterproof clothing, shoes, and wagon covers.

Charles Goodyear, at the age of 31, became interested in finding a treatment for rubber that would enable it to resist extremes of heat and cold. If such a treatment were possible, rubber could be used in a number of products.

In 1839, he accidently dropped some rubber mixed with sulfur on a hot stove, and found a solution to his problem. The rubber-sulfur mixture was "fired" on the hot stove. This process was called "vulcanization" after Vulcan, the Roman god of fire. Goodyear's chance discovery gave birth to a new industry!

Goodyear proved to be a better inventor than a business man. When he died in 1860, he was out of work and penniless. Although his discovery made millions of dollars for others, Goodyear left a debt of $200,000.

"Life," Goodyear once said, "should not be estimated exclusively by the standard of dollars and cents. I am not disposed to complain that I have planted and others have gathered the fruits. A man has cause for regret only when he sows and no one reaps."

summary

1. Natural rubber is a hydrocarbon made from the sap of rubber trees.
2. A rubber molecule is a polymer of C_5H_8 monomers linked together in a twisted zig-zag chain.
3. Raw rubber is soft and sticky, but when heated with sulfur, it loses its stickiness and becomes more elastic.
4. Neoprene, the first successful synthetic rubber, resists the harmful action of oils and greases better than natural rubber.
5. Thermoplastic articles become soft when heated and harden when cooled.
6. Thermosetting plastics take a permanent shape when heat and pressure are applied.
7. Cellulose, vinyl, acrylic, polyethylene, nylon, and fluorocarbons are families of thermoplastics.
8. Thermosetting families include phenolics, melamines, polyesters, epoxies, and polyurethane plastics.

review

Match each item in the left column with the best response in the right column. *Do not write in this book.*

1. latex
2. polymers
3. vulcanization
4. carbon black

a. filler for tires to give extra wear
b. first commercial synthetic rubber
c. phenolic plastic family
d. laminated articles

5. thermoplastic
6. Bakelite
7. epoxy
8. neoprene

e. softened by heat
f. rubber and plastics composition
g. ammonium carbonate
h. product from the rubber tree
i. heating raw rubber with sulfur
j. used as a bonding material

questions

Group A
1. What are five properties of rubber?
2. What is the simplest formula for rubber?
3. What conditions are needed for the proper growth of rubber trees?
4. Why is carbon black added to car tires?
5. Why is the vulcanization process so important in the making of rubber?
6. How is foam rubber made?
7. What is neoprene? What is the main difference in the structural formula of a natural rubber molecule and that of a neoprene molecule?
8. In what respect is neoprene better than natural rubber?
9. What synthetic rubber is mostly used in making car tires?
10. What raw materials are used in making SBR rubber?
11. What were the first plastics ever used? What property did they have that limited their use?
12. What does the word *plastic* mean as used by chemists?
13. What raw materials are used in making celluloid?
14. List four properties of plastics.
15. Describe the difference between thermoplastic and thermosetting plastics.
16. List several uses of the acrylic plastics. of vinyl plastics.
17. What are some of the uses of polyurethane plastics?

Group B
18. How is latex processed to obtain crude rubber?
19. Describe the difference between a monomer and a polymer.
20. What is the effect of adding reclaimed rubber to a batch of new rubber?
21. Tell how a plastic article is made by compression molding.
22. Describe how thermoplastic articles are made.
23. Discuss the properties of Teflon. What are some of its uses?
24. What chemicals are used in making Bakelite?
25. How are laminated gears made from Bakelite and cotton?
26. What type of plastic would you recommend for (a) housing for a computing machine? (b) garden hose? (c) adhesive bonding? (d) frying and baking pans? (e) squeeze bottles? (f) dinnerware? (g) radio and TV consoles? (h) recording tape? (i) fishing lines?

further reading

Asimov, Isaac, *Asimov on Chemistry*, Garden City, N. Y.: Doubleday, 1974.

Dean, H., *Manufacturing Industry and Careers*. Englewood Cliffs, N. J.: Prentice-Hall, 1975.

Jambro, D., *Manufacturing Processes: Plastics*. Englewood Cliffs, N. J.: Prentice-Hall, 1976.

Kaufman, M., *Giant Molecules: The Technology of Plastics, Fibers and Rubber*. Garden City, N. Y.: Doubleday, 1968.

Lappin, A. R., *Plastics and Techniques*. Bloomington, Ill.: McKnight, 1965.

Le Bras, J., *Introduction to Rubber*. New York: Hart, 1970.

Newman, T. R., *Plastics as an Art Form*. Radnor, Pa.: Chilton, 1972.

natural and synthetic fibers

15

You have seen that fuel, rubber, and plastics contain organic compounds. In this chapter, you will learn about natural and synthetic fibers, which also contain organic compounds. Fabrics come in four basic groups depending on their origin: animal, plant, mineral, and synthetic. All animal fibers are proteins: that is, compounds that contain carbon, hydrogen, nitrogen, and oxygen. Plant fibers are mostly cellulose, a carbohydrate composed of carbon, hydrogen, and oxygen. Coal and petroleum are the basic raw materials for making synthetic fibers. Over 70% of the fibers in use today are synthetic.

objectives:

☐ To develop an understanding of the chemistry of natural and synthetic fibers.

☐ To identify wool, cotton, linen, and silk fibers with the aid of a microscope.

☐ To describe how yarn is made from natural and synthetic fibers.

☐ To describe bleaching and dyeing processes of woven and formed fabrics.

natural fibers

Wool comes from many animals. Wool comes from the fleece of sheep, llama, alpaca, goat, or camel. The curly, kinky hairs of these animals mat together, trapping "dead air." This insulating property makes wool an ideal cold-

15-1 A worker shearing wool from a sheep. What is the next step in making wool cloth?

15-2 The carding machine forms a flimsy sheet of fibers. The fibers are made into a ropelike strand and coiled onto a large cylinder. What preparation of the wool is necessary before carding?

weather fiber because it prevents the loss of body heat. The use of wool dates back several thousand years. See Fig. 15-1.

Wool is used mostly for making cloth. Australia is the leader in wool production, supplying over 35% of the world's total. The United States ranks sixth, producing about 4% of the world's supply.

At the woolen mills, skilled workers sort the fleeces into groups of differing qualities. An "all wool" label on a garment does not tell the quality of the wool.

Federal law requires that labels on clothing tell the percentage of wool and other fibers mixed into the cloth. Wool that has never before been woven into cloth is called new, or *virgin wool*. Some wool, however, is not new. Scraps of unused woolen cloth are sometimes processed by machines back into fibers. This shredded fiber gives a low-quality wool that must be labeled *reprocessed wool*. Woolen rags from worn clothing can be reworked into fibers. This kind of wool is labeled *reused wool*. Reprocessed and reused wools do not wear as well as virgin wool.

Wool fibers are made into yarn and then into cloth. Wool, as it comes from an animal, is dirty and greasy. Before the wool can be made into yarn and woven into cloth, it must be washed with a detergent and a mild alkali in warm water to remove the grime, perspiration salts, and natural wax. After washing, it goes to the *carding* machine, where the tangled mass is spread into a flimsy, web-like sheet. Carding machines have large turning drums covered with short, steel teeth to comb the wool into a flat web. Carding blends the various fibers together. See Fig. 15-2.

Next, the rope-like strand from the carding machine must be drawn and twisted into thinner strands on the *roving frame*. This process gives the fibers the strength needed to begin the spinning process, making the strand into yarn. The yarn is then woven into cloth. See Fig. 15-3.

Wool is made up of polymers, molecules having a long structure. Chemically, wool is very much like hair, feather, and horn. Wool is easily dissolved by strong bases, such as sodium and potassium hydroxides. It burns poorly, with an unpleasant odor, and leaves a black charred mass as residue.

If you look at wool fibers through a microscope, you can see that they have tiny, scale-like plates, such as those shown in Fig. 15-4. These plates overlap each other like shingles on a roof. The scale-like fibers hook onto each other, causing wool to mat together. The fibers of wool tend to spring back into their original shape after they are bent or folded. This is the reason why your woolen clothes hold their shape better than do cotton, rayon, or linen garments.

Since heat weakens woolen fibers, hot irons should never be used on woolen cloth. Hot water should also be avoided because it may cause wool to shrink. Dry-cleaning solvents do not have this effect on woolen cloth. Cool water and a mild detergent are best for washing woolen garments.

15-3 Wool is made into yarn on a roving frame (left). The yarn is then loaded into a warper (right) and is ready to be woven.

15-4 Compare the scalelike wool fibers (left) with smooth cotton fibers (right) as seen through a microscope. Why do wool clothes hold their shape better than cotton, rayon, or linen?

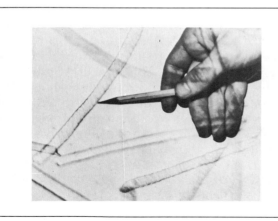

chapter 15　natural and synthetic fibers　　　261

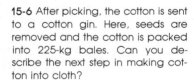

15-5 Cotton forms the foundation for the employment of more than 9 million Americans. Where is most cotton grown in this country?

15-6 After picking, the cotton is sent to a cotton gin. Here, seeds are removed and the cotton is packed into 225-kg bales. Can you describe the next step in making cotton into cloth?

Today, some wools are made shrink-resistant. Special treatment of wool fabrics permits machine washing with little or no shrinking. Wool can also be treated to make it moth-proof.

Cotton comes from a plant. The so-called "cotton belt," the region where cotton is grown in this country, includes 18 southern states, stretching from Virginia to California. Cotton plants are grown from seeds that are planted yearly. The length, color, and cleanliness of cotton fibers vary with soil and weather conditions. Most of the world's cotton is grown in the United States, Russia, China, and India. See Fig. 15-5.

Under the microscope, as shown in Fig. 15-4, cotton fibers look much like flattened, twisted tubes, quite unlike wool. Alkalis are not as harmful to cotton as they are to wool, but acids are. Can you suggest a way to find the percentage of wool contained in a sample of mixed wool and cotton cloth?

Body moisture is readily absorbed by cotton fibers and then lost through evaporation. This makes cotton garments comfortable for summer clothing. However, perspiration in contact with cotton fiber produces a weak acid reaction. This is one reason why cotton clothing wears out in places where perspiration is most active. Like other natural fibers, cotton is often used in blends with synthetic fibers.

Cotton is nearly pure cellulose, having a formula of $(C_6H_{10}O_5)_x$. The "x" in this giant molecule is not known, but it is believed to be as high as 500,000.

Cotton is made into cloth. Cotton arrives at the mills in large bales, as shown in Fig. 15-6. A machine called a *picker* rips the cotton from the bale into a fluffy mass. Dirt and seed particles are removed during this process. See Fig. 15-7. The cotton then goes to the carding machine where it is changed to a flat, web-like sheet. The sheets are gathered into loose strands, or roves. The roves are then made into cotton yarn by drawing and twisting. This process lines up the fibers side by side to give them added strength. See Fig. 15-8.

Garments made from cotton tend to shrink when they are first washed. To prevent this shrinkage, a process called *Sanforizing* is used. Sanforizing is a trade name for the process of placing the cloth in a steam bath while it is stretched, thus "preshrinking" the finished cloth.

To give cotton a glossy finish and make it stronger, cotton fibers are sometimes *mercerized* as well. In this process, cotton cloth is stretched and dipped into a solution of sodium hydroxide. This process causes the fibers to swell, and to become rounded, more lustrous, and stronger.

Specially treated cotton fibers have the "wash-and-wear" or wrinkle-proof properties of synthetic fibers. Some of these fibers resist water and stain. Clothing made from such treated cotton fibers requires less frequent laundering and keeps a freshly pressed look. Other cotton fibers can be treated to make them fire resistant. See Fig. 15-9.

The cotton seeds that are separated from the fibers are used to make oil for cooking and for making margarine. The solid that remains makes a protein-rich feed for cattle or flour for low-starch diets.

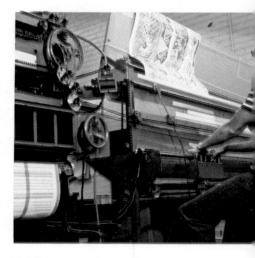

15-7 A picker removes dirt and seeds from the cotton. What is the next step in making cotton cloth?

15-8 This woman is recording a pattern on punched cards. The cards are then fed into the loom and the pattern is reproduced in the cloth.

15-9 Can you tell the difference between the treated and untreated fabrics after both have been washed?

15-10 Silkworm cocoons. Can you describe what happens inside the cocoons?

Silk comes from a caterpillar. Silk, like wool, comes from an animal. The cocoons of the mulberry silkworm provide the source of silk fibers. The caterpillars, or larvae, feed on mulberry tree leaves and spin cocoons to cover and protect themselves while they slowly change into moths. See Fig. 15-10.

Silk starts as a liquid that comes from tiny glands below the mouth of the caterpillar. When the fluid comes in contact with air, it changes into a solid and becomes a strand of silk. The caterpillar builds its cocoon by winding this strand around itself again and again. If laid in a straight line, the strand may reach over 1000 m in length.

After about three weeks, the insect bores a hole in one end of the cocoon and comes out as a moth. However, silk obtained from these cocoons is of little value once it is punctured by the insect. Steam or hot air is used to kill the insect before it comes out of the cocoon.

The cocoons are then treated with heat to soften the gummy material on the outside. The fibers are unwound and joined with fibers from other cocoons to form a single strand of silk. This process takes a long time. As many as 3000 to 5000 cocoons are needed to produce a kilogram of raw silk. For this reason, silk is much more costly than any other natural fiber.

A molecule of silk is a large protein polymer. Silk is the strongest natural fiber. It is elastic and wrinkle resistant. Silk is weakened by exposure to sunlight and perspiration, and tends to yellow with age. Silk is seldom washable; most silk garments must be dry cleaned. Silk feels smooth to the touch because the fiber is free from joints or scales. Alkalis and acids will damage and may even destroy this fiber. Japan is the leading producer of raw silk fiber.

Linen comes from the flax plant. Linen is the oldest fiber known, dating back many thousands of years. Linen is much smoother, more lustrous, and more soil resistant than cotton. Linen is not wrinkle resistant, but most cloth made from linen today has a wrinkle-resistant finish added. Cloth made from linen readily absorbs moisture, making it popular for warm-weather clothing.

The flax plant provides two main products: linen fiber from the stems, and seeds from the flowers. To obtain linen fiber, the crop must be cut before the seeds mature.

If flaxseed is the desired product, the plant is allowed to grow until the seeds mature. However, the stem fibers of a mature plant can no longer be used for linen.

The flax plants to be made into linen are pulled up and spread out in the sun to dry. The stems have a thin covering of bark, which must be removed. Linen fibers are nearly a meter in length, very soft, and yellow in color. These fibers, which are much stronger than cotton, tend to lie parallel. Linen, like cotton, is nearly pure cellulose, with similar chemical properties.

Linen cloth is made in a manner much like cotton. Spinning draws and twists the fibers. The cloth can then be woven on looms from the spun yarn. Linen, the most permanent natural fiber, outwears cotton and wool. Unlike cotton, linen fibers do not shed lint, making them ideal for dish towels, napkins, and tablecloths.

Although Ireland is the world's linen center, flax is also grown in Belgium and the Netherlands. In the United States, flax is grown for seeds that are used to make linseed oil.

activity

Burning properties of natural fibers. Hold a sample of wool fibers with tweezers in a Bunsen burner flame, or burn them with a match. Note the rate of burning, the nature of the residue, and the odor. Repeat this process using cotton, silk, and linen samples. Can you distinguish between the properties of animal and plant fibers?

Leather is made from the hides of animals. Leather is a substance that is strong, soft and warm. It can be made into footwear, clothing, luggage, furniture coverings, sporting goods, and many other items.

In the United States, leather is made chiefly from the hides of cattle that are raised for beef. The fresh hides are preserved by "salting" with coarse grains of sodium chloride. This process keeps the skins from rotting while awaiting shipment to the tannery for further processing.

At the tannery, the hides are soaked in cold water to remove the salt and any dirt. Since salt has a drying effect on the hides, the soaking also restores some of the moisture. The hides are then placed into a bath of lime and sodium sulfide to remove the hair.

Bating is the next step in the making of leather. Certain enzymes are mixed with wood "flour" to digest hair roots and other protein matter sticking to the hide.

The hides are now treated with various plant bark and wood extracts or chemical tanning agents. These agents cause physical and chemical changes in the hide that turn it into leather. Sodium bichromate or potassium bichromate, with small amounts of alum, are the most commonly used chemical tanning agents.

About 5 m² of leather can be obtained from an average-sized cow. Deer, pig, reptile, sheep, and elk are other animals that are used for making leather. Since there are few farm horses left in this country today, that source of leather has declined in recent years.

Paper is made from cellulose. The raw material for making paper is wood. Paper fibers are nonwoven fabrics, called "formed fabrics." This term means that the fibers are bonded together by methods that include heat, adhesion, solvents, and pressure.

In making paper, the fibers of wood are first separated from one another by chopping, shredding, and beating. The fibers are then mixed with a great deal of water to form a thin "soup." This soup is forced through a narrow slit down onto a vibrating wire screen. As the water drains through the screen, the fibers become tangled together in a matted sheet. These sheets are passed between heavy rollers that squeeze out the remaining water, and then through other steam-heated rollers that iron the surface to a smooth finish. See Fig. 15-11. Of course, there are many refinements used in this process. However, paper is basically a tangled mass of cellulose fibers.

Sheets of tangled wood fibers from logs of poplar, spruce, and certain kinds of pine are termed *mechanical pulp.* Newspapers are printed on bleached mechanical pulp paper because it is less costly than chemically processed paper. See Fig. 15-12.

"Kraft" paper, made from southern pines, is made from *chemical pulp.* Chemical pulp comes from wood chips boiled under pressure. "Kraft" is a Swedish word meaning "strength," a property that is needed for paper cartons, brown grocery bags, and wrapping paper.

15-11 Water is drained from the "soup" of shredded wood fibers. Can you describe the next step in making paper?

For the highest quality paper, cotton and linen rags are beaten for the pulp. Paper used for currency is made of rag paper. Colored silk and nylon fibers are added for strength and to make counterfeiting more difficult. Parchment paper is given a hard, translucent finish by running it through a tank of sulfuric acid and then quickly washing off the acid. Waxed paper is made by passing paper through a vat of melted paraffin. Roofing papers contain tar and asphalt.

A glossy paper texture is obtained by adding a coating of rosin, casein (a milk protein), or gelatin to the surface. White paper, as you might suspect, is produced by a bleaching process.

Some fibers are called mineral fibers. Mineral fibers are made from inorganic substances, including some metals. Two important mineral fibers in common use are *asbestos* and *fiber glass.*

Asbestos is a silky, fibrous mineral that can be spun into threads and woven into cloth. These fibers are fireproof and are unaffected by most common acids. Asbestos products are used as insulation for hot-water and steam pipes, fireproof clothing, and in brake linings for cars. See Fig. 15-13.

When a blast of high-pressure steam is directed against molten glass, forcing it through tiny holes, fiber glass is formed. This fluffy mass of very fine fibers can be spun into yarn and woven into cloth. Fibers of this type are not used for clothing but are used for draperies and curtains. Fiber glass cloth is easily washed and needs no ironing. Glass fibers mixed with plastic form a material that is very strong, tough, and light. This material is used in making fiber glass boats and car bodies.

Fiber glass is sometimes called glass wool. It does not rot or burn, it is vermin-proof, and is not affected by heat and light. Fiber glass is often used as a heat insulator for ovens, home freezers, and hot-water heaters.

synthetic fibers

Most synthetic fibers are made by forcing liquids through small holes. Synthetic fibers can be placed into two broad groups: (1) Those fibers that are made from cellulose; the most common examples are *rayon* and

15-12 Newsprint is cut, then wound into rolls. An average roll contains about 8 km of paper. What name is given to this kind of paper?

15-13 Asbestos is a mineral fiber made of magnesium silicate. What properties does this fiber have that make it useful in fire proof articles?

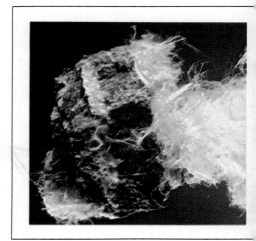

acetate; and (2) those fibers that are made by joining together monomers (small molecules) to form polymers (large molecules). Common examples in this group of fibers are *nylon, acrylic,* and *polyester.*

Each synthetic fiber has two names. One is *generic,* or the group name, like those mentioned above. The other is the *brand name,* used by the maker of the fiber as a copyrighted trade mark. Some of these brands that may be familiar to you are Dacron, Kodel, Orlon, Acrilan, and Qiana.

Most synthetic fibers are made by forcing liquids through tiny holes in a metal plate and allowing them to harden. A wide range of liquids produces a great variety of fibers.

Gold or platinum alloy is used in the plates because these metals are not affected by most chemicals. The plates, called *spinnerets,* are about the size of a sewing thimble and have from 10 to 150 small openings, depending on the thickness of the strand wanted. Large mills have hundreds of these spinnerets arranged in rows. See Fig. 15-14.

When a new fiber is developed, the United States Federal Trade Commission assigns it a generic name. In order to receive such a name, the new product must be different in chemical composition from all other fibers. The product must also have different properties of importance to the consumer.

Rayon and acetate are the oldest of the synthetic fibers. The basic material for the making of rayon fiber is cellulose from purified wood pulp. The pulp is dissolved through a series of chemical reactions and forced through the openings of a spinneret. The long filament then passes through a dilute acid solution for hardening.

Rayon absorbs moisture, making it suitable for clothing. However, rayon has two weaknesses: perspiration weakens rayon fibers, and they lose strength when wet.

Acetate is another well-known fiber made from wood pulp. The reaction between cellulose and acetic acid is the basis for this process. Acetate is made of fibers that do not wrinkle or shrink as much as rayon. Acetate is the only fiber used in cigarette filters because it is an efficient smoke remover.

Acetate fibers have two important weaknesses: (1) a very hot pressing iron will destroy the fibers, and (2)

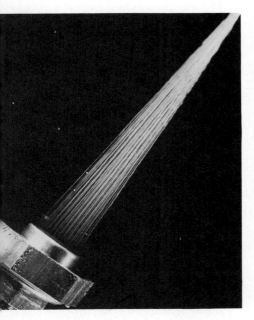

15-14 The spinneret is used in making long strands of synthetic fiber. Can you describe how a spinneret works?

some dry-cleaning solvents will dissolve the fibers. Acetate melts when burned.

Nylon, acrylic, and polyester are later inventions. The basic materials for making nylon are coal and petroleum. This polymer is squirted through spinneret holes to form nylon threads. The strands are then stretched to about four times their starting length. This stretching forces the molecules to line up, giving nylon an increased strength and making it more elastic. This elastic property has made nylon a popular fiber for socks and stockings, and other wearing apparel. See Fig. 15-15.

Acrylic and polyester were invented in the early 1950's. These fibers are easy to launder, are light in weight, dry quickly, and resist wrinkles. They are difficult to dye, do not absorb moisture well, and produce static electricity. Like all noncellulose fibers, they are thermoplastic. Thermoplastic fibers require low to moderate ironing heats. These fibers blend well with natural fibers in making cloth. Polyesters are used in making shatter-proof soft drink bottles. These bottles can be later recycled into fabrics, insulation materials, and auto parts.

15-15 Nylon ropes are strong and water-resistant. What raw materials are used in making nylon?

bleaches and dyes

Some chemicals are used to bleach cloth. All natural fibers have a yellow tinge. These yellow fibers must be *bleached* to produce white yarn or cloth. Sunlight is effective for bleaching moist plant fibers, such as cotton and linen. However, sunlight tends to "yellow" animal fibers, such as wool and silk. Most synthetic fibers do not need bleaching.

Sulfur dioxide or hydrogen peroxide are used to bleach wool and silk. Chlorine cannot be used because it destroys these fibers. However, chlorine works well for bleaching cotton. A solution of sodium hypochlorite is used for home-bleaching of cotton fabrics. This solution sells under the name of Clorox or Purex. See Fig. 15-16.

Chlorine is a poisonous, greenish-yellow gas. It reacts with many metals and with hydrogen. In the bleaching process, chlorine reacts slowly with water, uniting with the hydrogen and setting oxygen free. This freed oxygen then reacts with some dyes in the cloth, causing a loss of

15-16 A sodium hypochlorite solution in the bowl bleaches a blue cotton shirt. Can you explain the bleaching action of this solution?

color. As shown in the following equations, the oxygen does the actual bleaching of the cloth:

1. $2 Cl_2 + 2 H_2O \rightarrow 4 HCl + O_2 \uparrow$
2. oxygen (O_2) + dye in fabric $\rightarrow$ colorless compound

Sulfur dioxide is a dense, colorless, choking gas. By burning a little sulfur under a chemical hood, or in a well-ventilated room, you can make sulfur dioxide gas. If you put a wet red carnation in a glass beaker with the gas, the flower will be bleached by the sulfur dioxide gas. Try this test with dried apricots that have been soaked in water. Examine the apricots after about 20 min and note any changes in color.

Sulfur dioxide gas unites with water to form sulfurous acid (H_2SO_3), a very weak acid. Sulfurous acid readily unites with atoms of oxygen to form sulfuric acid (H_2SO_4). In the bleaching process, the sulfurous acid removes the color from some materials by removing oxygen. In other words, sulfur dioxide bleaches by removing oxygen, and chlorine bleaches by adding free oxygen.

Sulfur dioxide is a milder bleach than chlorine. Thus, sulfur dioxide is widely used for bleaching wool, silk, paper, and straw. White wool sweaters and white silk cloth often yellow with age because sulfur dioxide bleach is not permanent. The oxygen of the air slowly changes the wool and silk fabrics to their original colors.

It is interesting to note that because of the use of sulfur compounds in the paper industry, most books and documents printed in the past 50 years are slowly weakening and breaking down. If prompt action is not taken, almost no printed record of this period in history will survive. Many important documents are now being put on microfilm to save their contents for future use.

Hydrogen peroxide is prepared in large volumes by passing an electric current through cold, dilute sulfuric acid. You have seen that hydrogen peroxide is an unstable liquid that breaks down readily, forming water and free oxygen:

$$2 H_2O_2 \rightarrow 2 H_2O + O_2 \uparrow$$

The oxygen that is set free can remove the color from some materials. Besides wool and silk, hydrogen peroxide is commonly used to bleach paper, feathers, and hair.

activity

The bleaching of cloth. Put about 5 g of bleaching powder in a 500-mL beaker. Pour 10 mL of concentrated hydrochloric acid on the powder. Greenish-yellow chlorine gas will be set free at once. CAUTION: *Chlorine gas is very poisonous. Be sure to use a chemical hood or a well-ventilated room. Do not inhale the gas.* Now lower strips of wet, colored cotton cloth into the gas-filled beaker. Cover the beaker and leave the strips in the chlorine for about 45 min, then examine the cloth. The cloth must be wet for bleaching to take place. Can you explain why?

Dyes are mostly complex organic compounds. In earlier days, dyes were obtained from roots, barks, leaves, flowers, and berries. The dyes in use today are made by chemists. The chemist has learned how to build up molecules of bright-colored substances. There are over 5000 dyes now in use, but probably over 100,000 have been made at one time or another. Inorganic compounds have been combined with organic compounds to produce many shades of color. Such chemicals make some dyes *fast*. A fast dye is one that clings to the fabric, does not lose its color, and does not wash away easily.

Coal tar is the raw material for most dyes. Coal tar is a black, sticky liquid obtained as a by-product from coke ovens. While the tar does not contain dyes, it is used to obtain benzene. Benzene is used to make a colorless liquid called aniline. Aniline is the starting point in the making of many synthetic dyes:

$$coal\ tar \rightarrow benzene \rightarrow aniline \rightarrow aniline\ dye$$

If a fiber takes and holds the dye with little help, the dye is called a *direct dye*. Direct dyes can be used for all natural fibers. Common salt added to the dye bath forces the dye to leave the solution and attach itself to the cloth. In general, animal fibers are easier to dye than plant fibers.

Acid dyes are sometimes used to help silk and wool hold bright colors. These dyes do not cling to linen and cotton. Fabrics that have been colored with acid dye must not be washed with alkalis because the cloth cannot hold the dye.

Certain dyes will not adhere to cotton unless a chemical, called a *mordant*, is used first. The mordant forms a solid within the cotton fibers. Then, when the cloth is dipped into the dye bath, this solid joins with the dye, making a fast color. Dyers often use different mordants with the same dye to make a variety of shades. Aluminum hydroxide is a widely used mordant.

Certain solvents can remove stains from cloth. Before you try to remove a stain from cloth, it helps to know what caused the stain. Also, it is good to know the kind of cloth that is stained. A little testing may be needed to find the right solvent. Soap and water are effective in taking out many kinds of stains. However, some stains need solvents other than soap and water. In all cases, you must be careful to use a solvent that will not harm the fabric. For instance, you already know that some dry-cleaning solvents destroy acetate fibers. Bleaches containing chlorine will ruin wool and silk. Dilute acids must be washed out of cotton and linen because they weaken the fiber. Bleaches may not only take out the stain; they may bleach the dye out of the cloth as well. It is necessary to find a solvent that will take out the stain in the cloth without affecting the material. Table 15-1 contains data on the removal of some common stains from many fabrics.

Table 15-1:

Removal of Stains from Clothing	
Stain	**Type of Cleaner**
Acids	ammonia; rinse with water
Alkalis	vinegar; rinse with water
Blood	cold water and salt or soap and cold water
Coffee	boiling water
Fruit juices	water
Grass	alcohol; wash with soap and water
Grease	dry-cleaning solvent, benzene, or kerosene
Ink, iron rust	detergent and water, oxalic acid, or lemon juice
Lipstick, rouge	glycerin, or soap and water
Paint (oil)	turpentine or mineral spirits

Careers in the Textile Industry

Egyptians were making fine cloth at least 4000 years ago with cotton spun by hand. Fabrics were made by hand for many generations up to the beginning of the industrial revolution about 200 years ago. Today, modern machines do in seconds what it took hand weavers days to do.

The textile industry, and closely related industries, is large and covers a wide range of job opportunities. The textile industry provides fibers, cloth, dyes, and the finished cloth products for every person in the country. For such a large task, over 3.4 million people are employed directly in the making of textiles. The industry involves over 700 companies with 7000 plants in 47 states producing a yearly payroll of over $7 billion. In the United States, the making of textiles and clothing accounts for one manufacturing job in every eight.

Among the many hundreds of types of jobs in this large industry are fashion designer, garment cutter, knitting machine operator, pattern maker, fashion artist, dressmaker, tailor, fashion model, and sewing machine operator.

The fashion artists are creative people who make sketches of garments. Their job is to sketch clothing to look so attractive that people will come to the store to buy the finished garments. They recommend the use of new fabrics that are easy to wear and to care for. Many work in large department stores and advertising agencies.

Fabric designers suggest the makeup of all types of fabrics to be used in clothing, draperies, carpets, upholstery materials, and household linens. They also suggest color combinations and instructions for making the cloth. Many men and women are employed as fabric designers.

Knitting machine operators inspect the material when it is made into cloth. The operators must know how to start and stop knitting machines, replace the yarn on the needles, and be alert to the work of making high-quality cloth for various consumer uses.

The pattern makers translate the sketches of the fashion artist into paper patterns for each part of the garment. Cutters then cut the cloth from the pattern. Sewing machine operators use machines to stitch the pieces of cut materials together to form the completed garment.

/ummary

1. Wool and silk fibers are proteins from animal sources.
2. Cotton and linen fibers are made of cellulose, a carbohydrate, obtained from plants.
3. Paper is a formed fiber made from wood.
4. Mineral fibers include asbestos and fiber glass.
5. Synthetic fibers are made by squirting certain liquids through tiny holes in a metal plate called a spinneret.
6. Nylon, acrylic, and polyester are made by linking together small molecules (monomers) into large molecules (polymers).
7. Chlorine bleaches cotton and linen fibers by adding oxygen to a dye to make the cloth colorless.
8. Sulfur dioxide bleaches wool and silk by removing oxygen from some dyes to make the cloth colorless.
9. Aniline, obtained from coal tar, is the starting point for the making of many dyes.

review

Match each item in the left column with the best response in the right column. *Do not write in this book.*

1. wool
2. cotton
3. mercerizing
4. linseed oil
5. nylon
6. fiber glass
7. bleaching powder
8. hydrogen peroxide
9. mordant

a. elastic synthetic fiber used for making clothes.
b. mineral fiber not affected by heat and light
c. cellulose
d. used in dyeing cotton fibers
e. breaks down into water and oxygen
f. leather goods
g. strengthens cotton fibers
h. natural protein fiber
i. used in making small amounts of chlorine
j. making of "Kraft" paper
k. made from flax seeds

questions

Group A
1. Why is wool a good heat insulator? Name four animals that furnish wool.
2. Why is the label "all wool" no guarantee of quality?
3. List four steps in the making of woolen cloth from raw wool.
4. What elements make up cellulose? Why is cotton considered a polymer?
5. What products are made from cotton seeds?
6. How is cotton mercerized? sanforized?
7. Why is linen a desirable material for dish towels?
8. List the steps involved in making leather. Why is each step needed?
9. Why is asbestos suitable for use as car brake linings?
10. What raw materials go into making newsprint?
11. Why are spinnerets plated with either gold or platinum?
12. What are the advantages and disadvantages of acetate fibers?
13. What basic raw materials are needed for making nylon? List four properties of acrylic and polyester fibers.
14. Name four properties of chlorine gas. Describe how you would make a small amount of this gas. What precautions would you take in the making of this gas?
15. How would you remove a coffee stain? grease? paint?

Group B
16. Describe what happens to wool blankets when they are washed in hot water or with alkaline soap.
17. Describe how wool and cotton fibers look under a microscope.
18. Describe how silk is made by the silkworm. Why is silk such a costly fiber?
19. How is linen obtained from the flax plant?

20. How is paper made from wood?
21. List three general groups of synthetic fibers. Give two examples of each group.
22. Is it the chlorine in bleaching powder that bleaches cotton? Explain.
23. Describe the bleaching action of hydrogen peroxide and sulfur dioxide.
24. Can any dye of suitable color be used for cotton? Explain.
25. What is the difference between acid dyes and direct dyes? Why is a mordant used in dyeing cotton cloth?

further reading

Addy, J., *The Textile Revolution*. New York: Longman, 1976.

Adrosko, R. J., *Natural Dyes and Home Dyeing*. New York: Dover, 1971.

Bikkie, J. A., *Careers in Marketing*. New York: McGraw-Hill, 1978.

Coplan, M. J., *Fiber Spinning and Drawing*. New York: John Wiley & Sons.

Grilli, E. R., *Future for Hard Fibers and Competition from Synthetics*. Baltimore: Johns Hopkins University Press, 1975.

Hollen, N., et al., *Textiles*. New York: Macmillan, 1973.

Thorstein, T., *Practical Leather Technology*. New York: Van Nostrand Reinhold.

unit 3
motion, forces, and energy

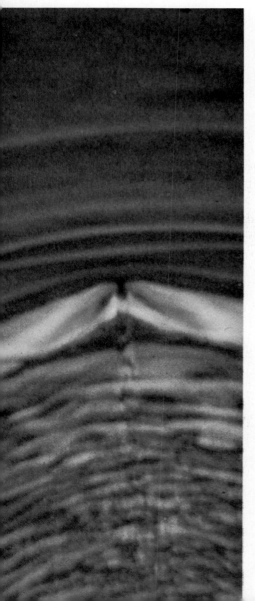

16

motion and its causes

Can you imagine a world with no movement? Think about all the movement going on around you. You blink your eyes, turn a page, cross your legs. The hands of the clock move as time passes. Perhaps a car drives past the window of your classroom. What is different about these motions? What is the same? How can you describe them? What causes these motions? In this chapter you will learn the answers to some of these questions.

objectives:

☐ To define motion, speed, velocity, and acceleration.

☐ To understand the difference between balanced and unbalanced forces.

☐ To describe forces including those of friction and gravity.

☐ To construct force diagrams.

☐ To solve problems dealing with torque and parallel forces.

☐ To develop an understanding of Newton's three laws of motion.

motion in our world

Motion is all around you. Think about the examples of motion in the above paragraph. Is there any definition that will fit all of these motions? Yes; **motion** *can be defined simply as a movement of an object from one place to another.*

Your eyelids move from one position to another and back again when you blink them. Your finger and the page of this book move from one place to another when you turn the page. The motion in each of these cases fits the above definition. Now suppose you wanted to describe motion to another person. If you said, "The car is in motion," the other person would not have a very good idea of what you meant. For instance, how fast was the car moving? The term *motion*, by itself, does not tell enough about the movement of an object. There are other ways to describe movement that give a better picture than just the term motion. For example, you might say, "The car was moving at 30 km/hr." When you say how fast something is moving, you are giving the **speed** of that object. *Speed is the distance an object moves in a certain amount of time.* In this example, the speed of the car was 30 km/hr.

In order to detect motion, you must compare the moving object with something that is not moving. The object that is not moving is called a *reference point*. Recall the examples of motion given above. What might be some reference points for (1) an eyelid blinking; (2) a page turning; (3) a car moving at 30 km/hr?

Velocity and acceleration describe motion. Can you think of one more important fact that is needed to describe the motion of the car in the above example? You know that the car is moving. You know how fast it is going. What else do you need to know? It would surely help to know in what direction the car is going.

The statement, "The car is going north at 30 km/hr," gives the car's **velocity.** *Velocity is speed in a definite direction.*

What if the speed and direction are not constant? A car starts at rest and is brought up to a suitable road speed. If the car comes to a stop sign, it is brought to a halt. If the road has a sharp curve, the car's speed is reduced and it is turned carefully around the bend of the road. A change in the speed or direction is called **acceleration.**

In common use, the term *acceleration* just means that an object speeds up. But in science, the term has a more accurate meaning. Acceleration means any change in speed or any change in direction. A car is accelerating even when it slows down or when it turns a corner. *Acceleration is a change in velocity per unit of time.* See Fig. 16-1.

16-1 Are the roller coaster cars being accelerated? Why?

sample problem

A car moving on a straight, level road changes speed from 35 km/hr to 45 km/hr in 5 sec. Find its acceleration.

solution:

Step 1. From the definition of acceleration, write the formula

$$a = \frac{v_f - v_i}{t}$$

where a is the acceleration, v_f is the final velocity, v_i is the initial velocity, and t is the elapsed time.

Step 2. Substituting data contained in the problem according to the above formula,

$$a = \frac{45 \text{ km/hr} - 35 \text{ km/hr}}{5 \text{ sec}}$$

$$= \frac{10 \text{ km/hr}}{5 \text{ sec}}$$

$$= 2 \text{ km/hr/sec}$$

Note that acceleration is expressed in speed (velocity) units per unit of time.

forces in our world

Forces cause motion. So far you have learned about motion and differences in motion. What causes these motions in the first place? What causes the car to start, or your eyelids to blink? What makes the pages of the book turn? In every case, a push or pull of some kind caused the motion. One set of muscles in your eyelids pushes the lids down. Another set of muscles pulls them up. Your finger lifts the page of the book and pushes the page to turn it over. The car engine pushes the car forward. Even though the pushes and pulls came from different sources, they are all called **forces**. A force is not something you can see. *A force is just a name given to any push or pull.*

unit 3 · motion, forces, and energy

There are many kinds of forces. The easiest examples of forces to understand are those in which you can actually see something pushing or pulling something else. Think of a baseball hitting a catcher's mitt, or a bowling ball hitting the pins. You can see the baseball push the mitt and the bowling ball push the pins. You can see the mitt and the pins move. Objects can also exert pulling forces. When a cowhand ropes a running steer, the steer pulls against the rope. Meanwhile, the cowhand and horse are pulling on the other end of the rope in the opposite direction.

It is easy to imagine the forces discussed above because you can see that they cause things to move. But what about an object that does not move when you push against it? Is there a force acting upon the object? There may not appear to be any, but the forces are there. Suppose two people pick up a rope and begin a tug-of-war. They tug on the rope but neither can "out-pull" the other. The forces are exactly equal and opposite, and the rope does not move. In other words, when they are combined, the forces acting on the rope exactly balance each other.

In this example, you can tell that forces are present because the people can be seen pulling on the rope. Even when the rope was lying on the ground, however, there were forces acting upon it. One such force was caused by the weight of the rope being pulled down against the ground. The other force was the ground pushing up against the rope.

Friction is a common force. Suppose you put a cardboard box on the gym floor and give it a push. You apply a force to the box and the box slides across the floor. What happens to the sliding box? Will it keep on sliding forever? Of course not. You know from experience that the box will come to a stop. Again, your experience tells you that **friction** caused the box to slow down and stop. *Resistance to motion, caused by one surface rubbing against another, is called friction.* To start the box moving again, you must overcome the friction between the box and the floor, called *static friction*. See Fig. 16-2 (p. 282). To keep it moving, you need to overcome *sliding friction*.

Friction is also affected by the smoothness of the sliding surfaces. If the surfaces are polished, the object will slide with less friction. For instance, a playground sliding

16-2 What kind of friction must the football player overcome to move the sled?

board has a very smooth surface so that children can slide down easily.

Another way to change the force of friction is to change sliding friction to *rolling friction*. If you attach wheels to the bottom of the box and again push it across the gym floor, how would it move? You would see that the box moves much farther with wheels than without them. Rolling friction is less than sliding friction.

Friction can also be decreased by using some form of *lubrication*. Oil is an excellent lubricant. It makes the surfaces very slippery. You can see how some of the factors affecting friction apply to the brick shown in Fig. 16-3.

Another kind of friction is *fluid friction*. (A fluid is either a liquid or a gas.) For example, the wind that pushes

16-3 How is the force of friction different on the three sides of the brick?

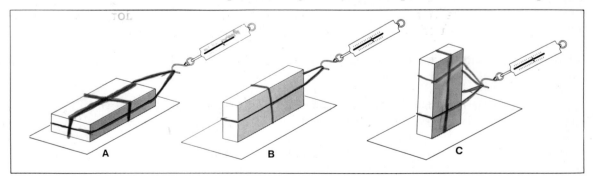

A   B   C

activity

Forces of friction. You will need the following items to do this activity: bricks, string, spring balance, board, lubricant.

Step 1. Attach a spring scale to a brick and place it on a board, as shown in Fig. 16-3A. Pull the brick across the board at an even rate. Repeat the activity with the brick on its side and then on its end as shown in Fig. 16-3B and C. Then put a second brick on top of the first, and finally a third brick on top of the other two. Be sure to keep the scale level.

In your notebook, record your readings in a table similar to the one shown below. *Do not write in this book.* Find the area of the brick's sliding surface in each trial.

	Area of Sliding Surface	Reading of the Spring Scale			
		No Lubrication		With Lubrication	
		To start	To pull	To start	To pull
1 brick (flat)					
1 brick (side)					
1 brick (end)					
2 bricks (flat)					
3 bricks (flat)					

Step 2. After completing the trials on a dry board, repeat them using a lubricant. If you find oils too messy, you can use soapy water or waxed paper as a lubricant. Record your readings in the table.

Questions:
1. Is there a link between an object's weight and the friction it has when pulled? Explain.
2. What effect does the lubricant have on friction? This activity should help you see the relationships between sliding area and friction, between weight and friction, and the effect of lubrication on friction.

against a moving car or airplane is fluid friction. In water, fluid friction tends to hold back the motion of a boat or submarine.

Sometimes, however, you want to increase friction instead of decrease it. For example, a car turning a corner

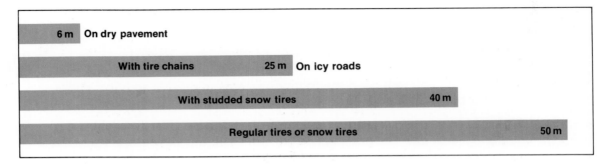

6 m	On dry pavement	
With tire chains	25 m	On icy roads
With studded snow tires	40 m	
Regular tires or snow tires	50 m	

16-4 Stopping distances of a car at 48 km/hr. Can you tell from the graph why you must be particularly careful when driving on ice?

needs a great deal of friction between the tires and the road. If this friction is not present, the car will skid off the road. Likewise, much friction is needed for a car to make a quick stop. See Fig. 16-4. Did you know that it is impossible to walk without friction and that nails will not hold without it? Can you explain why?

Gravity is another common force. You know that when a ball is thrown straight up in the air, it does not keep going up forever. It stops and comes back down again. Think about the up and down motion of the ball carefully. You applied a pushing force in an upward direction to start the ball moving. What force slowed and stopped the ball's upward movement? Then, what caused the ball to come back down? The force that caused these actions is the **force of gravity.** It is the force of earth's gravity pulling on objects at or near the earth's surface that gives the objects the feeling we call *weight.*

Sir Isaac Newton, the brilliant English scientist, lived during the seventeenth century. He worked out the *law of universal gravitation, which says that every mass in the universe attracts every other mass.* The weight of objects on the earth is just one example of this force.

Newton found the greater the masses, the greater the force between them. A 1000-kg mass on the surface of the earth is attracted toward the center of the earth by twice as much force as a 500-kg mass. Also, if the earth had twice its present mass but remained the same size, all objects on it would weigh twice as much.

Finally, the closer the two masses are, the greater the force of attraction between them. Tests showed that *gravity force varies inversely as the square of the distance between the centers of the masses.* If the distance is reduced to half as much, the force becomes four times as great;

while if the two masses are only one third as far apart, the force is nine times as great.

The inverse-square rule applies if the distance is measured from the center of the earth, not from the surface. On the surface, you are already 6400 km above the center of the earth. If you are 6400 km above the surface of the earth, you are already 12,800 km above the center of the earth. Thus, at 12,800 km you would be twice as far away from the center of the earth as you are now, and so you would weigh only one fourth as much. See Fig. 16-5.

Gravity causes uniform acceleration of free falling bodies. Galileo, the great Italian scientist, was one of the first to study acceleration. He timed a metal ball as it rolled down a groove in a sloping plank. Galileo found that the ball gained the same amount of added speed each second that it rolled. In other words, as the ball rolled down, it was accelerating at an even (or uniform) rate.

Instead of using a plank, scientists are now able to study the motion of objects as they are freely falling. These studies have shown that at the end of 1 sec, objects attain a speed of 9.8 m/sec (if we ignore air resistance). At the end of 2 sec, they fall at a speed of 19.6 m/sec. This means that the acceleration of a freely falling object is 9.8 m/sec, per second. (The statement is to be taken literally, "per second, *per second.*") In other words, speed of a freely falling object increases at a rate of 9.8 m/sec each second. This value can also be written 9.8 m/sec/sec or 9.8 m/sec^2.

This acceleration, caused by the gravity of the earth, is called *1 g*. For objects near the earth's surface, *g* equals 9.8 m/sec/sec. An acceleration of 5 *g*'s, therefore, means a change in velocity, or acceleration, of 49 m/sec/sec.

Since the downward speed of a falling object begins at zero and is up to 9.8 m/sec after 1 sec, the *average speed* during the first second of fall must be (0 + 9.8) ÷ 2, or 4.9 m/sec. (How do you find the average of two test grades of 70 and 80?) Since the object moved at an average speed of 4.9 m/sec for 1 sec, it must have fallen 4.9 m. If the object falls for 2 sec, its velocity at the end of 2 sec will be 19.6 m/sec. (Remember, the object gains 9.8 m/sec each second.) The average of its initial speed (0) and its final speed (19.6) is (0 + 19.6) ÷ 2 = 9.8 m/sec. If an object moves at an average speed of 9.8 m/sec for 2 sec, it will go a distance of (2 × 9.8), or 19.6 m.

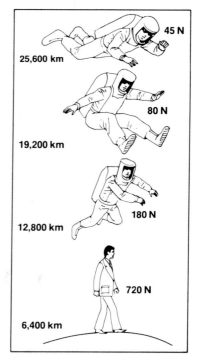

16-5 How does the inverse-square rule apply in this diagram? What would be the force of gravity on the astronaut at 32,000 km above the earth?

Notice that an object falls four times as far in 2 sec as it does in 1 sec (19.6 m ÷ 4.9 m = 4). This means that the change in distance is proportional to the square of the change in time. That is, if the time is doubled, the distance becomes four times as great.

Galileo noticed that if the time of fall were tripled (3x), the distance covered would be nine times as great. He expressed this as a law, stating that *the distance covered by an object that is uniformly accelerated depends upon the square of the time elapsed*. Thus, to find how far an object will fall, square the number of seconds that it falls and multiply by one half the acceleration owing to the force of gravity. The formula for this motion is

$$s = \frac{at^2}{2}$$

where s is the distance, a is the acceleration, and t is the time.

sample problem

An object starts from rest and falls freely under the force of gravity. (a) What is its velocity at the end of 5 sec? (b) What is its average velocity during the fall? (c) What distance has it fallen during this time?

solution (a):

Step 1. Given: acceleration owing to the force of gravity = 9.8 m/sec/sec; initial velocity (v_i) = 0; time = 5 sec; and final velocity (v_f) = ?

Step 2. The above quantities are related by the formula

$$a = \frac{v_f - v_i}{t}$$

Step 3. Substituting given quantities and solving for v_f,

$$9.8 \text{ m/sec/sec} = \frac{v_f - 0}{5 \text{ sec}}$$
$$v_f = 9.8 \text{ m/sec/sec} \times 5 \text{ sec}$$
$$= 49 \text{ m/sec}$$

solution (b):

Step 1. Given: $v_f = 49$ m/sec; $v_i = 0$

Step 2. Average velocity (v_a) is

$$v_a = \frac{v_f + v_i}{2}$$

Step 3. Substituting,

$$v_a = \frac{49 \text{ m/sec} + 0 \text{ m/sec}}{2}$$

$$= 24.5 \text{ m/sec}$$

solution (c):

Step 1. Use the general formula for distance

$$s = \frac{at^2}{2}$$

Step 2. Substitute known or given values

$$s = \frac{1/2 \times (9.8 \text{ m/sec}^2) \times (5 \text{ sec})^2}{2}$$

$$= 122.5 \text{ m}$$

Note: Check this result by noting that in 5 sec an object with an average speed of 24.5 m/sec will move 122.5 m.

Air gives resistance to freely falling bodies. In the discussion about free-falling bodies, we assumed that there was no air resistance to slow down the objects. In other words, we assumed that the objects were falling in a vacuum where they all accelerate at the same rate regardless of their size, shape, or density. See Fig. 16-6.

You know, however, that any objects you have ever thrown or dropped have fallen in the presence of air. You know that air exerts friction to resist any falling object. Yet, objects are affected differently. For example, a stone falls faster than a pillow. The reason, of course, is that air resistance slows down the light-weight pillow more than it slows the stone.

How does air resistance affect a sky diver? Suppose the diver jumps from a high-flying plane and delays opening the parachute. The falling speed will increase quickly. However, as the speed increases, the friction of air resistance also increases. Finally, the force of friction of

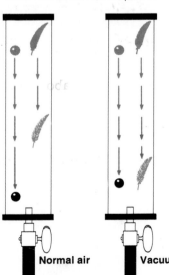

16-6 Can you think of any reason why a feather and a steel ball would fall at the same speed?

Normal air   Vacuum

the air becomes equal to the downward force of gravity; that is, it is equal to the diver's own weight. At this point there is no *unbalanced* force and no further acceleration. The sky diver will continue to fall at a constant velocity. This speed may be as high as 250 km/hr. When the parachute opens, it increases the air resistance. This increase in resistance reduces the rate of fall to a point that will permit a safe landing. See Fig. 16-7.

Torque tends to cause an object to rotate. As a child, you learned how to balance on a seesaw, which is a kind of *lever*. You learned that the balance point was affected by two factors: (1) how much you weighed, and (2) how far you sat from the point of support, or *fulcrum*. A child with a weight of 200 N can balance a child weighing 400 N by sitting twice as far from the fulcrum. [The weight of an object on the earth's surface is found by multiplying its mass in kilograms by 9.8 m/sec^2. The resulting weight is measured in *newtons* (N).]

In Fig. 16-8, you will find that if you multiply the force on the left by the distance from the fulcrum, you get 800 newton-meters (N-m). *The force times the distance (from the fulcrum) is called the* **torque.** The torque to the left of the fulcrum tends to turn the seesaw in a counterclockwise motion. Note that torque is always measured in force units (newtons) multiplied by distance (meters) from the fulcrum (or N-m).

If a seesaw is to balance, the clockwise torque must equal the counterclockwise torque. A force of 400 N placed 2 m to the right of the fulcrum will give a torque of 800 N-m. The seesaw, or lever, will then be balanced.

If there are two people on the left side of the seesaw, each one will create a counterclockwise torque. By adding

16-7 What force slows down the speed of this space capsule and its parachutes?

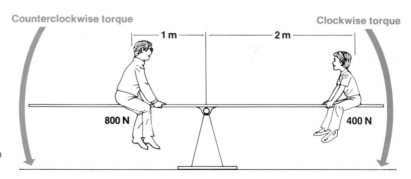

Counterclockwise torque   Clockwise torque

1 m   2 m

800 N   400 N

16-8 What would happen if the man moved away from the fulcrum?

their two torques together, you can find their total. The torque needed on the right side of the seesaw to balance it must equal the total torques on the left side.

In order to prevent rotation of a body or change in rotation, *the sum of the clockwise torques about any axis must be equal to the sum of the counterclockwise torques about the same axis.*

sample problem

A weight of 5 N hangs from a 100-cm uniform lever, 20 cm to the right of its midpoint fulcrum. Twenty centimeters to the left of the fulcrum is a weight of 2.5 N, and 50 cm to the left of the fulcrum is a weight of 1 N. Is the lever balanced? Explain.

solution:

Step 1. Make a drawing of the lever, showing the forces and torques acting on it.

Step 2. Clockwise torque:

$$5 \text{ N} \times 20 \text{ cm} = 100 \text{ N-cm}$$

Step 3. Counterclockwise torques:

$$2.5 \text{ N} \times 20 \text{ cm} = \underline{50 \text{ N-cm}}$$
$$1 \text{ N} \times 50 \text{ cm} = \underline{50 \text{ N-cm}}$$
$$\text{Total} \quad 100 \text{ N-cm}$$

Since the clockwise torque equals the sum of the counterclockwise torques, the lever is balanced. You may wish to check your results in the science lab, using a meter stick as a lever. Place the fulcrum at its midpoint, and hang known weights from it.

Parallel forces act together. Two or more forces acting in either the same or in the opposite direction are called *parallel forces.* Think of a bridge that is held up by two piers. The piers exert parallel upward forces. Objects on the bridge exert parallel downward forces.

Suppose a truck weighing 10 kilonewtons (kN) is resting on the middle of a 25-m long bridge. If the truck is exactly in the middle, its weight will be supported equally by the two piers. However, if the truck moves to a spot 15

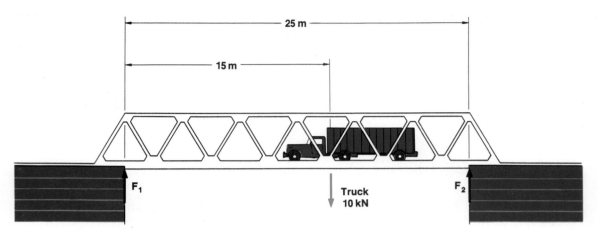

25 m

15 m

F₁

Truck
10 kN

F₂

16-9 How much of the truck's weight
is supported by each pier?

m from one end, as shown in Fig. 16-9, how much weight will each pier then support? (In this example, do not count the weight of the bridge itself. Also assume that all the weight of the truck is not located at the wheels but at its *center of gravity*. The center of gravity is the point where all the weight of the truck may be considered to be concentrated.)

The problem can be worked out if you think of the bridge as a lever, with one pier (F_1) acting as the fulcrum, and the other pier (F_2) pushing upward. Thus, 10 kN × 15 m = 150 kN-m of clockwise torque. Pier F_2 at the end of a 25-m lever must therefore give an equal counterclockwise torque of 150 kN-m. Since F_2 is 25 m from F_1, then 150 ÷ 25 = 6 kN. Therefore, the second pier (F_2) supports 6 kN of the truck's weight. The remaining 4 kN must be supported by the first pier (F_1).

Forces can be diagramed. One way you can describe a force is by its size or amount. The size of the force is measured in newtons (N). For example, you can say you pushed on a box with a force of 10 N. However, a force cannot be fully described by its size alone. The direction in which the force acts is also important. *Any quantity that has both size and direction is called a vector quantity.* Since forces have both size and direction, forces are vectors. A vector quantity may be shown as an arrow. The head of the arrow shows the direction. The length of the arrow shows the size. Suppose you want to use a vector to show a force of 10 N pushing toward the east. As a scale, let 1 cm equal 2 N. Thus, a line 5 cm long should be drawn

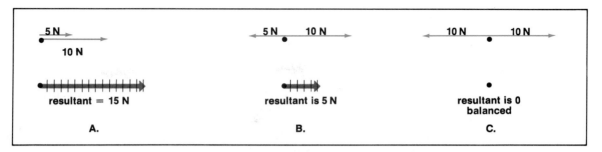

A. 5 N 10 N resultant = 15 N

B. 5 N 10 N resultant is 5 N

C. 10 N 10 N resultant is 0 balanced

with the arrowhead pointing toward the right. This arrow would show a force of 10 N in an eastward direction.

Forces can be added or subtracted. You know that when you push on something, it tends to move in the direction you push it. But what happens when you apply two forces in different directions on the same object? You can use force vectors to answer this question.

When two or more forces act together, their effect on an object is a combination of the forces. This combined effect is called the *sum* or *resultant* of the forces.

Suppose, for example, that one person is pulling on a rope with a force of 5 N. Another person is pulling on a second rope (attached to the same object and pulling in the same direction) with a force of 10 N. The resultant force is the same as if one person were pulling on a single rope with a force of 15 N. This force is drawn in **A** of Fig. 16-10. This example can be shown in the classroom using spring scales and string.

Now one of the people goes to the opposite side of the object, and they both pull again with the same forces as before. What is the resultant? In what direction will the object move? See part **B** of Fig. 16-10. Now suppose that the people are still pulling in opposite directions but with equal forces of 10 N each. The forces cancel each other out and the resultant is zero. See part **C**. When two or more vectors are acting at the same time on a point, they can be combined into a single resultant vector. The effect on the point is as if the resultant force alone were acting on the point.

What happens when two forces are not acting in the same line? In Fig. 16-11, two strings are shown pulling on a spring scale fastened to a board. The strings are pulled at an angle of 90° to each other. Other spring scales show that the force along OD is 6 N; the force along OB is 8 N.

16-10 Can you tell by studying the diagrams which points are acted on by balanced forces? unbalanced forces?

16-11 When the vectors are added, what is the value of the resultant? The protractor shows angle BOR.

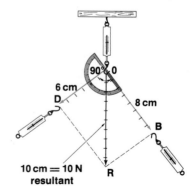

90° 0
6 cm
D
8 cm
B
10 cm = 10 N
resultant
R

You can show the 6-N pull by making OD 6 cm long, with a scale of 1 cm for each newton. Then make OB 8 cm in length to show the 8-N pull. Complete the drawing as shown in Fig. 16-11. You will have a four-sided figure in which each pair of opposite sides is parallel. Use a protractor to find angle DOR and BOR. Measure line OR of the diagram. What is its length? Its value gives the size and direction of the resultant of the two forces.

In this example, the resultant of the forces OD and OB is 10 N. The spring balance at O should read the same as the force obtained by measuring line OR. The single resultant force has the same effect and can be used in place of the 6-N and 8-N forces at 90° to each other.

Can you picture the forces acting on a corner post of a wire fence? What might happen to the corner fence post if it were pulled too hard by wires at right angles to each other? Where should a supporting wire be placed?

The size and direction of the resultant of any two forces can always be found by drawing a force diagram. Sometimes three or more forces act at different angles upon the same point. Can you think of a way to find the resultant of three or more forces acting at the same time upon the same point?

sample problem

If each of two wires at 90° to each other exerts a 100-N pull on a fence post, find the size and direction of the resultant force.

solution:

By scaled vector diagram. Select a convenient scale, where, for example, 1 cm equals 10 N. On a sheet of paper lay out two vectors 10 cm in length, at a 90° angle to each other. Complete the drawing and measure the diagonal. It should be slightly over 14 cm long. What force does this vector represent? A protractor will show that the angle made by the resultant with each force vector is about 45°.

the laws of motion

Newton's first law of motion describes "inertia." You have learned to describe how objects move in terms of speed, velocity, and acceleration. You have also learned that all motion and changes in motion are caused by forces. But just how are forces and motion related? Does the amount of force have anything to do with the amount of motion an object has? How does an object behave if no forces act on it? These are some of the questions about motion Sir Isaac Newton tried to answer.

Let us take the question of "no forces" first. Remember in the discussion of friction, you saw that if you put a box on the gym floor and gave it a push, it would start moving. However, the box would not move forever. It would soon come to a stop. The force of friction between the box and the ground caused this change in motion. Now suppose there were no friction forces acting on the box. Would the box slow down? Would it ever stop? Newton did many tests to find the answer to this question. He concluded that if there is no force acting on it, an object will continue to do whatever it has been doing. If an object is moving in a certain direction at a certain speed, it will continue to do so. If the object is not moving, it will not start moving.

Newton's conclusion is known as *Newton's first law of motion*. It is stated as follows: *If an object is at rest, it tends to stay at rest; if it is moving, it tends to keep on moving at the same speed and in the same direction*. Newton used the word **inertia** to describe this tendency to resist changes in motion. See Fig. 16-12.

You can see how the law applies in some examples. If you are standing in a bus and it starts suddenly, what happens? The bus and your feet begin to move, but the rest of your body is still at rest. You are almost toppled over! There is an opposite effect if the bus stops too fast. Your body tends to keep moving even though the bus and your feet have already stopped. Both cases show Newton's first law.

Inertia is also a factor in a moving car. The faster a car travels, the greater is its tendency to stay in motion. Compare the stopping distances of cars traveling at different rates of speed in Fig. 16-13 on p. 294. Note that at twice the speed, *four* times the stopping distance is needed. At a speed of 96 km/hr, or three times as fast as 32 km/hr, it takes *nine* times the distance to stop. Remember this when you are tempted to drive too fast or too close to the car ahead.

It is inertia that tends to throw a person off a fast-turning merry-go-round. Likewise, in baseball a base

16-12 What happens to the coin when you flick away the card? What would happen to the coin if you pulled the card away slowly?

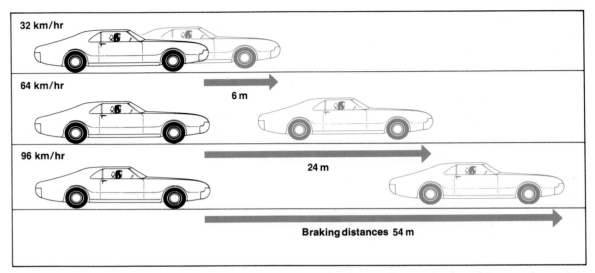

32 km/hr

64 km/hr

6 m

96 km/hr

24 m

Braking distances 54 m

16-13 Does kinetic energy build up in direct ratio to speed? How can you tell?

runner "rounds" the bases instead of making a sharp 90-degree turn at each base. Inertia makes it hard for the runner to change direction.

Even the turning earth shows the effect of inertia acting on it. The effect is so great that the earth has a slight bulge at the equator. In fact, if the earth turned 18 times as fast as it does now (or fast enough to give us a day of only 90 min instead of 24 hr), loose objects at the equator would be thrown into space.

Tests have shown that the mass and the inertia of a body are exactly proportional. That is, if one object has twice the mass of another, it also has twice the inertia. Therefore, scientists can measure the inertia of an object in order to find its mass. They cannot find the mass of an object by weighing it because its weight is not the same all over the universe. On the moon an object weighs about one sixth as much as it does on the earth, but its inertia and mass remain unchanged.

Balanced forces cause no net change. In the discussion of force vectors, you saw that unbalanced forces cause nonmoving objects to move in the direction of the resultant. If the forces are *balanced*, that is, if their resultant is zero, there is no motion. Although forces are being applied, they are canceling each other out. In such cases, the *net force* is zero, and the object is balanced.

An object in uniform motion is also balanced. Imagine an airplane traveling a perfectly straight and level course eastward at 200 km/hr. This plane's velocity is constant and there is no change in speed or direction. The only way the plane can keep this uniform motion is if all the forces on it are balanced in such a way that they cancel each other out. If any of the forces acting on the plane change in size or direction, the forces will be unbalanced and the velocity of the plane will change.

This idea of balanced forces is really just another way of looking at Newton's first law of motion. (See p. 292.) The two ideas are basically the same.

Net force changes motion: Newton's second law of motion. The second law of motion shows how force, mass, and acceleration are related to one another. Newton found that the size of the acceleration for a given mass depends upon the size of the unbalanced force applied to it. In order to throw the ball, a baseball pitcher applies a force on the ball. To throw the ball faster, a greater force must be applied.

It is also easy to see that acceleration is related to the mass of an object. The greater the mass of an object, the greater its inertia. Thus, the greater the mass, the greater the force needed to change the acceleration. In other words, the same force gives a greater acceleration to a baseball than it does to a bowling ball.

For example, a car built for racing has both a powerful engine and a light-weight body. See Fig. 16-14. The same engine accelerates a car more than it would a truck.

16-14 Can you see why a racing car has such a light front end and such a large engine?

To summarize the statements above, *the acceleration of a body is directly proportional to the net force acting on the body and inversely proportional to its mass.* This is a statement of *Newton's second law of motion.* Stated as a formula,

$$a = \frac{F}{m} \text{ or } F = m \times a$$

where F is measured in newtons, m in kilograms, and a in meters/second^2.

sample problem

What force would your arm have to exert to accelerate a bowling ball with a mass of 10 kg down an alley at a rate of 5 m/sec^2?

solution:

$F = ma$
$\quad = 10 \text{ kg} \times 5 \text{ m/sec}^2$
$\quad = 50 \text{ kg-m/sec}^2, \text{ or } 50 \text{ N}$

16-15 The net force is the force that exceeds the balanced force (difference between pushing force and friction force). How much force is needed to balance friction in this picture?

Net force, 6N

Pushing force, 10N

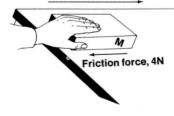

M

Friction force, 4N

Only the net force is used in finding the acceleration of a body. No matter how hard you push against a standing railroad train, you will not be able to move it. The pushing force of your body would all be lost by friction. Therefore, no net or unbalanced force is left to accelerate the train. Study the example shown in Fig. 16-15. Before a body will accelerate, all opposing forces, including the force of friction, must first be balanced. Beyond that, any added force will cause the body to move. It is important to note that net force always causes a change in motion, that is, an acceleration.

Newton's third law: Action causes reaction. If you place a heavy book on the table, the weight of the book exerts a downward force on the table. From the opposite direction, the tabletop exerts an upward and equal force on the book. Therefore, no net force is acting on the book and the book does not move.

Does it seem strange to think of the table exerting an upward force on the book? Perhaps it becomes more clear if you can imagine trying to support the book on a stretched sheet of very thin tissue paper. The paper is too weak to push up on the book with a force equal to the

unit 3 motion, forces, and energy

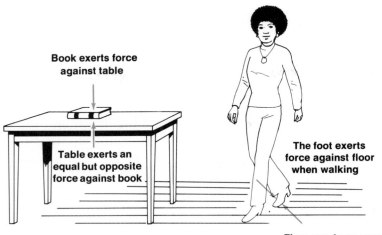

Book exerts force against table

Table exerts an equal but opposite force against book

The foot exerts force against floor when walking

Floor exerts an equal but opposite force against foot

16-16 Action and reaction are pairs of forces that apply to two bodies. How can motion occur if action and reaction are equal and opposite?

book's weight. Since there is a greater force downward than there is upward, the book will tear the paper and fall to the floor.

In walking across the classroom, your feet exert a backward push against the floor, while the floor exerts an equal and opposing force (friction) against your feet. See Fig. 16-16. What would happen if you tried to walk on a floor that was as smooth and slippery as ice? Why?

In these cases, two distinct objects are involved in each pair of forces. In the first case, the objects are the book and the table; in the second, the objects are your feet and the floor. In both examples, one force is the action and the equal and opposite force is the reaction. *When one object exerts a force on a second object (action) the second object exerts an equal and opposite force upon the first (reaction).* This is a statement of *Newton's third law of motion.*

Mass times velocity is momentum. Let us explore Newton's third law in greater detail. Suppose a man and a boy are facing each other, each standing on a pair of roller skates as shown in Fig. 16-17 (p. 298). They push each other, and each one moves away in the opposite direction. Each person has the same force acting on him, but one moves away faster than the other. The boy, having a smaller mass, moves away with a greater velocity.

In this example, suppose the boy has a mass of 40 kg and the man 100 kg, and the man moves backwards at a

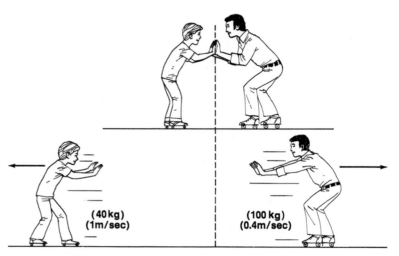

16-17 Why does the boy move away from the starting point faster than the man?

(40 kg)
(1m/sec)

(100 kg)
(0.4m/sec)

velocity of 0.4 m/sec. The boy will move off in the opposite direction at a velocity of 1 m/sec. Notice that when you multiply the mass and velocity of the boy (40 kg × 1 m/sec), you get the same product as when you multiply the mass and velocity of the man (100 kg × 0.4 m/sec). *The product of the mass and velocity of a body is called its* **momentum.**

In any reaction between two bodies, the change in momentum produced in one body is equal and opposite to the change in momentum in the other. In the example given, $M \times V$ (man) = $M \times V$ (boy).

Action-reaction explains a gun's kick. The sharp recoil when a gun is fired is another instance of Newton's third law of motion. The action and reaction are between the bullet and the gun. The burning powder pushes just as hard on the gun as it does on the bullet. This push gives the bullet momentum in one direction and the gun equal momentum in the other direction. In other words, the momentum of the bullet ($M_b \times V_b$) equals the momentum of the gun ($M_g \times V_g$). If any three of these four values are known, you can find the fourth value.

Look at the gun and bullet in another way. Since the gun and the bullet were at rest with respect to each other before the shot was fired, the total momentum of the gun and the bullet was zero. In other words, the velocity of each object was zero. Mass times zero velocity gives zero momentum. What were the values after the shot was fired? If you think of velocities in one direction as *positive*

298   unit 3   motion, forces, and energy

and those in the opposite direction as *negative*, then the total momentum after firing again adds up to zero. The total momentum is the same both before and after the firing. This example shows the *law of conservation of momentum*, which states that in any reaction between masses, *there is no net change in total momentum*.

sample problem

A gun having a mass of 5000 g fires a 20-g bullet at a speed of 500 m/sec. Find the recoil velocity of the gun.

solution:

Step 1. Substitute the given quantities in the equation, $M_g \times V_g = -M_b \times V_b$. (The minus sign means opposite direction.)

$$M_g = 5000 \text{ g}; M_b = 20 \text{ g}$$
$$V_g = ?; V_b = 500 \text{ m/sec}$$
$$5000 \text{ g} \times V_g = -(20 \text{ g} \times 500 \text{ m/sec})$$

Step 2. Solve for V_g in this equation.

$$V_g = \frac{-(20 \text{ g} \times 500 \text{ m/sec})}{5000 \text{ g}}$$
$$= -2 \text{ m/sec}$$

(The gun "kicks" in the opposite direction.)

There are many examples of the reaction principle. Jets and rockets are other examples of Newton's third law. Remember that action and reaction must take place between two different masses. The jet plane or rocket is one mass, and the exhaust gases are the other mass. Think of the rocket as a gun firing bullets of high-speed molecules. The molecules go one way and the rocket goes the other way. All movement in water, air, or space rests upon the reaction principle.

You have seen other examples of the reaction principle. In a rowboat, the oars push against a mass of water. This gives the boat a push in one direction and the water a push in the opposite direction. If you leap from the front of a rowboat, the boat will be kicked back with as much momentum as your body develops jumping forward. If you jump from a light-weight boat, you will likely land in

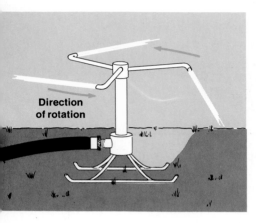

16-18 If the rate of flow of water through the nozzle increases, what happens to the speed of the sprinkler head? Why?

the water! What would happen if you were to leap for shore from a large ship? Why?

The rotary water sprinkler shows another good example of the reaction principle. The jet of water squirted out by the nozzle gives a certain momentum. As a result, the nozzle receives the same "kick" in the opposite direction. The reaction causes the sprinkler to move in a circular pattern. See Fig. 16-18.

two worlds of work: GOODS and SERVICES

Have you ever thought about what kind of job you will have when you are an adult? There are hundreds and hundreds of different jobs available. How can a person choose from among them?

Perhaps it may help to know that all jobs can be put into two groups called **goods** and **services.** If you understand these two groups, then you will have a better idea of the kind of work you may wish to do.

What is meant by "goods"? Goods include the food you eat, the clothes you wear, and the home you live in. Goods also include cars, TV sets, pencils, and tennis shoes. Goods include anything that is grown or made that people can use. The production of goods comes from four major areas: mining, manufacturing, agriculture, and construction.

The other world of work is service. Service is doing something for someone. For example, the farmer may grow the food, but how does it get to your table? Someone transports the food. Transportation is an important service function. Other service functions include providing people with telephones, water, natural gas, electric power, and sewer lines. Still other services include police and fire protection, banking, and real estate.

Even though the United States produces a great many goods, there are more people employed in service jobs than in jobs that produce goods. In fact, for many years, jobs that provide services have been growing faster than the jobs that produce goods.

When you think about what you may want to do when you are an adult, perhaps it will help to know that some jobs produce goods and others provide services.

summary

1. Speed is distance traveled in a given unit of time.
2. Velocity is speed in a given direction.
3. Acceleration is a change in velocity per unit time.
4. Force is a push or pull.
5. There are many kinds of forces, including friction, gravity, and parallel.
6. Newton's first law states that objects in motion tend to stay in motion and objects at rest tend to stay at rest.
7. Newton's second law states that a net force is needed to cause a change in motion.
8. Newton's third law states that for every action there is an equal and opposite reaction.
9. Mass times velocity is momentum.

review

Match each item in the left column with the best response in the right column. *Do not write in this book.*

1. acceleration
2. force
3. gravity
4. inertia
5. momentum
6. speed
7. torque
8. velocity
9. uniform motion
10. friction

a. a push or pull
b. force of attraction between two masses
c. change in velocity per unit time
d. faster than the speed of sound
e. not speeding up, slowing down, or turning
f. product of mass and velocity
g. resistance owing to surfaces rubbing together
h. tends to produce rotation
i. speed in a given direction
j. tendency of a mass to stay at rest or to keep moving
k. unit of mass
l. rate of motion

questions

Group A

1. What is the difference between speed and velocity?
2. What forces are usually acting on a body at rest?
3. Forces on an object in uniform motion and an object at rest are both in balance. Why?

4. How does friction affect the efficiency of a machine?
5. Which has less resistance, rolling or sliding friction? Give examples.
6. What is gravity?
7. What is torque?
8. Define inertia and give three examples.
9. What is acceleration? What causes it?
10. State Newton's second law of motion. Give an example of this law.

Group B

11. A car starts from a rest position, is accelerated to 60 km/hr, and then is held at 60 km/hr. At what steps in this sequence of events are the forces on the car in balance?
12. Is friction ever desirable? Explain.
13. Why would a person weigh less on the moon than on the earth?
14. If you doubled the net force acting on a moving object, how would its acceleration be affected?
15. If twice the net force results in twice the acceleration, why doesn't a 10-kg stone fall twice as fast as a 5-kg stone?
16. What happens to the distance needed for stopping a car if its speed is increased from 10 km/hr to 40 km/hr?
17. What happens to acceleration if the mass is tripled and the force remains the same?
18. Why does a slowly moving freight train have more momentum than a high-speed bullet?
19. Would the acceleration of a falling object 1000 km above the earth's surface be more or less than 9.8 m/sec/sec? Explain.
20. Newton's third law of motion involves two objects and two forces. Explain what this means in terms of action and reaction.

problems

1. A car is slowed down from 50 km/hr to 20 km/hr in 6 sec. What is the deceleration in kilometers per hour per second?
2. An object is dropped from a high building. How far will the object fall in 5 sec? Neglecting air resistance, what would its velocity be at the end of this time? What would its *average* velocity be during the fall? If it took 7 sec for the object to hit the ground, find the height of the building.
3. A man weighing 800 N sits on one side of a seesaw, a distance of 2 m from the fulcrum. How far on the other side of the fulcrum must a 500-N boy sit to balance the seesaw?
4. A truck weighing 12 kN is located 20 m from the east end of a 60-m bridge. Find the weight supported by each of the two piers. (Neglect the weight of the bridge itself.)

5. Find the resultant of two forces, one a force of 20 N acting east, and the other of 5 N acting on the same point in the same direction. What would be the resultant if they acted in opposite directions?
6. There is a 10-N pull on an object in an easterly direction. On the same object, another pull of 15 N acts to the south. What is the size and direction of the resultant force?
7. What force is needed to accelerate a 1000-kg car at 3 m/sec²?
8. If a 5-kg gun fires a 25-g bullet at a speed of 500 m/sec, what is the recoil speed of the gun?

further reading

Branley, Franklyn M., *Weight and Weightlessness.* New York: Thomas Y. Crowell, 1972.

Jacobs, Richard D., *Matter and Motion.* New York: Cambridge Book, 1976.

Lefkowitz, R. J., *Forces in the Earth: A Book About Gravity and Magnetism.* New York: Parents' Magazine Press, 1976.

Legunn, J., *Motion.* Mankato, Minn.: Creative Education, 1971.

Liss, Howard, *Friction.* New York: Coward, McCann & Geoghegan, 1968.

Schneider, Herman, *Scientists Find Out About Matter, Time, Space, and Energy.* New York: McGraw-Hill, 1976.

17

using force and motion

In the last chapter, you saw what causes motion and learned about the laws of motion. In this chapter, you will see how motion is affected by machines to increase force, change direction, or increase speed. You will also see how balanced and unbalanced forces affect flight through air and space.

objectives:

☐ To describe how forces operate in simple machines.

☐ To describe how force and motion are related in flight through the air.

☐ To explain the forces and motions that make space flight possible.

☐ To describe the forces and motions that keep satellites in orbit.

machines

A machine can change forces three ways. Machines make things easier to do. A machine is a device that affects a force in any of three ways. A machine can change (1) the *size* of the force, (2) the *direction* of the force, or (3) the *speed* of the force. Look at each of these ways more closely. (1) If you pull on the rope of a block and tackle (which is a system of pulleys), you can lift a heavy weight with little effort. In this case you are changing the size of the force. (2) If you push down on one end of a seesaw, you can lift someone at the other end. In this example,

unit 3 motion, forces, and energy

you are changing the direction of the force. (3) If you slowly turn the crank of a hand-driven eggbeater, the blades turn much faster than the crank. Here you are changing the speed of the force.

No doubt you could think of other ways in which machines make it easier to do things. For example, is there more force, greater speed, or a change in direction if (a) you push down on a bicycle pedal; (b) you press down on the brake pedal of a car; (c) you push down on the handle of a jack that is lifting a car?

There are two main kinds of machines. Often you call a simple device, such as a knife or a screwdriver, a tool. Maybe you think of a machine as something more complex, such as a typewriter, an airplane, or a car. In science, however, the word *machine* includes all these things. A knife and a screwdriver are as much machines as are cars.

All the machines ever invented are classed as either *simple* or *complex*. Simple machines include the *lever, pulley, inclined plane, wedge, screw,* and *wheel and axle.* Not all of these machines are really different from one another. The screw and wedge are just forms of the inclined plane. The wheel and axle and pulley are really kinds of levers. Gear wheels and hydraulic devices are other forms of the above simple machines.

Complex machines are devices made of two or more simple machines. Regardless how complex a machine is, it is still made up only of simple machines. See Fig. 17-1.

A machine can give a mechanical advantage. The mechanical advantage (MA) of a machine tells how much the machine increases the effort you put into a job. Mechanical advantage is sometimes expressed as the ratio of two distances. For example, suppose that you use a rope and pulleys to lift a weight. You pull the rope 2 m and the weight is lifted 1 m. If you divide the distance the rope moved by the distance the weight moved, you have a ratio of 2 to 1 (2:1), or a mechanical advantage of 2. This number is called the ideal mechanical advantage (IMA). The IMA tells you how many times the machine can increase the force if there is no friction. Since friction is present in all machines, the actual mechanical advantage (AMA) is always less than the IMA.

17-1 How many kinds of simple machines can you find in this picture?

The AMA can be found by dividing the weight of the lifted object by the effort put into the machine. The number you get tells you just how much a machine multiplies the force you put into it. For example, a long bar is used to lift a weight of 1000 N with an effort of 40 N. The bar has a mechanical advantage of 1000/40, or MA = 25. This number means that for every newton of force that is applied, this machine produces a force of 25 N.

All machines waste effort. Some people are surprised to learn that a machine does not increase the amount of energy that is put into it. As was stated earlier, a machine can change the size, speed, or direction of a force, but it cannot increase the amount of energy. In fact, the energy that comes out of a machine is always less than the energy that goes into it.

Why is there a loss of energy? No machine can function without friction. Therefore, some of the energy that goes into the machine is used to overcome friction. The loss of energy varies with the kind of machine used. If the machine is efficient, it loses little energy.

The efficiency of a machine can be found by comparing the energy that goes into a machine (called *input work*) with the energy that comes out (called *output work*) as follows:

$$\text{efficiency} = \frac{\text{output work}}{\text{input work}}$$

For example, if you put 4000 N-m into a machine and you get only 3000 N-m out of it, the machine has an efficiency of 75% (3000/4000 = 0.75). See Fig. 17-2.

You can also find the efficiency of a machine by comparing the actual and ideal mechanical advantage as follows:

$$\text{efficiency} = \frac{\text{AMA}}{\text{IMA}}$$

For example, if the actual mechanical advantage of a machine is 2 while the ideal mechanical advantage is 2.5, the efficiency is again 80% (2/2.5 = 0.8).

If a machine were 100% efficient, the input work would be equal to the output work. In addition, the AMA would be equal to the IMA.

Pulleys are one type of simple machine. It is much easier to use a pulley at the top of a flagpole to raise the flag than

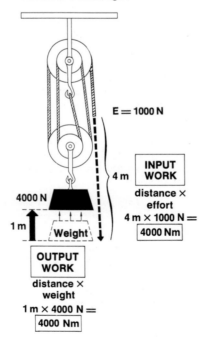

17-2 An effort of 1000 N, moving through a distance of 4 m, is needed to lift the 4000 N weight a distance of 1 m. What is the actual mechanical advantage of this block and tackle? the ideal mechanical advantage?

E = 1000 N

4 m

INPUT WORK
distance × effort
4 m × 1000 N =
4000 Nm

4000 N

1 m   Weight

OUTPUT WORK
distance × weight
1 m × 4000 N =
4000 Nm

it is to climb the pole to raise it. With a single *fixed* pulley there is no increase in the effort force or in the distance through which the effort moves. However, by using a fixed pulley, you can change the direction of the effort force. Changing the direction means you can stand on the ground while pulling the flag to the top of the pole.

Compare a fixed pulley with a single *movable* pulley as shown in Fig. 17-3. With the movable pulley, the force is applied in the same direction as the effort. There is no change in direction, but there is a decrease in the effort needed to lift the weight.

Piano movers may use a *block and tackle* to raise a piano to a second-story window. A block and tackle is a system of pulleys used to raise a heavy object with a small amount of effort. Although a piano may have a weight of about 3000 N, one person can raise it by pulling down on the rope.

In the pulley system shown in Fig. 17-2, the 3000 N weight is the load to be raised. (Of course, the lower, movable block also has some weight, and it too is part of the load. Its weight is not included in the example.)

Note that the weight and pulley hang from four strands of rope. The total upward force must equal the total downward force of 3000 N. Therefore, each of the four strands must lift 750 N and the force on the effort rope must be 750 N. If friction is added, a pull of *more* than 750 N will be needed.

There are three classes of levers. Now that you know what is meant by mechanical advantage, input work, and output work, you can see how these ideas apply to other simple machines. The *lever,* for instance, is just a rigid bar that is free to turn about its point of support, which is called a *fulcrum.* The torques and parallel forces discussed in Chapter 16 can be applied to any lever.

The loaded wheelbarrow shown in Fig. 17-4 is a form of lever. A load of 1000 N is placed so that the center of gravity for the wheelbarrow and its load is 0.8 m from the fulcrum (F). How much *effort force (E)* must be applied upward at the handles in order to lift the wheelbarrow?

Look at the torques about F of the parallel forces acting on the wheelbarrow. The clockwise torque produced by W (the weight of the wheelbarrow and its load) is equal to 1000 N × 0.8 m = 800 N-m. The force of effort E needed to overcome W can be found by using the counterclockwise

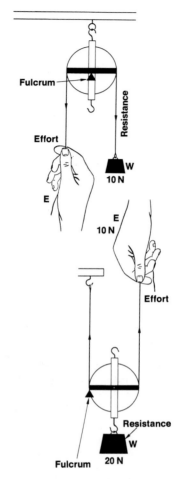

17-3 How does the mechanical advantage of the single fixed pulley compare with that of the movable pulley?

17-4 How does a wheelbarrow use the principles of a simple machine? How much effort is needed to lift the wheelbarrow?

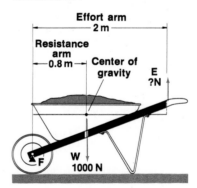

A.
First class lever
The MA increases as the fulcrum is moved toward the weight. It may have an MA greater or less than one.

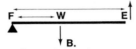

B.
Second class lever
The MA increases as the weight is moved toward the fulcrum. Its MA is always greater than one.

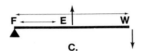

C.
Third class lever
The MA is always less than one. It gains speed at the cost of lowered MA.

17-5 How does the effort arm compare with the resistance arm in each lever?

torque of E about the same fulcrum ($E \times 2$ m). Equating the two torques,

$$E \times 2 \text{ m} = 1000 \text{ N} \times 0.8 \text{ m}$$
$$E = 400 \text{ N}$$

Note that even though the bottom of the wheelbarrow is sloping, all distances are measured at right angles to the forces of lift.

The wheelbarrow is an example of a second-class lever. Figure 17-5**B** shows that in a second-class lever, the weight W is between the fulcrum and effort. The mechanical advantage (MA) is always greater than 1. The MA can be increased by moving the weight toward the fulcrum or by lengthening the effort arm.

Figure 17-5**A** is a diagram of a first-class lever. The fulcrum is located between the weight and the effort. The MA may be greater or less than 1. To increase the MA, the fulcrum must be moved nearer to the weight.

The third-class lever in Fig. 17-5**C** always has an MA of less than 1. A third-class lever is used to increase speed.

With your elbow on the table, raise this book in your hand. You have just used your forearm as a lever. Levers are so common in daily life that you do not notice how often you use them. Things like shovels, brooms, scissors, nutcrackers, and piano keys are types of levers. Even bottle openers are levers. In fact, a typical bottle opener can be used to show two types of levers. See the student activity below.

activity

A bottle opener as a lever.
Step 1. Place the opener over a bottle cap as shown in Fig. 17-6**A**. This is an example of a first-class lever.
    1. What is the length of the effort arm?
    2. What is the length of the resistance arm?
    3. What is the mechanical advantage of the opener?

Step 2. Now place the opener over the bottle as shown in Fig. 17-6**B**. In this example, the opener is used as a second-class lever.
    4. What is the length of the effort arm?
    5. What is the length of the resistance arm?
    6. What is the mechanical advantage?

Questions:
1. With which method should the bottle be easier to open?
2. Have you found the AMA or IMA in the above activity? Explain.

The wheel and axle is a lever. Have you ever tried to open a door by turning the rod with your fingers when the doorknob was missing? If you have, you can see how a doorknob makes the job easier. The doorknob acts like a *wheel and axle*. The steering wheel of a car is another good example of the wheel and axle. Just a little effort on the rim of the steering wheel is enough to change the direction of the front wheels of the car. The screwdriver and the hand drill are other common examples of the wheel and axle. From the above examples, you can see that although the "wheel" and the "axle" are of different radii (or diameters), they are rigidly attached to each other and move as one piece.

From a study of Fig. 17-7 and a knowledge of torques, you see that $W \times$ line AF is equal to $E \times$ line BF. By rearranging this equation, the actual mechanical advantage of this machine becomes $W/E = BF/AF$. (Remember that if friction is not counted, the IMA and the AMA are the same.) Therefore, the ideal mechanical advantage of the wheel and axle is simply the ratio of the radius (diameter) of the wheel to radius (diameter) of the axle. You can make the IMA as great as you wish by using a very large wheel coupled with a very small axle. A winch and a windlass are examples of the wheel and axle. Note that the distances around the outsides of the wheel and the axle may also be used to find the IMA. Can you explain why?

The inclined plane increases force. The *inclined plane* is another machine that is often used to raise heavy objects. Look at Fig. 17-8 on p. 310. Here, if you do not count friction, a force of 250 N is enough to roll a weight of 1000 N up the plank. Therefore, the mechanical advantage, $W/E = 1000 \text{ N}/250 \text{ N} = 4$. You can also see that you have to use a plank 8 m long in order to raise the load a height of 2 m. Thus, you see that the four-times gain in force is obtained by a four-times increase in distance. In other words, if you raise the 1000-N weight, W, a height of

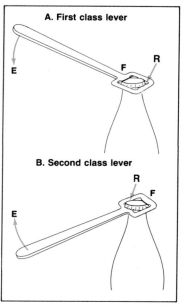

17-6

A. First class lever

B. Second class lever

17-7 How is the wheel and axle like a lever? How is it different from a pulley?

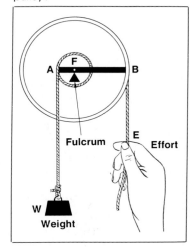

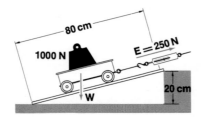

17-8 In what way can you increase the mechanical advantage of the inclined plane?

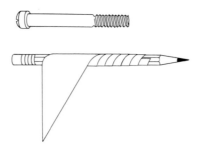

17-9 Can you see how a screw is like an inclined plane?

2 m, your 250-N effort must move a distance of 8 m. Can you find the input work? the output work?

The IMA of an inclined plane can be found by dividing the length of the plank by its height above the ground, 8 m/2 m = 4. Of course, when friction is counted, the effort E must be greater than 250 N.

If a road rises 4 m along a stretch of 100 m, then it is an inclined plane, with an IMA of 100 m/4 m = 25. This means that a car engine must exert a force equal to 1/25 of the car's weight to climb the grade at constant speed. In addition, the car must also have enough force to overcome friction.

The wedge and screw are forms of the inclined plane. It would be easy to split a log with a thin *wedge* if you struck the wedge with a heavy hammer. A wedge is very much like an inclined plane. The thinner the wedge, the easier it is to split a log. Likewise, the thinner the wedge, the higher the mechanical advantage.

Wedges are not very efficient because much of the effort is used up in friction. A knife is a form of wedge. The sharper the knife, the less force is needed to cut.

It would be hard to lift a car without the help of a machine. Some car jacks are examples of a simple machine called the *screw*. From a study of Fig. 17-9, you can see that the screw is just an inclined plane, wrapped in a spiral around a cylinder.

The screw is like an inclined plane that is very long compared to its height. The IMA of the screw is found by dividing the distance the effort moves in one full turn by the *pitch*, or distance, from one thread to the next. Can you show why this is so?

The wood screw, threaded bolt, and the bench-vise are examples of the screw that have high mechanical advantages. However, for these items, there is so much friction to overcome that the screw has a very low efficiency.

flight through the air

Thrust gets the plane moving. Have you ever stood behind a propeller-driven plane when the engine was started? If you have, you know that air is driven back-

unit 3 motion, forces, and energy

ward. Recalling Newton's third law, you know that this backward force on the air will produce a forward force on the airplane. The backward force on the air is the action, and the forward force on the airplane is the reaction.

The force pulling the airplane forward is called *thrust*. When the thrust is great enough to overcome the inertia, the plane begins to move down the runway. As it gathers speed, another force called *drag* begins to act on the plane. Drag is caused by the friction of the air that is pushed aside by the moving plane. At first, when the plane is going slowly, the drag does not amount to much. Later, when the plane is flying steadily, the drag is equal to the thrust.

Lift gets the plane off the ground. As the plane goes faster and faster down the runway, another force begins to take effect. Air pressure builds up under the wing until it lifts the plane off the runway. How does the wing act to cause the lift? To understand what happens, think of holding out your hand, palm down, from the window of a moving car. What happens when you turn your hand slightly so that the air strikes the underside of the hand? The air pushes your hand upward. That is one way that moving air acts to push the airplane up. This push is called the *kite* effect.

Lift is also produced by air going over the wing. How can this happen? Look closely at a cross section of a wing in Fig. 17-10. The wing is shaped so that it is somewhat rounded on top. When the airstream hits the front edge of the wing, it divides. Part of the air goes above the wing and part of it goes below, as you would expect. The air above the wing speeds up as it goes over the raised curve. Surprising as it may seem, when the speed of air increases, the pressure that it has goes down. This rather startling fact was discovered by Daniel Bernoulli during the early eighteenth century. This effect is now known as the **Bernoulli** (ber-*noo*-lee) **principle.** This principle states that *when the speed of a fluid (liquid or gas) is increased, its internal pressure is decreased.*

To review what happens, just think of three steps. First, the air speeds up as it goes over the raised curve of the wing. Second, as the air speeds up, the air's pressure goes down. Third, the normal air pressure under the wing is still present and pushes up. The combination of these effects produces the lift to get the plane off the ground.

17-10 Where is the air moving fastest? How does air speed affect lift?

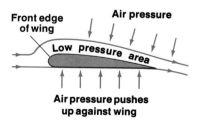

Front edge of wing

Air pressure

Low pressure area

Air pressure pushes up against wing

activity

Bernoulli's Puzzle.
1. Hold a sheet of paper in front of you and blow against the lower half of it. What happens?
2. Predict what would happen if you blew on and slightly over the rounded top curve of the paper as shown in Fig. 17-11. Now try it and compare the result with your prediction. Was your prediction correct?

17-11

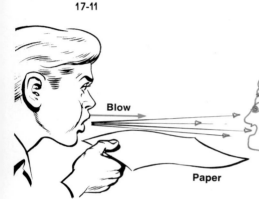

Blow

Paper

Forces keep the plane steady. An airplane is designed so that it is steady while in flight. It will even correct itself if something is done to upset the plane's balance. It is more like the flight of an arrow than that of a stick. A stick will just tumble in its path if you throw it. However, when an arrow is shot from a bow, its tail feathers act to keep it steady in flight. An airplane needs something like tail feathers too. It has horizontal and vertical *stabilizers* to keep the plane steady in flight.

The wings of many airplanes are angled upward. That is, the tip of the wing is higher than the base. This is called an *upsweep*. This upsweep tends to keep the airplane from rolling and swinging from side to side.

Forces change motion. An airplane has controls that cause it to change speed, turn left or right, move up or down, and roll clockwise or counterclockwise. Stated in another way, an airplane has controls that cause the forces of flight to become unbalanced. Thus, the speed or direction or both can be changed.

The thrust or forward speed of the airplane comes from the propeller (or from the exhaust gases if it is a jet plane). Up and down movements of the airplane's nose are controlled by adjusting the *elevators*. To turn the nose downward, the elevators are turned downward. Air striking the bottom surface of the elevators pushes the tail up, thus tilting the nose down.

When the pilot pulls back on the control wheel, or control stick, the elevators tilt up. The air blows against their top surfaces, forcing the tail down and the nose of the plane up. With extra power, the plane is able to climb.

The *rudder* is used to move the nose of the plane to the left or right. When the rudder swings to the right, the airstream pushes the tail to the left. The plane swings about, with its nose pointing to the right. However, just turning the rudder is not enough to make a smooth turn. Although the plane is facing a new direction, it is still moving on its original path. The plane is said to be skidding. It is somewhat like a car skidding on an icy curve. Therefore, turning the rudder is not a good way to turn an airplane to the left or right. Instead the pilot must bank the plane, that is, roll it slightly on its side. The pilot turns the control wheel or the control stick to adjust the *ailerons* (*ay*-ler-ons). One wing goes up and the other goes down. The plane now turns just as smoothly as a car on a banked highway curve. See Fig. 17-12.

Most planes also have control surfaces called *flaps*. Two flaps are hinged to the rear edge of the wing. Since the flaps on each wing turn downward at the same time, the plane does not bank, as it would if the flaps operated like ailerons. In order to slow down for a landing, the flaps are used as airbrakes to increase the drag. The flaps also increase the lift, so that less speed is needed to keep the plane from falling. Therefore, landings are possible at safer speeds and on shorter runways than would be needed without the flaps. Flaps are also used in takeoffs.

Keep in mind that all of these controls are simply ways of changing the forces acting upon a plane. For a plane to start to climb, the lifting force must be greater than the plane's weight. To increase the plane's speed, the thrust must be greater than the drag.

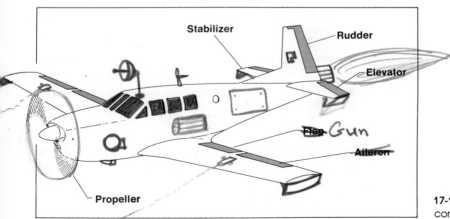

17-12 What is the function of each control surface on this light plane?

activity

Testing controls. Build a paper airplane and test its flying ability a number of times. Make adjustments so that its flight is as stable as possible. Then cut slits into the back edges of the wings to make elevators. Also make a rudder. See Fig. 17-13. Make small adjustments in the angles of the elevators and rudder and see how this affects the flight of your plane. Record your findings in your notebook.

Questions:
1. How did the plane fly when
   a. the elevators were raised slightly?
   b. the elevators were lowered slightly?
   c. one elevator was raised and the other lowered?
   d. the rudder was turned slightly?

2. What combination of adjustments produced the best flight?

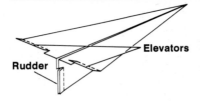

17-13 If the controls are set as shown, in which direction will the paper airplane turn?

17-14 Where is a supersonic airplane when the sonic boom is heard by someone on the ground?

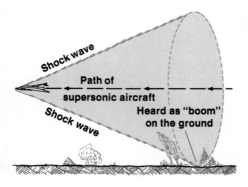

Some jet planes fly faster than the speed of sound. You have seen that the way air moves over the surfaces of an aircraft is very important. If a plane is flying at less than the speed of sound, the air tends to flow more or less smoothly over the surfaces of the plane. When the plane goes faster than the speed of sound, however, air flow changes. At these speeds, the air piles up along the front edges of the aircraft to form a shock wave. The shock wave, in turn, spreads out behind the aircraft much like water waves spread out from a moving boat. If the edges of the shock wave were visible, the shock wave would look like a cone behind the aircraft. See Fig. 17-14.

Sonic boom. What happens when the shock wave of a supersonic aircraft touches the ground? If the shock wave is strong enough, it can be heard as a *sonic boom*. The boom is caused by a jump in air pressure as the shock wave hits the ground. The greater the pressure of the wave, the louder the boom. A pressure jump of $15 \ N/m^2$ is heard as a distant boom that would hardly be noticed. A pressure jump of up to $50 \ N/m^2$ is generally not harmful. However, an increase of over $50 \ N/m^2$ sounds like a loud thunderclap and is enough to cause some damage to windows.

The loudness of the sonic boom is affected by the speed, size, and height of the plane above the ground. A large airliner flying at twice the speed of sound at a height of 25 km would cause a pressure jump on the ground of about 20 N/m². If it were flying at 16 km, the pressure would jump by 50 N/m². At a height of 10 km, the pressure jump would be 120 N/m², and at a height of 3 km the figure would be 420 N/m². What would the total force be on a large store window measuring 3 by 6 m in each of these cases? Can you see why there is some objection to low-level supersonic flights across land areas?

flight through space

Getting off the launching pad takes a great deal of force. Which is a better engine to boost a 5000-kN spaceship into orbit: one that gives 1000 kN thrust for 1 hr, or one that gives 10,000 kN of thrust for 3 min?

Many people would choose the engine that gives the lower thrust because it is on for a whole hour. However, they would be wrong. Why? Just think how high you can lift a 5000-kN weight (the spacecraft) with a 1000-kN thrust. Even though the thrust is applied for a full hour, the spaceship would not even get off the launching pad. Only rocket engines that give "brute force" can be used for lifting a spaceship off the pad.

"Stages" are needed for many space flights. Suppose you were all alone 300 km in the heart of a desert with a truck, a jeep, and a motorcycle. Suppose also that each vehicle has a sealed gas tank with enough gas to go 100 km. You cannot transfer gas from one vehicle to another. How would you travel the 300 km out of the desert to safety?

The answer: Put the motorcycle into the jeep and the jeep into the truck and drive the truck 100 km until it runs out of gas. Then continue on with the jeep and motorcycle until the jeep runs out of gas. Finally, go the last 100 km with the motorcycle. This is a costly way to take a trip, but the "payload" (the driver) gets out of the desert.

The desert problem can be compared to how "stages" are used to launch objects into orbit. So much energy is needed to put a payload into orbit that launchers are built in steps, or stages, each riding "piggyback" on the other

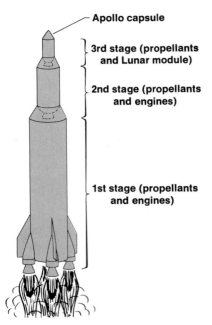

Apollo capsule

3rd stage (propellants and Lunar module)

2nd stage (propellants and engines)

1st stage (propellants and engines)

17-15 Why is the first stage of a three-stage rocket so large in comparison to the other stages?

17-16 One satellite has a high orbit, the other a low orbit. Which one travels faster?

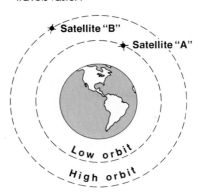

Satellite "B"

Satellite "A"

Low orbit

High orbit

as shown in Fig. 17-15. The first stage gets the spacecraft off the pad. Then when the first stage burns out its fuel, it drops off and the second stage ignites. The second stage, in turn, pushes the spacecraft farther into space, and when it burns out it too drops off. Finally, the small third stage finishes the job of getting the payload to its target.

The Saturn V rocket, used to send astronauts to the moon, was a three-stage rocket. If you were to lay a Saturn V rocket flat on a football field, the tip would reach beyond the goal post on one end of the field and the engines would reach beyond the goal post on the other end of the field.

Gravity affects satellites in orbit. When a satellite is in orbit around the earth, it is affected by inertia and gravity. Remember Newton's first law: an object in motion tends to stay in motion. That motion would take the satellite in a straight line into space. What pulls it into a curved path around the earth?

The satellite is not only affected by inertia. It is also affected by the force of gravity. Gravity is always pulling the satellite toward the earth. In an orbiting body, the force of gravity pulling the body toward the earth is balanced by inertia, which tends to keep the body going in a straight line.

At any given height there is one given speed that will permit a satellite to follow a circular path around the earth. The lower the orbit, the greater the speed. Why? Gravity is stronger closer to the earth, so the inertia must be greater to balance the gravity. The greater inertia comes from greater speed. In an orbit 200 km above the earth, the satellite must go at a speed of 29,000 km/hr. In an orbit 36,000 km above the earth, a satellite needs to go only 5500 km/hr.

At still greater distances from the earth, satellite speeds are slower yet. For instance, the moon, 384,000 km from the earth, travels at only 3500 km/hr. See Fig. 17-16.

Flight to the moon and planets is difficult. There are many problems in trying to go to the moon. You have already seen that just getting into earth orbit takes a great deal of force using three-stage rockets. If the spaceship gets into earth orbit, it then coasts for a while until it is lined up for a flight to the moon. When it is in just the right place, the engines are fired again and the extra force

316   unit 3 motion, forces, and energy

pushes the spaceship away from the earth's gravity. The spaceship follows a long curved path through space. It finally reaches a place where the gravity of the moon is stronger than the gravity of the earth.

If the spaceship is properly aimed, it will swing into an orbit around the moon. Its engines are fired again to slow the ship enough so that it can be "captured" by the gravity of the moon. Further careful steps are needed to send a "moon lander" to the moon's surface, get it back to the spaceship, and to get the spaceship back to earth.

Flights to the planets are even more complex. The distances are much greater, and the trips take much longer. Furthermore, when the spaceship leaves the earth-moon area, the sun's gravity has a greater influence on the spaceship than the gravity of the earth or moon.

To get to another planet, a spaceship must (1) leave the gravity field of the earth, (2) become a satellite of the sun, (3) enter the gravity field of the target planet.

Surprising as it may seem, it does not take very much more speed and fuel to go to the planets than to go to the moon. Most of the fuel is needed just to get off the ground. Once in orbit, somewhat more fuel is needed to get to the moon, but very little extra is needed to go beyond. See Table 17-1.

Table 17-1:

Minimum Launch Velocities and Time Needed to Reach Objectives		
Objective	Length of Trip	Minimum Velocity (km/hr)
Earth orbit	30 min	28,000
Moon	63 days	40,000
Mercury	110 days	48,000
Venus	150 days	42,000
Mars	260 days	42,000
Jupiter	2.7 years	50,000
Saturn	6 years	53,600
Uranus	16 years	55,600
Neptune	31 years	56,800
Pluto	46 years	57,600

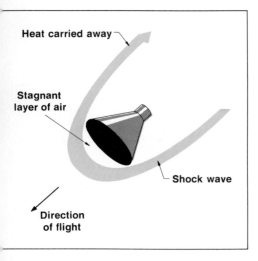

17-17 Why does a re-entry space capsule have such a blunt shape?

The launch velocities are the minimums needed for the listed trips. By increasing the launch velocities, the trip time can be reduced.

Re-entry capsules become very hot. How is a spaceship brought back to earth? First of all, its rocket engines are fired to slow down the spaceship. With less speed, the ship also has less of a tendency to travel in a straight line. Yet, gravity is still pulling the ship down into the atmosphere. When the spaceship hits the air, it slows down because of friction. See Fig. 17-17.

The friction of the air causes a great deal of heat. There is so much heat that if it all entered the spaceship, the ship would melt. To solve the heat problem, the re-entry capsule has a broad, blunt shape. This blunt shape causes a strong shock wave to build up in the air. The shock wave carries off most of the white-hot air so that the heat does not enter the capsule. Even so, the capsule gets so hot that it melts away some of the ceramic covering.

Finally, when the speed is reduced enough by the air, a parachute is opened and the capsule descends gently to the ocean. Here it floats until picked up by a ship.

Johannes Kepler
Astronomer

Johannes Kepler was a German scientist who found the key to the movements of the planets. He studied heavenly bodies millions of kilometers away with nothing more than pencil and paper. Kepler's own eyesight was too poor to permit him to make many useful sightings. However, Kepler had access to many carefully recorded sightings by the Danish astronomer, Tycho Brahe.

Astronomers at that time could not account for the slight differences that were noted in the orbits and speeds of the planets. Kepler tried all the accepted theories, but none seemed to fit the sightings. Kepler, like many great scientists before and after him, had great persistence. He worked for four years on the orbit of Mars alone. He worked many more years on the other planets. Finally, Kepler found that the paths of planets are not circular, but oval-shaped, and that there is a relationship between a planet's speed and its distance from the sun.

During these years, Kepler received some small grants to do his research, but a greater share of his income came from doing a most unscientific task. Kepler served as the official astrologer for kings and royalty! This is a rare example of astrology aiding the cause of science.

summary

1. Machines change the direction, size, or speed of forces.
2. The ideal mechanical advantage of a machine is a ratio of the effort distance compared to the weight distance.
3. The actual mechanical advantage of a machine is a ratio of weight to effort.
4. All machines waste effort through friction.
5. The ratio of output to input work gives the efficiency rating of a machine.
6. Levers (including pulleys and wheel and axles) and inclined planes (including screws and wedges) are simple machines.
7. The forces on a flying airplane are thrust, drag, lift, and gravity.
8. Flight can be controlled by using a rudder, elevators, ailerons, and flaps.
9. Brute force and staging are needed to get a spaceship into orbit.
10. Inertia and gravity affect an orbiting satellite.

review

Match each item in the left column with the best response in the right column. *Do not write in this book.*

1. lift
2. thrust
3. drag
4. kite effect
5. supersonic
6. rudder
7. AMA
8. IMA
9. machine
10. inertia and gravity

a. point in a satellite's orbit farthest from the earth
b. a device that changes the amount, speed, or direction of a motion
c. an upward force on an airplane
d. performance of a frictionless machine
e. control surface of a plane that increases lift
f. ratio of resistance to effort
g. faster than the speed of sound
h. control surface on a plane that causes a change in direction
i. that which keeps a satellite in orbit
j. forward force produced by a propeller
k. lift caused by air striking the underside of the wing
l. force that slows the forward motion of an airplane

questions

Group A

1. What are three purposes of a simple machine?
2. Explain the difference between high mechanical advantage and high efficiency.
3. How do you find the ideal mechanical advantage of a pulley system? inclined plane? wheel and axle?
4. How are the single pulley and the wheel and axle related to the lever?
5. Name four parts of a plane used in controlling its flight.
6. Explain the action of a propeller.
7. What is meant by "staging" a spacecraft?
8. What keeps a satellite from flying off into space?
9. What keeps a satellite from crashing into the earth?

Group B

10. Explain how you would determine the efficiency of a pulley system.
11. Diagram a pulley system that you could use to lift a 3000-N load by pulling down with a 500-N force (assume no friction).
12. Of the machines discussed in this chapter, which do you think is the most efficient? least efficient? Why?
13. Into what two groups can the six simple machines be placed?
14. Do you think it is possible to make a perpetual motion machine? Explain.
15. List four main forces acting on a plane in flight.
16. Describe two ways to increase the lift of an airplane.
17. Why is a satellite traveling faster at the low point of its orbit than at the high point?
18. What causes space capsules to become hot when they re-enter the earth's atmosphere?

problems

1. What is the AMA of a machine that lifts a weight of 100 N with an effort of 20 N?
2. What is the efficiency of a machine that does 160 N-m of output work with 200 N-m of input work?
3. The movable pulley block of an oil rig is supported by eight strands of wire cable. (a) What is the ideal mechanical advantage? (b) Neglecting friction, how much weight could be supported by an effort of 500 N? Refer to Fig. 17-2, if necessary.
4. What is the efficiency of a block and tackle that lifts a 500-N weight a height of 1 m when an effort of 80 N moves through a distance of 10 m? What is the input work? output work?

5. What is the IMA of an inclined plane 20 m long and 2 m high?
6. A 500-N cart is rolled up a 20-m plank to a platform 5 m off the ground by an effort of 150 N parallel to the plank. Find (a) the ideal mechanical advantage, (b) the actual mechanical advantage, (c) the input work, (d) the output work, and (e) the efficiency.
7. What is the weight that could be lifted by a second-class lever 60 cm long if the fulcrum is at one end and a 15-N upward force is applied at the other? Assume that the weight is placed 20 cm from the fulcrum. What downward effort would be needed if the fulcrum and the weight were to exchange places?
8. What weight is lifted by a lever with a 3-m resistance arm when 24 N is applied to the 5-m arm?
9. A plane has a wing area of 64 m². The average lift on the wings is 800 N/m². Find the total lift.
10. A satellite is orbiting quite close to the earth at a speed of 28,000 km/hr About how long will it take to make one revolution? (Assume the orbit diameter is 12,800 km.)

further reading

Clarke, Arthur C., and Silverberg, R., *Into Space: A Young Person's Guide to Space.* New York: Holt, Rinehart and Winston, 1971.

Helman, H., *Lever and the Pulley.* New York: M. Evans, 1971.

Hulton, K. G., *Machines Seen at the End of the Mechanical Age.* New York: Holt, Rinehart and Winston, 1972.

Madison, *Aviation Careers.* New York: Franklin Watts, 1977.

Meyer, Jerome S., *Machines.* Cleveland, Ohio: World, 1972.

Munson, Kenneth G., *Famous Aircraft of All Time.* New York: Arco, 1977.

Lomask, Milton, *Robert H. Goddard: Space Pioneer.* Champaign, Ill.: Garrard, 1972.

Sharp, David, *Looking Inside Machines on the Move.* Chicago: Rand McNally, 1977.

Simon, Seymour, *The Paper Airplane Book.* New York: Viking, 1971.

Taylor, John W., *Janes Pocket Book of Record Breaking Aircraft.* New York: MacMillan, 1977.

18

forces in solids, liquids, and gases

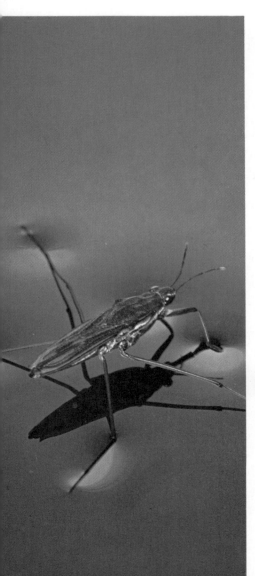

In Chapters 16 and 17, you saw how exterior forces affected matter, that is, how matter was affected by such forces as friction, gravity, and inertia. You also saw how forces were affected by machines. In this chapter, you will look more at the *interior* forces of matter. In other words, you will study some of the forces that exist among molecules. You will also see how forces are transferred within solids, liquids, and gases.

objectives:

☐ To describe some of the useful properties of matter.

☐ To compare cohesive and adhesive forces in solids and liquids.

☐ To calculate the pressures on solids and in liquids when given the forces and areas.

☐ To compare buoyancy with specific gravity.

☐ To understand the relationship between pressure and volume of confined gases.

forces affect shape

Some forms of matter are elastic. Why is steel used in car springs? Why is steel used in a watch spring? Why does a rubber band snap? How do the properties of matter affect its uses?

When the ends of a coiled steel spring are pulled apart, the applied forces cause the spring to stretch. If the forces are increased, the spring will stretch farther. When the forces are released, the spring snaps back to its original shape. A rubber band acts in the same way. It can be stretched, and when released, it returns to its former shape. The property of matter that allows it to return to its original shape after being distorted by a force is called *elasticity*. However, if a spring is stretched too far, it will not return to its former shape. The spring has been stretched beyond its *elastic limit*.

Because steel is very elastic, it returns to its beginning shape when the stretching force is removed. Tests show that steel is better than rubber in this respect. Steel is more elastic than rubber because it takes a greater force to stretch steel. Steel also returns to its beginning shape with more force. That is why steel is used in such different places as car springs and watch springs. However, in another way, rubber is better than steel. Rubber is more *resilient* (ree-*zil*-yent) than steel. Rubber can be stretched farther without going beyond its elastic limit. Substances such as putty or modeling clay are called *inelastic*. They do not regain their original shape.

Spring scales make use of the elastic properties of matter. For instance, when a given weight is hung on a spiral spring, the spring will stretch by a certain amount. When another equal weight is added to the first weight, the spring scale will stretch twice as far. If three equal weights are used, the spring will stretch three times as far, and so on. However, this step-by-step increase will not continue forever. At a certain point, the spring scale will be pulled beyond its elastic limit. See Table 18-1.

Table 18-1:

Elastic Limit of a Spring	
Weight on Spring (newtons)	Stretch of Spring (centimeters)
0	0.0
100	1.5
200	3.0
300	4.5
400	6.0
500	8.1
600	10.7

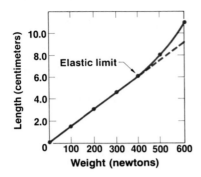

18-1 Why is it easier to find the elastic limit on this graph than in Table 18-1?

You can see that up to the 400-N level, each 100-N unit had an equal effect on the spring. However, when 500-N and 600-N weights were added, the elastic limit of the spring was exceeded. At these points, the spring could no longer be returned to its original shape. Instead, the spring was *permanently* stretched. The data in Table 18-1 can also be seen as a graph in Fig. 18-1.

The elastic property of the coil spring and rubber band also explains the compression of bridge piers, the bending of steel building beams, and the twisting of drive shafts.

Cohesion holds like molecules together. Why doesn't a steel beam fall apart? Why doesn't a concrete wall crumble? What holds solids together? The property of matter that holds it together is called **cohesion** (co-*hee*-zhun). *Cohesion is a tendency for like particles of matter to stick together.* Of course, it is easy to see that cohesion is quite strong in a steel beam.

What about liquids? Do they also have a cohesive force? If you observe water carefully, you will see that water does have a cohesive force. Other liquids also have cohesion, but the strength varies with the substance. Mercury, a liquid metal, has a strong cohesive force as compared to other liquids, but it is still weak compared to solids.

activity

Water drops and floating corks. You will need waxed paper, glass plate, corks, drinking glass, toothpick, and soap solution.

Part one: Place a few water drops on the waxed paper. Start with a small drop and make each succeeding drop larger until the largest is about 10 mm across.

Questions:
Study the drops carefully. Make a diagram in your notebook showing a side view of each drop. Then dip a toothpick into a soap solution and touch the tip to each drop. Again, draw side views of the water drops.
1. Which water drop is the highest?
2. Which water drop is the roundest?
3. What effect did the soap solution have on the water drops?

Part two: Place several drops of water on a clean piece of glass. How do the drops compare with the drops on the waxed paper?

Part three: Place a cork into a drinking glass half full of water. Try to get the cork to stay in the middle. What happens? Why?

Adhesion holds unlike molecules together. *The attraction between molecules of two different substances is called* **adhesion.** A good example of adhesion is the attraction between adhesive tape and skin. Glue and paste show adhesion when they stick to wood and paper.

After you have finished washing your hands, some of the water still sticks to your skin by adhesion. Yet if you dip your finger in mercury, no mercury sticks to it. This is because the force of attraction between the molecules of mercury (cohesion) is greater than the force of attraction between mercury and your skin (adhesion).

If you look carefully at a glass or test tube filled with water, you will see that the water is not level where it touches the sides of the glass. The attraction between water and glass is strong enough to pull some of the water slightly up the side of the glass. Mercury, on the other hand, is pulled down from the walls of the glass. The cohesive force of mercury is stronger than the attraction of mercury and glass. See Fig. 18-2.

There are special forms of cohesion and adhesion. Cohesion between molecules on the surface of a liquid produces an interesting effect. This effect does not exist below the surface because there all the cohesive forces on a molecule are balanced. At the surface, however, molecules are attracted only by the molecules below and at the sides. This attraction is shown by the force vectors in Fig. 18-3. The attraction causes the surface to act as if it were covered by a thin elastic film. This effect is called *surface tension.* You can see the effect of surface tension in a number of ways. For instance, you can pour water carefully into a glass until the liquid rises above the rim. The surface tension of the water is strong enough to keep the water from overflowing. However, if more water is added, the surface tension will break and the water spills over.

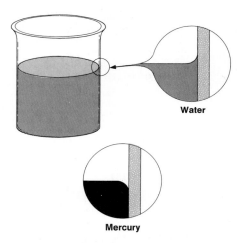
Water

Mercury

18-2 Why does water climb up the side of the glass while mercury draws away from it?

18-3 In which of the three enlarged molecules are the forces unbalanced? How do these unbalanced forces affect the liquid?

Molecule at surface is under unequal tension

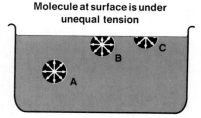

Capillary tubes

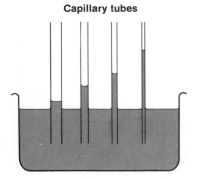

18-4 Why does the water rise higher as the tubes get thinner?

Mercury is depressed

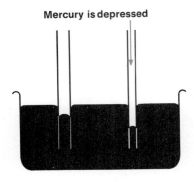

18-5 What force holds down the mercury inside the tubes? Why is the surface of the liquid curved down where the mercury touches the glass?

Mercury shows **this property** even more clearly. Mercury may pile up as much as 0.6 cm above the rim of a vessel before it spills over the edge.

Have you ever wondered why blotting paper soaks up ink, or a bath towel soaks up water? Perhaps you have seen water rise in glass tubes with a very small opening. It does this by *capillary action.* The word "capillary" comes from the Latin word for "hair." Capillary action takes place in fine hair-like tubes and is a form of adhesion.

If you take a glass tube with a small inside opening and dip it into water, how high will the water rise inside the tube?

The height of the water level in capillary tubes varies inversely with the tube's inside diameter. That is, the smaller the inside of the tube, the higher the water level will rise. See Fig. 18-4.

In some liquids, such as mercury, with stronger cohesive than adhesive forces, the reverse happens. Instead of rising inside the tube, the liquid is pulled down below the level of the liquid outside the tube. The surface tension is so great that the liquid is kept from rising very far up the tube. See Fig. 18-5.

Capillary action occurs in places other than in fine glass tubes. For instance, blotting paper and paper towels soak up water quite well. This is due partly to the tiny, hair-like passages that exist between the fibers of the paper. Capillary action also plays a small part in the growth of plants by lifting soil water up to the roots and by moving the sap up to the stem and leaves.

If soil is tightly packed, moisture will be brought to the surface and the ground will dry up. If the soil is tilled, so that it is crumbly and loose, the moisture will be conserved. The spaces between particles of loose soil are much larger than in firm soil. Therefore, capillary action does not occur in loose soil.

activity

Soaking sugar. Place a sugar cube into a container holding a small amount of ink. Pour some powdered sugar on top of the cube. What happens to the sugar cube? the powdered sugar? How is this activity like capillary action in tightly and loosely packed soils mentioned above?

pressure: force on a unit area

Solids exert pressure. It is much easier to cut meat with a sharp knife than with the side of a fork. When the same amount of force is applied in both cases, the **pressure** exerted by the knife is much greater than the pressure exerted by the fork. All of the force on the knife is focused on a small area of meat beneath the thin edge of the blade. When the same force is applied to the fork, it is spread out over a larger area of meat because the fork has a larger surface area touching the meat than the knife edge. *Force acting upon a unit area of surface is called pressure.* To measure the pressure against any surface, the force is divided by the area upon which the force acts. Study the sample problem.

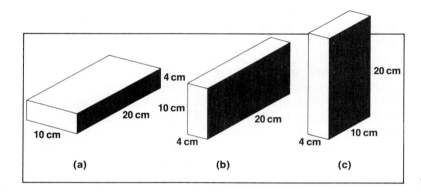

(a)   (b)   (c)

18-6

sample problem

A rectangular block measuring 20 cm by 10 cm by 4 cm with a weight of 40 N is placed on a table. Find the pressure exerted on the tabletop when the block is placed in each position A, B, and C shown in Fig. 18-6.

solution:

Case A. The area of the surface of the block in contact with the table is

$$10 \text{ cm} \times 20 \text{ cm} = 200 \text{ cm}^2$$

Since pressure is force acting on a unit area, substitute the values given.

$$\text{pressure} = \frac{\text{force}}{\text{area}} = \frac{40 \text{ N}}{200 \text{ cm}^2} = 0.2 \text{ N/cm}^2$$

(continued)

Case B. If the block is turned sideways, the area of contact is

$$20 \text{ cm} \times 4 \text{ cm} = 80 \text{ cm}^2$$

The pressure is then

$$\frac{40 \text{ N}}{80 \text{ cm}^2} = 0.5 \text{ N/cm}^2$$

Case C. The area of contact is

$$10 \text{ cm} \times 4 \text{ cm} = 40 \text{ cm}^2$$

Therefore, the pressure is

$$\frac{40 \text{ N}}{40 \text{ cm}^2} = 1.0 \text{ N/cm}^2$$

The problem shows that although the force against the tabletop remains the same in each position, the pressure is five times as great in position C as in position A. You can see that pressure can be increased by decreasing the area of contact, or by increasing the force.

Liquids exert pressure. Because it has weight, water pushes down on the bottom of a river bed or a water tank. The deeper the water, the greater the pressure. Many cities have water tanks to provide the needed pressure for their water supply systems. We will look at a method of finding the pressure of water at any depth in a tank. Suppose that a city has a standpipe 30 m high with a bottom area of 50 m². The tank holds (50 m²) × (30 m) = 1500 m³ of water.

The weight of fresh water is 10,000 N/m³ (or 10 kN/m³). Therefore, the water in the tank will weigh (1500 m³) × (10 kN/m³), or 15,000 kN. Of course, this weight or force is spread evenly over the entire bottom of the tank. Since the bottom has an area of 50 m², the pressure on the bottom is force/area, or 15,000 kN ÷ 50 m² = 300 kN/m². In other words, each square meter on the bottom of the 30-m-high tank has a force of 300 kN pressing upon it.

The pressure exerted by a liquid can be found in another way. The pressure is merely the product of its depth and its weight/volume. Since the tank in the example is a column of water 30 m deep,

$$\text{pressure} = \text{depth} \times \text{weight/m}^3$$
$$= 30 \text{ m} \times 10 \text{ kN/m}^3$$
$$= 300 \text{ kN/m}^2$$

What is the pressure in newtons per square centimeter (N/cm²) in the above problem? Remember that there are 10,000 cm² in 1 m².

A water tank is often put on top of a hill to give it more height. For the same reason, the tank is kept nearly full at all times.

The pressure at any given tap or faucet in the city depends on the difference between the height of the water in the standpipe and the height of the faucet. The lower a faucet is in a building, the greater is its water pressure. A faucet in the basement delivers water with more force than one on the top floor.

Pressure acts in all directions. Liquids push not only against the bottom of a tank but against the sides as well. See Fig. 18-7. For instance, if you pulled the plug out of a small hole in the side of a barrel filled with water, the water would spurt out because of pressure. The pressure at any given point on the side of the tank is equal to the downward pressure at that point. The sides of deep tanks are made stronger near the bottom because the pressure is greater there than near the top.

If you push an empty glass down into water, you feel an upward push of the water against the bottom of the glass. The deeper you push the glass, the greater is the upward force. This shows that the upward pressure at any depth equals the downward pressure.

Liquid pressure does not depend on the shape of the container. Figure 18-8 shows four various-shaped vessels filled with water and connected by a level pipe. The diagram shows the truth of the old saying, "Water seeks its own level." You will now see why this is so.

Liquids flow in a pipe because of a difference in pressure. The flow is always from a point of high pressure to one of low pressure. If the liquid is not flowing along the tube, the pressure throughout its length must be the same.

The height of the water in each vessel is the same. You already know that the pressure depends on the depth of the liquid. If the water level in one vessel were lower than in the others, the pressure in that vessel would also be less. Water would then flow from the other vessels through the pipe toward the low-pressure point until all levels were the same.

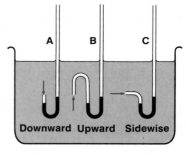

18-7 How does the upward and sidewise pressure of a liquid compare to the downward pressure at any given point?

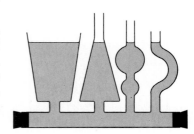

18-8 What is the relationship between the shape of the container and the pressure at the bottom of the container?

chapter 18   forces in solids, liquids, and gases   329

Before building a dam, engineers find the total force of water that will press against its wall. They begin by finding the pressure half-way down from the top. The pressure at this point is the *average pressure*. The average pressure multiplied by the area of the whole wall of the dam gives the total force exerted by the water. Dams and other structures are always built to resist a force several times greater than what is expected. Even so, dams sometimes break because of excess pressure or other causes.

Surprising as it may seem, the length of the lake backed by the dam has no effect on the force on the dam. The force is the same whether the lake is 100 km long or only 100 m long. The push against the dam depends only on the size of the dam and the depth of the water that touches the dam.

Pressure is multiplied in a confined liquid. An increase in depth increases the pressure of a liquid in an open vessel. Now consider the effect of extra pressure applied to a liquid in a closed space.

Figure 18-9 shows how easy it is to smash a liquid-filled glass jug by striking a sharp blow on its stopper. The principle that explains why the jug cracks so easily was found over 300 years ago by the French scientist, Blaise Pascal. *Pascal's law states that pressure applied to a confined fluid acts equally in all directions.*

If a pressure of 100 N/cm² is applied to the stopper of the confined liquid shown in Fig. 18-9, each square centimeter of the jug would be pushed out with an added force of 100 N. If the area of the surface of the jug is 1600 cm², a force of 100 N/cm² × 1600 cm², or 160 kN, pushes against its inner surface. This force may be enough to shatter the glass jug.

Hydraulic devices make use of Pascal's law. The word *hydraulic* refers to any device that works by putting pressure on a liquid, just as *pneumatic* (nu-*mat*-ik) devices are operated by a compressed gas. Hydraulic and pneumatic machines can multiply forces greatly.

A hydraulic press has two pistons. The pressure, or force acting on each square centimeter, is the same for both pistons. However, one piston is much larger than the other. Since *total force* is equal to the area times the pressure, the total force acting on the larger piston will be much greater than that acting on the smaller piston.

18-9 How does the pressure on the stopper affect the force against the sides of the bottle?

100 N/cm²

100 N/cm²

unit 3 motion, forces, and energy

The smaller piston shown in Fig. 18-10 has an area of 1 cm², and the larger piston has an area of 100 cm². A downward force of 1 N on the smaller piston would produce an upward force of 100 N on the larger piston (1 N/cm² × 100 cm² = 100 N). Now suppose a 100-N push is applied to the smaller piston. A force of 10,000 N would act against the larger piston (100 N/cm² × 100 cm² = 10,000 N). Without friction, this device could support a force of 10,000 N by exerting a force of only 100 N. What is the formula for ideal mechanical advantage (IMA) of this machine? Would the kind of liquid used (water, oil, or alcohol) make any difference in the IMA or in the actual mechanical advantage (AMA)?

Hydraulic machines are used in many industries where large forces are needed. Hydraulic presses are used for baling cotton, squeezing oils from seeds, shaping car bodies, punching holes in steel plates, and molding products from plastics. See Fig. 18-11. Other uses of Pascal's law include hydraulic brakes on cars and planes, automobile lifts used in service stations, and barber chairs.

Fluids exert a buoyant force. If a piece of cork is dropped into water, some upward force pushes it to the surface. *This upward force that fluids exert on objects immersed in them is called* **buoyancy.** Remember that by fluids, we mean both liquids and gases.

Area = 1 cm²   **Area = 100 cm²**
100 N
1 N

18-10 What is the relationship between the force applied to the small piston and the weight lifted by the larger piston?

18-11 How are hydraulic forces being used by the machine shown in this picture?

18-12 What is the relationship between the object in the water and the water in the catch bucket?

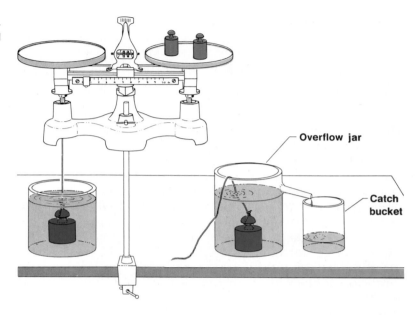

Overflow jar

Catch bucket

Maybe you have noticed that it is easier to lift a heavy object when it is underwater than when it is in the air. This is due to the upward buoyant force of the water.

You can measure this upward force by trying a simple test. First weigh an object on a scale. Then attach a string from the scale to the object. Lower the object into a beaker of water as shown in Fig. 18-12. Now weigh the object again while it is underwater. Suppose the object weighs 10 N in the air but only 7.5 N underwater. By subtracting, you can see that the apparent loss of weight is 2.5 N. This is equal to the buoyant force.

Next, carefully lower the object into an overflow jar that is full of water. Weigh the overflow of water in the catch bucket. If this test is done carefully, an amount of water weighing 2.5 N will overflow.

This experiment shows that *the apparent loss of weight of the object is equal to the weight of the liquid displaced.*

Similar tests were done by the Greek scientist Archimedes (ahr-kih-*mee*-deez) more than 2000 years ago. He found that an object immersed in a fluid is pushed up with a force that equals the weight of the displaced fluid. *Buoyancy is the force that pushes an object up and makes it seem to lose weight in a fluid.*

unit 3 motion, forces, and energy

Buoyancy causes some objects to float. Suppose the object in Fig. 18-12 were made of wood instead of metal. Suppose also that the wooden object weighed only 2 N instead of 10 N. Because the water can supply an upward force, in this case of up to 2.5 N, the object will float instead of sink.

If the density of the object is less than that of the fluid, the object floats; if the density of the object is greater than that of the fluid, the object sinks.

Equal-sized cubes of balsa, cork, maple, and ice will all float in water. See Fig. 18-13. Notice, however, that some cubes float higher in the water than others. The greater the density of the floating object, the more liquid it displaces, and the deeper it sinks into the liquid. A floating object sinks deep enough to displace a weight of liquid equal to its own weight, as shown by the following experiment.

Weigh a wooden block on a platform balance and then lower it into an overflow jar. This test is like the experiment shown in Fig. 18-12. Weigh the water that overflows into the catch bucket. Note that the weight of the block is equal to the weight of the water that overflows. Can you explain what will happen to a block if its density is the same as that of the liquid into which it is placed?

If you repeat the above test using a less dense liquid such as kerosene or alcohol, the block will sink deeper and cause more liquid to overflow. However, the weight of this displaced liquid still equals the weight of the block. Remember that since the liquid is less dense, it takes more of it to equal the weight of the block.

Buoyant forces are exerted by gases as well as by liquids. Since fluids include both gases and liquids, buoyant forces are exerted by air as well as water. For instance, an inflated balloon will rise in the air if the weight of the balloon is less than the weight of the air it displaces. The displaced air exerts a net upward force, causing the balloon to rise. For this reason, balloons are filled with light gases such as helium. Weights are used to hold the balloon down until it is ready for flight. When the weights are released, the balloon rises.

To return the balloon to earth, the gases inside

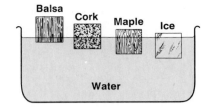

18-13 What is the relationship between the weight of each block and the weight of the water each displaces?

the balloon are released or compressed into a tank. This compression reduces the volume of the gas. The balloon decreases in size and displaces less air, so the buoyancy decreases.

A submarine works on the same principle. In order to dive, the submarine takes on seawater and becomes heavier. It rises by forcing out the water, becoming lighter. A ship floating in the water is another example of Archimedes' principle. If a ship weighs 40,000 kN, it displaces 40,000 kN of water.

Density can be expressed as specific gravity. It is often useful to compare the density of a solid or liquid with that of water. To make this comparison, scientists use the concept of **specific gravity.** Specific gravity is found by dividing the weight of an object by the weight of an equal volume of water. *The specific gravity of a substance is a number that compares the density of a substance with the density of water.*

Table 18-2 gives the specific gravities of some common substances.

Using the data given in the table, can you tell which substances will float in water? Which will sink in water? Give reasons for your answers.

The easiest way to find the specific gravity of a liquid is to use a floating device called a *hydrometer* (hy-*drom*-uh-ter). A hydrometer is a hollow glass tube weighted at the lower end so that it floats upright. A hydrometer will sink into a liquid until it displaces its own weight. Of course, a hydrometer sinks deeper in liquids of low density than in liquids of high density. The specific gravity is read directly from the scale on the glass tube. See Fig. 18-14.

Many industries, such as those making gasoline, salt, sugar, and soap, use specific gravity readings to control the quality of their products. A chemist may use it to help identify a substance or to learn how pure it is. A gas station attendant uses a hydrometer to check the antifreeze in the cooling system and the acid in the battery.

Gases exert pressure. We live at the bottom of an ocean of air made up of a mixture of gases. The air in an average classroom weighs several thousand new-

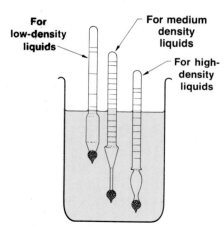

18-14 Which hydrometer is heaviest? Which has the greatest density?

For low-density liquids

For medium density liquids

For high-density liquids

Table 18-2:

Specific Gravities of Some Common Substances			
Butter	0.87	Ice	0.92
Copper	8.9	Iron, steel	7.79
Cork	0.25	Lead	11.3
Diamond	3.5	Mercury	13.6
Gasoline	0.7	Water	1.00
Gold	19.3	Wood (oak)	0.85
Human body (lungs full)	0.98-1.01	Wood (pine)	0.5

tons and all of the air around the earth weighs many billions of kilonewtons. This weight of air exerts a pressure of 10 N/cm^2 at sea level. The following experiment demonstrates the effects of air pressure.

A thin rubber balloon is tied over a glass bell jar. The bell jar has an opening at the bottom for a tube from a vacuum pump. As air is pumped out of the jar, the air from the outside is forced into the jar. The air pushes the rubber film farther and farther down into the glass. The pressure may be enough to break the balloon.

At first, the pressure of the air inside the glass jar is equal to the pressure of the air on the outside. Then, when the air inside the bell jar is removed, the inside pressure decreases. The outside pressure, however, remains unchanged. This outside air pressure pushes the balloon in. This principle, as shown in Fig. 18-15, explains how you can drink through a soda straw. In this case, you lower the air pressure inside the straw by sucking out the air. The outside air pressure then pushes the liquid in to fill the low-pressure area in the straw.

Air pressure is measured with an instrument called a barometer. An Italian scientist named Torricelli invented a *barometer* over 300 years ago.

Torricelli took a glass tube about 1 m long, closed one end, and filled it with mercury. Placing his thumb over the open end, he turned the tube over, dipping

Air pushes down on the liquid surface with one atmosphere pressure

Air pressure is reduced here

18-15 How does air pressure make it possible for you to drink through a straw?

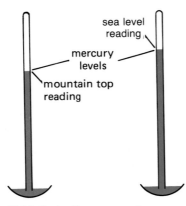

18-16 Torricelli compared a mercury barometer reading at sea level with a reading taken on a mountain top. Why was there a difference in the two readings?

its open end below the surface of mercury in a dish. When Torricelli took his thumb away, some mercury ran out, but a column about 760 mm high stayed in the tube. What kept the rest of the mercury in the tube?

During the 1600's people could do no better than guess that "nature abhors a vacuum." Yet, even this principle could not explain why nature did not abhor a vacuum above the 760-mm level in the tube. The mercury would rise no higher.

Torricelli looked at the problem in another way. Why didn't *all* of the mercury run out? He reasoned that it was the air pressure pushing on the mercury in the dish that supported the mercury in the tube. Otherwise, the rest of the mercury would surely run out of the tube. The mercury column exerts a pressure that is equal to the pressure of the air at a given place.

To test Torricelli's theory, Pascal took the mercury and tube to the top of a mountain. What do you suppose he found? The height of the column of mercury was lower on the mountain top than on the ground. Why? It was obvious that the air pressure was lower at that altitude. As a result, it could not support as high a column of mercury. See Fig. 18-16.

The air pressure varies from day to day at the same place. Pressure changes indicate changes in weather. The barometer, an instrument that measures air pressure, can be used to help predict weather.

Gases follow Boyle's law. How does the volume of a confined gas change when you change its pressure? Suppose a gas is placed in a cylinder with a movable piston, such as shown in Fig. 18-17. Call the initial volume of the gas V_1 and its pressure P_1. When the pressure on the piston is increased to a new pressure, P_2, its volume is decreased to a new value, V_2. If you kept increasing the pressure, you would get a set of readings such as those shown in Table 18-3. These readings are plotted on a graph in Fig. 18-18. **Boyle's law** gives the relationship between volume and pressure in a confined gas.

Robert Boyle was an English scientist who did many tests with confined gases. *Boyle's law states that for any gas, the pressure multiplied by its volume is a constant, as long as the temperature remains unchanged.*

18-17 The volume of a gas decreases as the pressure increases, providing the temperature remains the same. How does the increase in pressure affect the density of the gas?

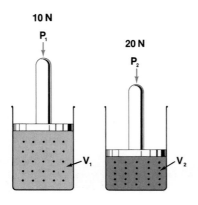

336   unit 3   motion, forces, and energy

Table 18-3:

Pressure-Volume Relationships		
Pressure P (N/cm^2)	Volume V (cm^3)	Pressure × Volume $P \times V$
15	12	180
30	6	180
45	4	180
60	3	180
90	2	180
180	1	180

The following formula expresses Boyle's law:

$$P_1 \times V_1 = P_2 \times V_2$$

If any three of the quantities in the formula are known, you can find the value of the fourth.

You know that when the pressure of a confined gas is increased, the gas molecules are crowded closer together, and the volume is decreased. If the pressure is doubled, for example, the volume of the gas is reduced to one-half of its former volume. What effect does the increase in pressure have on the density of the gas?

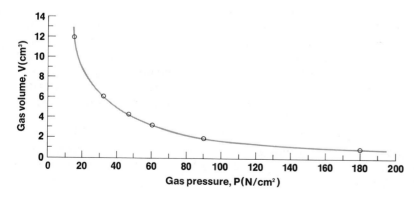

18-18 What is the relationship between volume and pressure?

chapter 18   forces in solids, liquids, and gases   337

sample problem

When 300 cm³ of a gas under a pressure of 15 N/cm² is compressed to a volume of 20 cm³, what will be the new pressure of the gas? Assume that there is no change in temperature.

solution:

Step 1. Given:

$$P_1 = 15 \text{ N/cm}^2 \quad P_2 = ?$$
$$V_1 = 300 \text{ cm}^3 \quad V_2 = 20 \text{ cm}^3$$

Step 2. From Boyle's law,

$$P_1 \times V_1 = P_2 \times V_2 \text{ or } P_2 = \frac{P_1 \times V_1}{V_2}$$

Step 3. Substituting,

$$P_2 = \frac{15 \text{ N/cm}^2 \times 300 \text{ cm}^3}{20 \text{ cm}^3}$$
$$= 225 \text{ N/cm}^2$$

(Note: Gases do not follow Boyle's law *exactly*, especially when they are almost cold enough to become liquid.)

Archimedes
Buoyancy and Bath Water

Archimedes (287 B.C.–212 B.C.), a Greek scientist, was trying to solve a problem for his king. The king had given a quantity of gold to a goldsmith to be made into a crown. When the crown was finished, the king felt that some of the gold had been stolen and replaced with copper. However, he did not know for sure. The crown *looked* like gold and it *weighed* the same as the gold given to the goldsmith. How could the king be sure? Archimedes was asked to help.

The answer came as a surprise. One day, while settling down into a full bathtub, Archimedes noticed that some water spilled over. Eureka! He had the answer! What had Archimedes learned? He realized that he could find the volume of the crown by lowering it into a full vessel of water and measuring the amount that spilled over. Archimedes knew that copper was only about half as dense as gold, so a given *weight* of copper would have twice the *volume* of gold. Thus, if copper had been used, it would displace *more* water than if an equal weight of gold had been used.

When the crown was lowered into water, it did indeed displace more water than an equal weight of gold. The goldsmith was a thief!

summary

1. Some substances are elastic; that is, they return to their original shape after being stretched.
2. Cohesion holds like molecules together.
3. Adhesion holds unlike molecules together.
4. Pressure is the force on a unit area.
5. Pascal's law states that pressure applied to a confined fluid acts equally in all directions.
6. Fluids exert a buoyant force on objects immersed in them.
7. Archimedes' principle states that an object immersed in a fluid is pushed up with a force that equals the weight of the displaced fluid.
8. Specific gravity of a substance is its density compared to the density of water.
9. Gases follow Boyle's law: $P_1 \times V_1 = P_2 \times V_2$

review

Match each item in the left column with the best response in the right column. *Do not write in this book.*

1. adhesion
2. cohesion
3. elastic
4. pressure
5. Pascal's law
6. Boyle's law
7. buoyancy
8. specific gravity

a. pressure times volume of a given gas is a constant
b. force acting on a unit area
c. Torricelli
d. upward force on objects in a fluid
e. force of attraction between unlike particles
f. pressure on a confined fluid acts equally in all directions
g. tendency of matter to return to its original shape
h. force of attraction between like particles
i. pneumatic
j. density of a substance compared to the density of water

questions

Group A
1. How resilient is steel compared with rubber?
2. What is elasticity? Is all matter elastic? Explain.
3. Both cohesion and adhesion are forces between particles. What is the difference between them?
4. Why does mercury not "wet" your finger?
5. How does the height to which water rises in a capillary tube vary with the diameter of the tube?
6. What happens to the level of mercury in a capillary tube?
7. Why does the tilling of topsoil tend to keep it from drying out?
8. Explain what is meant by the following statement: Liquid pressure does not depend on the shape of the container.
9. How could you show that there is a pressure against the side of a barrel filled with a liquid?
10. Why does a piece of iron float on mercury but sink in water?
11. What is an easy way of finding the specific gravity of a liquid?
12. If a solid substance has a specific gravity of 0.9, will it sink or float in water? Why?
13. If the pressure of a confined gas at a constant temperature is tripled, what will happen to the volume?

Group B
14. Explain why the bristles of a camel's-hair brush spread apart when dipped into water but come together when the brush is removed from the water.
15. Explain how kerosene is drawn up a wick in a kerosene lamp.
16. Why is it possible for a steel needle or paper clip to float on the surface of water, even though the density of steel is much greater than that of water?
17. Which exerts more pressure at the bottom of a tank, gasoline or water? Explain.
18. How can a small force on one piston of a hydraulic press produce a large force on the other piston?
19. Without tasting them, how could you find out which of two sugar solutions was more concentrated?
20. Does Archimedes' principle apply to both gases and liquids? Give examples.
21. Describe how you would find the specific gravity of lead.
22. Determine from Table 18-2 whether lead will float in mercury.
23. Explain why balloons can float at specific levels in air, while anything that sinks in water goes all the way down.
24. How can you "weigh" a ship too massive to be placed on a scale?
25. Why doesn't the mercury run out of the open lower end of a barometer tube?
26. Does Boyle's law apply to liquids as well as gases? Explain.

problems

1. A coil spring hanging from a hook stretches 3 cm when a 2-N weight is attached. (a) How much would a 1-N weight stretch the spring? (b) a 3-N weight? (c) What weight would be required to lengthen the spring a total of 7.5 cm? (d) What assumption did you make in finding your answers?
2. A block measuring 10 cm × 5 cm × 2 cm and weighing 100 N is placed on a table. Find the pressure exerted on the table when the block is placed on its base, its side, and its end.
3. One piston of a hydraulic press has an area of 1 cm^2. The other piston has an area of 25 cm^2. If a force of 150 N is applied on the smaller piston, what will be the total force on the larger piston?
4. When the deep-sea vessel *Trieste* reached a depth of 11,000 m, what was the water pressure in newtons per square centimeter?
5. A ship displaces 25,000 m^3 of seawater. What is its weight? (Note: The weight of seawater is 10 kN/m^3.)
6. A 30-kN tank weighs 54 kN when filled with kerosene and 60 kN when filled with the same volume of water. Compute the specific gravity of the kerosene.
7. A pine raft is 4 m long, 2 m wide, and 20 cm thick. How much of a load can this raft carry when placed on a lake? (Hint: Use Table 18-2.)
8. If the pressure on 1 L of a gas is decreased from 105 N/cm^2 to 15 N/cm^2, what will be its new volume? Assume that the temperature remains the same during the process.
9. What is the total amount of force exerted by air pressure on the floor of your classroom? Why doesn't the floor cave in?

further reading

Arnold, Ned, and Arnold, Lois, *The Great Science Magic Show.* New York: Franklin Watts, 1979.

Asimov, Isaac, *Building Blocks of the Universe.* New York: Abelard-Schuman, 1974.

Cobb, Vicki, *Gases.* New York: Franklin Watts, 1970.

Hutchins, Donald, ed., *Late Seventeenth Century Scientists.* New York: Pergamon Press, 1969.

Ruchlis, Hy, *Bathtub Physics.* New York: Harcourt Brace Jovanovich, 1967.

Schneider, Herman, and Schneider, N., *How Scientists Find Out: About Matter, Time, Space, and Energy.* New York: McGraw-Hill, 1976.

work, energy, and power

The concept of work was discussed briefly in Chapter 17 when you learned about the use of machines. You will recall that work is the product of force times distance. In this chapter, you will see how work is related to energy and power. You will learn that work cannot be done unless energy is applied, and that energy exists as kinetic or potential energy.

objectives:

☐ To explain and give examples of the concepts of kinetic and potential energy.

☐ To describe how potential energy can be changed to kinetic energy.

☐ To show how energy can be changed to forms such as chemical, electrical, and heat energy.

☐ To define work, energy, and power, and to show how they are related.

energy and work

The joule is the basic unit of energy in the metric system. In this chapter, you will learn about various forms of energy, all of which have the ability to do work. The metric unit that identifies the work done by energy is the **joule.** *A joule (J) is the work done when one newton (N) of force acts through a distance of one meter (m) or 1 J = 1 N-m.* In later chapters, you will see that the joule is also used to measure heat energy and electrical energy.

Kinetic energy is energy of motion. You know that a huge boulder rolling down a steep hill has a great deal of energy. A flying airplane, a speeding train, or even a rolling bowling ball all have a certain amount of energy.

How much energy do these moving objects have? What kind of energy do they have? These moving objects all have **kinetic energy,** *which is energy of motion.* The amount of energy they have depends on two things: (1) the mass of the object and (2) the speed of the object.

Suppose two cars are moving in a straight line down the road. If their speeds are the same, the car having the greater mass will have the greater amount of kinetic energy. If these cars were to strike an object, the larger car would cause more damage to the object than the smaller car.

Now suppose that two cars having the same mass are moving at different speeds. The car moving at the greater speed will have the greater amount of kinetic energy. The kinetic energy increases as the *square* of the speed. For example, at three times the speed, a car has nine times as much kinetic energy. When solving problems with kinetic energy, the term *velocity* is used instead of speed. As you learned in an earlier chapter, velocity is speed in a definite direction. Therefore, the kinetic energy of an object depends upon its mass (m) and also on the square of its speed (v^2).

Potential energy is stored energy. You have seen that when an object like a huge boulder is rolling down a hill, it has a large amount of energy. But did the boulder have any energy before it began rolling?

You also know that it takes energy to lift a box from the floor to a table top. Does the box have any energy as it rests on the table? What happens to the energy used to lift the box? That energy is stored by the box as **potential energy.** *Any energy stored and called upon for later use is called potential energy.* This potential energy can do work. The water held back by a dam, for example, can be used to produce electricity as the water falls through the pipes. As another ex-

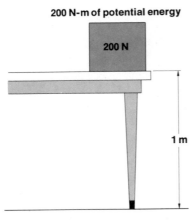

200 N·m of potential energy

200 N

1 m

19-1 The unit of energy is the joule. It is equal to the work done when a force of one newton acts through a distance of one meter.

ample, the wound-up mainspring of a clock uncoils and the stored or potential energy in the spring is released to work on the gears of the clock.

The potential energy of a body is equal to the work that it can do. In the case of a lifted object, the potential energy is equal to the work done on the body to lift it. See Fig. 19-1. You can find the work done to raise an object against the force of gravity by multiplying its weight (force) by the height (distance) it was lifted (work = weight × height).

Potential energy is stored energy, or energy that an object has because of its position or condition. Thus, a stick of dynamite has potential energy because it can do work when it explodes. A lump of coal has potential energy because it can give off heat when it burns. In other words, *chemical energy* is also a form of potential energy.

sample problem

What is the potential energy acquired by a 55-N block when it is placed on a ladder 4 m above the ground?

solution:

Step 1. Potential energy = weight × height

Step 2. Substituting given values in the above formula

Potential energy = 55 N × 4 m
= 220 N-m or 220 joules (J)

Energy can be changed from one form to another. The water behind a dam has potential energy. The potential energy changes to kinetic energy as the water gathers speed in its fall to the river below.

Suppose a student throws a ball straight up into the air. The ball goes up, stops, and comes back down. In this up-and-down path, there is a constant change of energy from kinetic to potential and back again to kinetic. The ball starts up with the kinetic energy given to it by the student's arm. Gravity reduces the ball's speed (and the kinetic energy) to zero as it reaches the highest point of the throw.

At this point, the ball has its greatest amount of potential energy. This energy, in turn, is changed back to kinetic energy as the ball falls back toward earth. The ball returns to the student's hand with almost the same speed and kinetic energy as it had when it was thrown. (Some of the ball's energy is changed into heat energy because of air resistance.)

Another example of the transfer of energy occurs in a swinging pendulum. Lift a pendulum bob to point A, as shown in Fig. 19-2. Because of its height, the bob now has its maximum amount of potential energy. When the bob is released, it falls to its lowest point at B. However, the bob does not stop there. It moves right past point B. Why? Although the bob has lost all of its potential energy at B, it now has gained its greatest amount of kinetic energy. The kinetic energy is enough to lift the bob to point C on the arc. By the time the bob reaches point C, all of its kinetic energy has been changed back into potential energy (assuming there are no friction losses).

These changes keep repeating as the bob swings back and forth. At the high points of the arc, the bob has its greatest amounts of potential energy but no kinetic energy. At the lowest point of the arc, the bob has its greatest amount of kinetic energy but no potential energy. Despite this constant transfer from one kind of energy to another, *the total amount of energy does not change*. The sums of the potential and kinetic energies are the same at any point of the arc.

Can you think of how the forms of energy change on a roller coaster? When is potential energy greatest? least? When is kinetic energy greatest? least?

Falling water provides energy. Have you ever seen water pouring over the top of a dam or a spillway? This water does no useful work. Only the water that goes through the power station does useful work. Why?

In the early years of human history, water was used to turn millstones, which ground grain into flour. Dams were built to hold back the water and make mill ponds. Water was led from the pond through a channel to a water wheel. The turning water wheel did the work needed to grind the grain. Today, falling water is used to make electricity.

Large volumes of falling water have large amounts

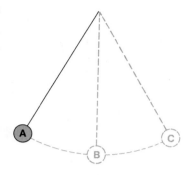

19-2 At what point in the swing of the pendulum is the kinetic energy greatest?

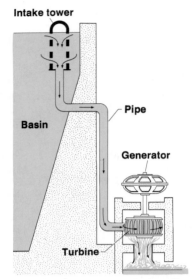

Intake tower

Pipe

Basin

Generator

Turbine

19-3 What kind of energy does the water have at the intake tower? at the turbine?

19-4 How does the size of the lake affect the potential energy available in the water behind the dam? How does the size of the lake affect the pressure against the face of the dam?

of energy. At Niagara Falls, for instance, part of the water goes over the falls, and part of it is channeled off to run the turbines to produce electric energy.

From points above the falls, some of the water is sent through a large channel to a basin located near the top and to the side of the falls. Next, the water drops about 60 m through large pipes to the water turbines as shown in Fig. 19-3. These turbines are modern versions of old-time water wheels.

The falling water in the pipe causes the turbine to spin. The shaft of the turbine is coupled to the rotor of an electric generator. Power obtained from falling water is called *hydroelectric power.*

Energy in the form of electricity is generated on both the American and Canadian sides of Niagara Falls. The cost of electric energy in this region is lower than in many other parts of the country because the waterfall is a natural one. Where there is no natural waterfall, people have had to build dams to create a source of water for hydroelectric power. The cost of the dam, therefore, increases the cost of the power that is produced.

The dam must be anchored on solid rock under the riverbed; otherwise, the water pressure at the base of the dam could cause water to seep under the dam and break through. See Fig. 19-4.

Energy cannot be made or destroyed. A moving car has kinetic energy. This energy comes from fuel that is burned in the car's engine. When fuel burns in the engine, chemical energy is changed into heat energy. The burning gases expand with great force against the pistons. In this way, heat energy is changed into mechanical energy, which moves the car. This change gives the car the kinetic energy of motion.

The **law of conservation of energy** *states that energy cannot be made or destroyed but can only be changed from one form to another.* In the case of a car, or falling water, or a ball thrown into the air, energy is not made or destroyed. It is only changed from one form to another.

This law did not seem to apply in the case of nuclear energy. There seemed to be too much energy coming from too little matter. But in 1905, Albert Einstein showed that matter itself can be converted into

unit 3 motion, forces, and energy

energy. This energy is released when atoms of matter are split or fused. See Chapter 6 for a fuller discussion of nuclear energy.

The matter–energy principle and the law of conservation of energy can be stated in another way. *The total amount of matter and energy in the universe does not change.* This far-reaching principle is one of the most useful in all of science. Many tests have confirmed this statement, and there are no known exceptions to this law.

work and power

There is no work without motion. Applying energy is not enough to produce work. The energy must cause something to move. For example, if you push hard against a box but it does not move, you have done no work! However, if you lift a 10-N weight a height of 1 m, you do 10 N-m of work. See Fig. 19-5. If you raise a 10-N weight a height of 2 m, you do 20 N-m (20 joules) of work. *Work is equal to force times the distance through which the force moves* as shown in the following equation:

$$W = f \times d$$

In this formula, W is the work done, f is the force applied, and d is the distance the object moves under the influence of this force. In addition, the force must be in the direction of motion. Work is not done by a force that acts at right angles to the direction of motion.

You have seen the formula for finding work. You may now have a better idea of the amount of work that can be done when potential energy is changed to kinetic energy. Think again of a lake being held back by a dam. When the water in the lake falls down through pipes to turn the turbine blades, it does a certain amount of work. The kinetic energy of falling water is changed to mechanical energy in the turbine. This mechanical energy, in turn, is changed to electrical energy. *Work always involves a transfer of energy.*

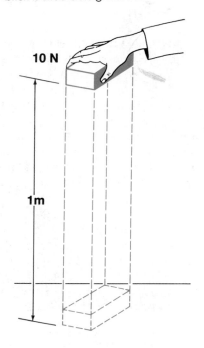

19-5 How much work is done if the brick is lifted a height of 1 m?

10 N

1m

19-6 How much work is done if 150 N of force are used to push the trunk 5 m? Does it make any difference how fast the trunk is pushed?

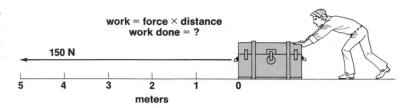

work = force × distance
work done = ?

150 N

5   4   3   2   1   0
meters

sample problem

A trunk weighing 500 N rests on the floor. How much work is done in pushing it a distance of 5 m using a force of 150 N? See Fig. 19-6.

solution:

Step 1. Remember that only the force acting in the same direction the trunk is moving is used in finding the work.

work = force × distance

Step 2. Substituting,

work = 150 N × 5 m
     = 750 N-m or 750 joules

What is the work done if the same trunk is lifted to a height of 5 m?

19-7 Which of the two diagrams represents more work being done? Which shows more power?

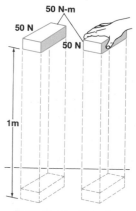

50 N-m

50 N

50 N

1m

Time=25 sec   Time=5 sec

Power is the rate of doing work. Knowing the amount of work done on an object does not tell you how long it took to do the job. What is the difference, for example, if a job is done in 20 sec or in 1 min? There is no difference in the amount of work done. However, there is a difference in the amount of **power** used.

The rate at which work is done is called power. The greater the amount of work to be done in a given time, the greater the power needed. When 50 N-m of work is done in 5 sec, the power is 10 N-m/sec (50 N-m ÷ 5 sec). If, on the other hand, the same work is done in 25 sec, the power needed is 2 N-m/sec. See Fig. 19-7.

Power depends on the following three factors: (1) the force applied, (2) the distance through which the force moves, and (3) the time during which the force is applied. The formula for power is

$$\text{power} = \frac{\text{work}}{\text{time}}$$

$$= \frac{f \times d}{t}$$

Go back to the sample problem on page 348 where you found the work done on a trunk moved across the floor. A force of 150 N moved the trunk 5 m. Now add the time factor. Suppose it took 5 sec to move the box. How much power was applied?

$$\text{power} = \frac{\text{work}}{\text{time}}$$

$$= \frac{750 \text{ N-m}}{5 \text{ sec}}$$

$$= 150 \text{ N-m/sec}$$

Horsepower and watts are units of power. You have probably heard the term **horsepower (HP)** used to rate the strength of engines. *One horsepower is 746 N-m (or 746 joules) of work being done per second.* An engine of 2 HP can do 1492 N-m (1492 joules) of work in 1 sec (2 × 746). How many newton-meters (joules) per *minute* are equal to 1 HP?

The units of power in the metric system are the *watt* (W), which is 1 N-m/sec, and the *kilowatt* (kW). One horsepower equals 746 W, or 0.746 kW (since 1 kW = 1000 W). It is useful to remember that one horsepower equals about 3/4 kW.

activity

As powerful as a horse. You can find your strength, in horsepower, by doing this activity. Select a set of stairs. With a stop watch, have another student measure the time in seconds it takes you to climb from the bottom to the top of the stairs. (You must begin with a standing start.) Make several trials and record the average time in your notebook. Measure and record the height climbed in meters. Also record your own weight in newtons. From these data, find your horsepower and compare it with the values obtained by other members of your class.

Jobs in Transportation

There is a wide range of jobs in the field of transportation. The most obvious job suggested by the title is that of driver. Drivers are needed for buses, taxicabs, and trucks, for either local or long-distance driving. For example, some truck drivers do local hauling such as delivering food to grocery stores, concrete to building sites, and trash to dumps. Other drivers are involved in long-distance hauling of furniture, steel, and other products across the country.

Railroads are another segment of the transportation industry. People are employed for on-train jobs such as conductors, engineers, and brake operators. Off-train jobs include track workers, station agents, and bridge and building workers. The railroad industry has been in a decline for many years, so jobs are limited.

The merchant marine is another area with limited job prospects. People are employed as sailors on ships, and clerks and dock workers off the ships.

The area of greatest growth in transportation is the airline industry. There is a much better-than-average growth in jobs such as pilots, flight attendants, flight engineers, and ticket agents.

The transportation industry is so vast that it offers job prospects on land, sea, and in the air.

summary

1. Kinetic energy depends on the mass of an object and the square of its velocity.
2. Potential energy is stored energy.
3. Energy can be changed from one form to another.
4. The law of conservation of energy states that energy cannot be made or destroyed.
5. Work is the result of a force acting through a distance.
6. Power is the rate of doing work.
7. Horsepower and watts are units of power.

review

Match each item in the left column with the best response in the right column. *Do not write in this book.*

1. work
2. power
3. horsepower
4. turbine
5. hydroelectric
6. potential energy
7. kinetic energy
8. law of conservation of energy

a. electric energy generated by falling water
b. the energy of motion
c. 746 N-m/sec
d. energy cannot be created or destroyed
e. pendulum
f. device turned by force of water
g. product of force and distance, with force parallel to distance
h. stored energy, or energy of position
i. rate of doing work
j. pipes that direct water to turbines

questions

Group A

1. Upon what factors does the kinetic energy of an object depend?
2. Which of the following two stones of equal weight has more potential energy: one on top of a mountain or one in a valley? Explain.
3. How can falling water do work?
4. Why are dams thicker at the bottom than at the top?
5. Arrange the following words in the order in which they are connected in a power plant: turbine, generator, basin, pipes.
6. How much work do you do if you push against a car but cannot move it? Explain.
7. Explain the difference between work and power.
8. What is measured by horsepower?

Group B

9. How is gasoline used to produce kinetic energy?
10. Describe how kinetic energy can be changed into potential energy.
11. Explain the kinetic and potential energy conditions of a baseball thrown straight up into the air. Describe conditions when the baseball (a) is released from the hand, (b) is at the peak of its flight, (c) returns to the hand.
12. Discuss the transfer of energy in the swing of a pendulum bob.
13. How do water turbines develop electric power?
14. List five forms of energy. How are they related?
15. Explain the meaning of the law of conservation of energy.
16. If a force acts on a body at a right angle to its motion, is any work done? Explain.

problems

1. How much work is done if you push a 200-N box across a floor with a force of 50 N for a distance of 20 m?
2. How much work is done if you lift a 20-N box up to a shelf 2 m high?
3. How much power is needed to carry a 100-N box up three flights of stairs with a vertical height of 12 m in 20 sec?
4. How much power is needed to push a box with a force of 200 N a distance of 10 m in 25 sec?
5. A 400-N student climbs a 3-m ladder in 4 sec. Find (a) the work done and (b) the power needed.
6. A force of 200 N is needed to drag a 1000-N box a distance of 25 m along a level roadway. (a) How much work is done? (b) If the box is lifted by a crane to a height of 25 m, how much work is done? (c) What is the kinetic energy at a height of 25 m? (d) What is the potential energy at 25 m?
7. Suppose you exerted a force of 60 N to push a 300-N box a distance of 12 m in 20 sec. You then lifted the box 1 m into a truck in 3 sec. (a) How much work did you do when you pushed the box? (b) when you lifted the box? (c) How much power did you exert when you pushed the box? (d) when you lifted the box?
8. If a 600-N student climbed a 4-m flight of stairs in 8 sec, what horsepower did that student generate?

further reading

Harmon, Margaret, *The Engineering Medicine Man: The New Pioneer*. Philadelphia: Westminster Press, 1975.

Hinkelbein, Al, *Energy and Power*. New York: Franklin Watts, 1971.

Hoffman, Banesh, and Dukas, Helen, *Albert Einstein: Creator and Rebel*. New York: Viking, 1972.

Hogben, Lancelot, *Maps, Mirrors, and Mechanics*. New York: St. Martin's Press, 1974.

Keen, Martin, *How It Works*. New York: Grosset & Dunlap, 1976.

Schneider, Herman, and Schneider, N., *How Scientists Find Out: About Matter, Time, Space, and Energy*. New York: McGraw-Hill, 1976.

Webb, Robert, *James Watt: Inventor of the Steam Engine*. New York: Franklin Watts, 1970.

Weiss, Harvey, *How to Be an Inventor*. Minneapolis: Thomas Y. Crowell, 1980.

heat energy

20

You have seen that matter can change from one form to another. Solids can change into liquids, liquids can change into gases, and one compound can change into another. When such changes occur, energy is either absorbed or given off. Usually, this energy is in the form of **heat.** *Heat is the energy possessed by a substance because of the motion of its molecules.*

objectives:

☐ To distinguish between heat and temperature.

☐ To describe how heat and temperature are measured.

☐ To demonstrate that matter expands when heated and contracts when cooled.

☐ To explain how heat is absorbed or given off during a change of phase.

☐ To compare the transfer of heat by conduction, convection, and radiation.

heat and temperature

Heat and temperature are not the same. The terms *heat* and *temperature* are often used in everyday life to mean the same thing. However, they are not the same. They have different meanings that you should be aware of. They will be used in several examples in this chapter.

One way of showing the difference in the terms is to compare a candle flame with a large tank of warm

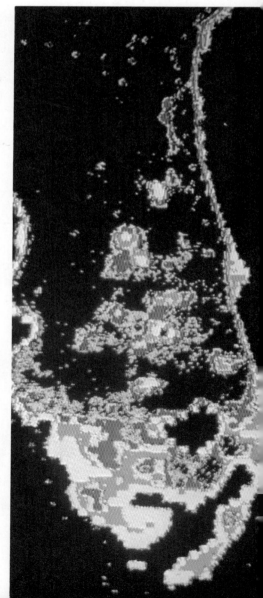

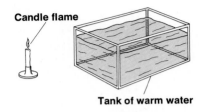

Candle flame

Tank of warm water

20-1 Does the candle flame or the tank of water have the higher temperature? Which one has more heat?

water. You can put your hand into one, but not the other. Why? The answer is obvious. The flame is too hot. It has a very high temperature.

Now answer another question. Which of the two examples has more heat energy, the candle flame or the tank of water? The answer is the tank of water. Do you know why? See Fig. 20-1.

In order to understand this answer, you must know what happens to molecules as a substance gets hotter or colder. When a substance is cold, its molecules vibrate slowly. When heated, however, the molecules speed up (the kinetic energy of the molecules increases). The hotter a substance becomes, the faster its molecules vibrate. **Temperature** *can be defined as the measure of the average motion (average kinetic energy) of the molecules.*

The faster the molecules vibrate, the higher the temperature. In the candle flame, the molecules vibrate very fast, so the flame has a high temperature.

Heat, on the other hand, depends not only on the

activity

Heat vs temperature. You will need two styrofoam coffee cups, two thermometers, a pan, a heat source, a large bundle of nails, and a single nail. Place the bundle of nails and the single nail into a pan of water and heat the water until it boils. Fill each of the two cups about two-thirds full of cool water. Measure the temperature of the water and record. Using tongs, put the bundle of nails into one cup and the single nail into the other cup. Put a thermometer into each cup and record the readings every 30 sec for 5 min.

Questions:
1. What was the temperature of the water in each of the two cups before the nails were added?
2. What was the temperature of the nails when they were taken from the pan? (Hint: You can tell even without using a thermometer.)
3. What was the change in temperature in the cup after adding the bundle of nails? the single nail?
4. Since the nails all had the same temperature when put into the cups, how do you account for the temperature changes?

average motion of the molecules, but also on the *number* of molecules involved. In summary, the amount of heat a substance has depends on the following two factors: (1) the average kinetic energy of the molecules (the temperature) and (2) the number of molecules present (the mass of the substance).

As a result, the candle flame has a much higher temperature than the tank of water; but the water, because there is so much of it, has more heat than the flame. Can you see why 100 g of water has ten times as much heat energy as 10 g of water at the same temperature?

Temperature is measured in degrees. A day in summer may be said to be hot and a day in winter cold, but these terms are not very exact. A *thermometer* gives a more exact reading of how hot or cold it is. It measures the temperature in *degrees*.

The *Celsius* (*Cel*-see-us) (C) scale is used throughout the world to measure temperature. On the Celsius temperature scale, water freezes at 0° and boils at 100°. Normal body temperature is 37°C, and a comfortable room is about 22°C. See Fig. 20-2.

Does "zero" on the scale refer to the lowest point that temperature can reach? No, it refers to the point where pure water freezes. In winter, the temperature goes well below zero in many parts of the country. Matter can get much colder than zero. As matter gets colder, the motion of its molecules slows down. The point at which the molecules stop moving is called *absolute zero*. Absolute zero is 273° below zero on the Celsius scale (−273°C).

Scientists have been able to study matter cooled down very close to absolute zero. They have found that at these temperatures, matter reacts in some surprising ways. For instance, liquid helium becomes a *superfluid*. That is, it becomes so adhesive that it creeps up the sides of its container. Some poor conductors of electric current become *superconductors* near absolute zero. A superconductor is a substance with little or no resistance to electric current.

At temperatures well above absolute zero, but still far below freezing, some common substances react strangely. Lead normally gives off a dull "thud" when struck, but when very cold, it rings like a bell. Mercury,

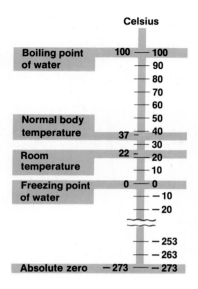

20-2 How many degrees are there between the boiling and freezing points of water?

20-3 This rubber ball has been cooled to near −273°C. What causes it to shatter when it is dropped?

normally a liquid, is a solid when it is very cold. Likewise, many gases change to liquids and then to solids when they are very cold. Rubber, which is normally soft and pliable, becomes hard and brittle when it is cooled to near absolute zero. See Fig. 20-3.

A second metric scale can also be used to measure temperature. It is the *Kelvin* (K) scale. The Kelvin scale is the same as the Celsius scale except that it has a different zero point. Its zero point is at absolute zero (−273°C). The freezing point of water is 273°K (0°C), and its boiling point is 373°K (100°C). In other words, the Kelvin scale is 273° higher than the Celsius scale.

Heat is measured in calories and joules. Heat cannot be measured with a meter stick or weighed on a balance. It can only be measured in an indirect way. Heat is measured by its effect on matter. Water is most often used as "standard matter" because it is so common.

The amount of heat required to raise the temperature of one gram of water one Celsius degree is defined as the **calorie.** Another unit, the *joule* (J), which is also used to measure work, can be used to measure heat. One calorie is equal to 4.185 joules.

Table 20-1:

Energy Values of Some Common Foods		
Food	100 grams (equal to)	kcal
Hamburger, cooked	1 medium	360
Tuna fish, canned	4/5 cup	200
French fries	20 pieces	390
Candy, chocolate	2 medium bars	500
Peanuts	3/4 cup	560
Milk, whole	3/8 cup	70
Bread, white	4 thin slices	280
Eggs, whole	2 medium	160
Apple, fresh	1 small	60
Tomato, fresh	1 small	20

Sometimes a larger unit called the *kilocalorie* (kcal), which is equal to 1000 calories, is used. A kilocalorie is a more useful unit to use to measure large numbers of calories such as the energy content of foods. The "calories" that you count when you are on a diet are actually kilocalories. A small baked potato of 100 grams, for example, has an energy value of 100 "calories" (actually 100 kcal). Table 20-1 lists the energy content of some other common foods.

Suppose you heat 50 g of water from 20°C to 90°C. If 1 cal raises 1 g of water 1 C°, then 50 cal are needed to raise 50 g of water 1 C°. To warm up all 50 g by 70 C° would take 50 × 70 = 3500 cal, or 3.5 kcal.

Users of coal need to know not only the price of coal, but also the number of calories of heat provided by the coal. Coal that supplies 8000 cal/g will give 6% more heat than coal with only 7500 cal/g.

Each substance has its own specific heat. Have you ever taken a frozen dinner from the oven and noticed that the foil cover is not hot? Why?

The answer has to do with how much heat it took to warm up the foil and food. Different substances need different amounts of heat to warm them up by the same number of degrees. *The number of calories needed to warm up 1 g of any substance by 1°C is called its* **specific heat.**

It takes one calorie of heat to warm up 1 g of water by 1 C°. Thus, the specific heat of water is an even 1.00 cal/g C°. Aluminum, with a specific heat of 0.21 cal/g C°, needs only 0.21 cal to warm it by 1 C°.

sample problem

How much heat is needed to warm up 750 g of iron from 10°C to 130°C?

solution:

Step 1. In Table 20-2, the specific heat of iron is listed as 0.11 cal/g C°. The temperature change is 130°C − 10°C = 120 C°.

Step 2. heat = (specific heat) × (mass) × (temperature change)
= 0.11 cal/g C° × 750 g × 120 C°
= 9900 cal

Table 20-2:

Specific Heats of Some Common Substances	
(cal/g C°)	
Aluminum	0.21
Brass	0.09
Copper	0.09
Ice	0.50
Iron	0.11
Lead	0.03
Steam	0.48
Water	1.00
Wood	0.42
Zinc	0.09

As a result, the thin foil warms up, or cools down, quickly compared to the moist food in a TV dinner. Table 20-2 gives the specific heats of some common substances.

Note that the specific heat of water is the highest of those substances listed in Table 20-2. In fact, water has one of the highest specific heats of all matter. Liquid ammonia (NH_3), with a specific heat of 1.2, is one of the few substances with a higher value than water.

Study Fig. 20-4 to see what happens when equal masses of some common metals are placed on a block of ice. They melt down to different levels in the ice depending on their specific heat values. (Be sure to remember that equal masses do not mean equal sizes or volumes.)

The specific heat of a substance is an important physical property because it tells if a substance can be used for a special purpose. For instance, aluminum pots and pans hold approximately twice as much heat as those made of equal weights of iron. You can see this by studying Table 20-2.

The high specific heat of water helps to explain why a large body of water is likely to have a milder climate than a land mass. The seasonal buildup of heat causes only a slow gradual increase in temperature in the body of water. Dry land, on the other hand, with a much lower specific heat, reaches a much higher temperature given the same amount of heat. Furthermore, soil is a poor conductor, keeping the heat from going deeply into the ground. Therefore, the heat causes a quick rise in temperature on land. Con-

20-4 These blocks have equal masses. Why are they different sizes? When heated to the same temperature, why do they melt the ice to different depths?

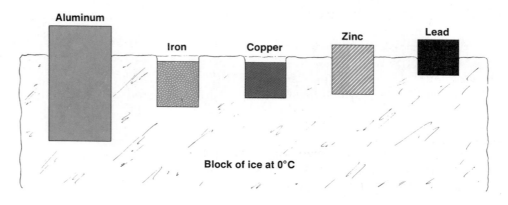

Aluminum   Iron   Copper   Zinc   Lead

Block of ice at 0°C

versely, land areas cool off much faster than do bodies of water.

some effects of heat

Heat causes matter to expand. Almost all gases, liquids, and solids expand when heated and shrink when cooled. The expansion is caused by the action of molecules. When heated, the molecules vibrate more rapidly and move farther apart. On the other hand, when matter is cooled, the molecules slow down and move closer together. This slowdown in the motion of the molecules reduces the overall volume of the substance, and the material shrinks.

You can see how gas expands by doing a simple experiment. Set up a flask and U-tube as shown in Fig. 20-5. Pour a little colored water into the bend of the tube. When you place your hand on the flask for 1–2 min, the air in the flask is slightly warmed and expands. As the air expands, it pushes the water toward the open end of the U-tube. If you now cool the flask, the liquid will move toward the flask. It is clear that air expands when heated and contracts when cooled. Other gases will act in the same way as air.

Liquids also expand when heated. Fill a flask with colored water and insert a rubber stopper and glass tube, as shown in Fig. 20-6. Mark the height of the colored water in the tube. Heat the flask by putting it into a pan of hot water. Observe that the colored water rises in the tube as it warms up. Water expands when it is heated. How is the size of the glass flask affected by the increase in warmth? Do you think that the flask expands more or less than the water? The difference in the rate of expansion of glass and mercury accounts for the way glass thermometers work. Mercury expands and contracts much more than glass. Therefore, mercury moves up and down the glass tube of a thermometer in response to changes in temperature.

The effect of heat on the size of solids can be shown by a ball-and-ring expansion device. See Fig. 20-7. At room temperature, the metal ball will just

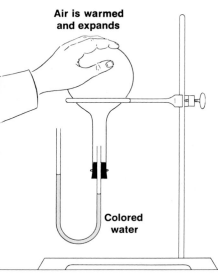

Air is warmed and expands

Colored water

20-5 What causes the colored water to move when you place your hand on top of the flask?

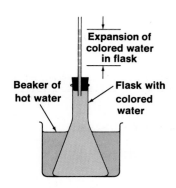

Expansion of colored water in flask

Beaker of hot water

Flask with colored water

20-6 In what direction will the water move in the glass tube when the water is heated?

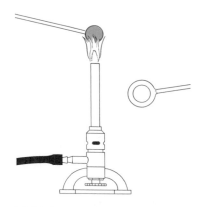

20-7 When the ball is heated, it no longer fits through the ring. Why? What happens if both the ball and ring are heated?

Table 20-3:

Coefficient of Linear Expansion	
(per Celsius degree)	
Pyrex	0.000003
Glass	0.000009
Iron	0.000011
Steel	0.000013
Copper	0.000017
Brass	0.000019
Aluminum	0.000023
Lead	0.000029

fit through the metal ring. When the ball alone is heated, it expands until it is too large to fit through the ring. If the ring is also heated, it too expands and the ball can again fit through the ring. You can see that heating the metal has caused it to expand.

All substances do not expand by the same amount. Studies show that gases expand more than liquids and liquids expand more than solids. In fact, for a given increase in temperature, gases expand more than 10 times as much as water and about 100 times as much as steel.

Different liquids expand by varying amounts for a given temperature change. Each solid also has its own rate of expansion that is different from other solids. However, *all gases expand and contract by the same amount for a given temperature change.*

Substances expand at different rates. When the temperature of a liquid is raised 1 C°, the actual increase in volume is a small fraction of the original volume. This fraction is called the *coefficient* (koh-eh-*fish*-ent) *of volume expansion.*

A 10° rise in temperature will produce ten times as much expansion as a rise of 1°. One hundred milliliters of a liquid will expand 100 times as much as 1 mL of a liquid. Thus, if you know the coefficient of volume expansion of a substance, the original volume, and the change in temperature, you can find the change in volume.

The increase is noticeable for large amounts of liquids. For example, a delivery truck holding 6000 L of gasoline at 32°C emptied its tank at a service station. Many hours later, when the gasoline had cooled to 10°C, it was found that the volume had shrunk by more than 80 L.

Although substances expand in all directions when heated, engineers are often concerned only with changes that occur in the length of solids. An increase in length is called *linear expansion.* The increase per unit length when a solid is warmed by 1 C° is called the *coefficient of linear expansion.* The rates of expansion for some common solids are shown in Table 20-3.

You can see from Table 20-3 that steel expands when heated. A long steel bridge can expand and contract by as much as 1 m between the hottest and coldest days of the year.

How do engineers solve the problem caused by expansion and contraction? One way of allowing for such expansion is to use finger-like joints at many points along the bridge roadway. The joints fit loosely into each other and slide back and forth as the bridge warms and cools. The next time you go over a long bridge, try to notice these joints. See Fig. 20-8.

sample problem _____

The steel center span of the Verrazano-Narrows Bridge, the world's longest suspension bridge, is 1420 m long. If the temperature changes 50 C° from winter to summer, how much will this bridge expand?

solution:

Step 1. From Table 20-3, the coefficient of linear expansion for steel is 0.000013/C°.

Step 2. expansion = (coefficient) × (length) × (temperature change)
= 0.000013/C° × 1420 m × 50 C°
= 0.9 m

Have you ever cracked a cold drinking glass by pouring hot water into it? The glass cracks because of uneven expansion of its inside and outside surfaces. Glass is a poor conductor of heat. As the inside surface is heated and expands, the outside surface is still cool and unchanged. The difference in expansion between the inner and outer surfaces causes the glass to crack. If the glass is thin, there is less chance that it will crack. Why?

Pyrex-type glass has become popular for lab use and for baking dishes in the home. This type of glass expands only one third as much as common glass. See Table 20-3. With the smaller rate of expansion, there is less strain when this type of glass is heated or cooled. Pyrex resists the strains caused by sudden heating and cooling.

Devices called *thermostats* control the burning of fuel in many furnaces. Many of these devices contain

20-8 Bridges undergo a great deal of expansion and shrinkage during the summer and winter months. How do engineers prevent changes in length from destroying the bridge?

Normal   **Iron**

Before heating   **Brass**

Hot   **Iron**

After heating   **Brass**

20-9 Does iron or brass expand more when heated? How can you tell?

a small bar made of two layers of metal. The two metals expand and contract at different rates when the temperature changes.

The bimetal bar shown in Fig. 20-9 is made of a brass strip riveted to a strip of iron. When it is heated, the brass expands more than the iron, causing the bar to bend.

In a thermostat, the strips touch a contact point if the temperature gets too cool. This contact switches on an electric current, starting the electric motor or opening the valve that controls the fuel supply to the furnace. The increased heat from the furnace causes the bar to bend away from the contact point, stopping the current. Thus, the temperature of your home stays within a few degrees of any thermostat setting.

Water expands when it freezes. Water, like almost any other substance, will shrink as it is cooled. However, unlike other substances, water is at its densest point at 4°C. Any further cooling causes it to *expand* instead of contract.

This behavior of water between 4°C and 0°C is of great value to all of us. Because of this odd property, lakes freeze on top instead of on the bottom. Otherwise, if the coldest water sank to the bottom, lakes would freeze from the bottom up. In the summer, warm water would stay near the top of the lake and not reach the ice. As a result, much of the ice would not melt. In time, all lakes and oceans would be mostly ice.

Fortunately, water expands near the freezing point. When cooled below 4°C, water becomes less dense and rises to the surface. When it freezes into ice, it expands still more. The volume of ice is about 1.1 times that of the water from which it was formed. In other words, the ice is less dense than the water, so it floats. The ice floating on the surface serves as a blanket to protect the rest of the water from freezing quickly.

Water close to the surface of an ice-covered lake is at 0°C. Water at the bottom of the lake is at 4°C. Water is at its greatest density at 4°C and thus stays at the bottom. If you have ever fished through ice in winter, you know that fish are found mostly at

20-10 Does warm water rise? Why is it warmer at the bottom of the lake than at the top?

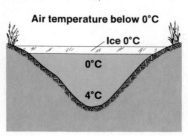

Air temperature below 0°C

Ice 0°C

0°C

4°C

the bottom of the lake where it is slightly warmer. See Fig. 20-10.

Water is an important *exception* to the rule that a substance contracts when it changes from a liquid to a solid.

heat transfer

Heat is exchanged during phase changes. Heat will cause a change of phase. A solid will change to a liquid and a liquid will change to a gas. For example, water will change to steam when heated.

Heat is always gained or lost in such a change of phase. Melting ice gains heat as it changes from ice to water, and water gains heat when it changes into steam. When the process is reversed, heat is lost. The same amount of heat is lost when steam condenses back to water and when water freezes into ice.

It takes 80 cal of heat to change 1 g of ice at 0°C to water at 0°C. Note that the added heat does *not* change the temperature. The heat is used just to change the ice to water. The heat that is absorbed in the melting process is called the *heat of fusion*.

Ice has a high heat of fusion compared with other

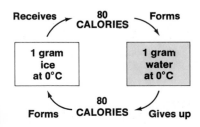

20-11 In this diagram showing heat of fusion, how much heat is needed to change 100 g of ice at 0°C to water at 0°C?

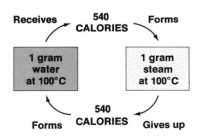

20-12 In this diagram showing heat of vaporization, what happens when water is changed to steam?

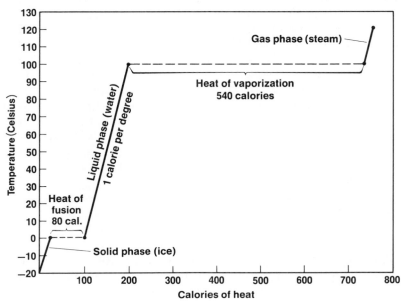

20-13 What causes the sharp increase in calories of heat at 0°C and again at 100°C, even though there are no increases in temperature at those points?

substances. The food in an ice chest stays cool because each gram of ice that melts absorbs 80 cal from its surroundings. See Fig. 20-11 on the previous page.

Water at 100°C absorbs 540 cal of heat for every gram that changes to steam. The heat needed to change a liquid into a gas is called the *heat of vaporization*. It takes 100 cal to raise 1 g of water from 0°C to 100°C. However, to change this hot water into steam, more than five times as much heat is needed (540 cal/g). See Fig. 20-12 on the previous page. Water has a very high heat of vaporization compared to other liquids. For example, the value for ammonia is 341 cal/g, and for mercury it is only 68 cal/g.

When steam condenses into hot water, the heat of vaporization is set free. Can you see why steam radiators are used to heat buildings?

Figure 20-13, on the previous page, shows how water changes from a solid to a liquid to a gas as heat is added to it. The dotted lines show that the temperature stays constant until enough heat has been added to bring about the change in phase.

A change of phase can have a cooling effect. The heat of vaporization of a substance can be used to produce a cooling effect. The cooling effect is similar to what happens to your hand if you wet it and then let it dry in a breeze. Your hand feels cool as the water dries. Heat is needed to evaporate the water. The heat comes from your hand, leaving it feeling cool.

Now suppose a scientist places some liquid ammonia in a tube under high pressure. If the liquid ammonia passes through a small opening into a low-pressure space, it will change to a gas. As ammonia changes to a vapor, it absorbs heat from its surroundings, causing a cooling effect. This is the process used by cold storage plants to make ice and to cool foods.

Study Fig. 20-14 to see how ammonia gas is used in ice-making. Ammonia gas takes heat from the brine (salt water) tank. The ammonia gas then goes to a pump where a piston compresses it back into a liquid. This change from a gas to a liquid gives off heat. Water running over the coils absorbs this heat. The liquid ammonia is then ready to pass through the valve and repeat the process.

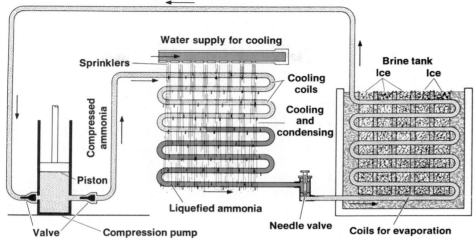

Ammonia gas returning to compressor

Water supply for cooling

Sprinklers

Cooling coils

Cooling and condensing

Brine tank

Ice   Ice

Compressed ammonia

Piston

Liquefied ammonia

Valve   Compression pump

Needle valve   Coils for evaporation

20-14 Can you trace the path of the ammonia through the pipes of an ice-making unit? Where are the hottest and coldest points in the system?

Ammonia has been largely replaced in modern electric refrigerators, air-conditioning units, and freezers by another coolant called *Freon*. Freon is an easily liquified gas. Air, instead of water, is often used to cool the coils. An electric motor drives the compressor pump, while a fan blows the air over the coils.

Would you be able to cool off the kitchen on a hot summer day by keeping the door of an electric freezer open? Can you explain your answer?

Heat travels in three ways. When two places have the same temperature, there is no movement of heat between them. When there is a temperature difference, however, heat moves to equalize the difference. Heat can move from one place to another in three ways: by **conduction, convection,** or **radiation.**

Conduction is the transfer of heat through a substance by direct molecular contact. Silver conducts heat very well. If you stir a hot liquid with a silver spoon, the handle of the spoon becomes hot almost at once. Heat from the liquid is transferred to the metal. Inside the metal the heat moves from molecule to molecule until the entire spoon is hot. On the other hand, if you stir the hot liquid with a plastic spoon, the plastic handle stays cool because plastic is a poor

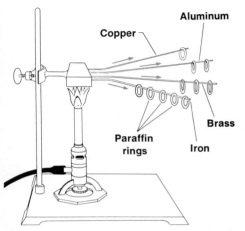

20-15 Metals differ in their ability to conduct heat and therefore melt the paraffin rings. Which metal is the best conductor of heat? Which is the poorest?

20-16 Water transmits heat very well by convection but poorly by conduction. Why hasn't the ice melted in this test tube? CAUTION: The test tube should not be pointed at anyone. Steam could explode from the tube if it is heated too rapidly.

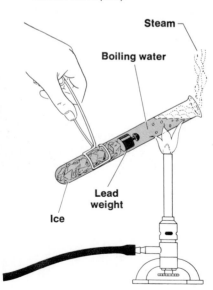

conductor. The ability of several common metals to conduct heat is shown in Fig. 20-15. Most metals conduct heat much better than nonmetals.

Solids are better conductors of heat than liquids, and liquids are better conductors than gases. This is true because molecules of most solids are closer together than the molecules of liquids. The molecules of liquids, in turn, are much closer together than those of gases. See Fig. 20-16.

Substances that are poor conductors of heat are called *insulators*. Cork, sawdust, and snow have enclosed "dead air" between their particles and thus are good insulators. Air that does not circulate is a good insulator. Clothes keep you warm in winter mostly because of the air in the spaces between the fibers.

During the last few years, much progress has been made in home insulation. Many fluffy substances, as well as boards of soft texture, are now widely used as heat insulators. When installed in the outside walls of a building and above ceilings, they keep heat inside the house in winter and outside in the summer. Good insulation saves fuel and increases comfort.

Convection is the transfer of heat by the movement of liquids and gases (fluids). Earlier in this chapter you learned that when a substance is heated, it expands. This expansion makes the substance less dense. Expansion causes gases and liquids to rise when heated. Conversely, cooler liquids and gases sink. Currents of liquids and gases that are set up by this unequal heating are called *convection currents*. Is convection likely to take place in solids?

Convection in a liquid can be shown by heating water in a beaker. Set the beaker so that just one side of it is over a burner. Drop a few crystals of dye or a few drops of food coloring on the surface of the water close to the side that is being heated. As the dye dissolves, you can see the movement of the colored water in the beaker. This movement shows how convection currents are formed in a heated liquid. See Fig. 20-17.

A common example of a convection current is the rise of heated gases up a chimney. This rise creates a "draft" of fresh air through the firebox. Some furnaces and heaters also transfer heat by convection.

unit 3 motion, forces, and energy

Most modern furnaces use blowers, however, rather than depending on convection currents to heat a house.

In a hot-water heating system, water moves through the pipes by convection. Heat is given up to the cold room, and the cooler, denser water flows back to the furnace, where it is reheated. In some systems, a pump is used to force the water through the radiators.

In some homes, the heating and cooling systems are combined into a single unit. A *heat pump* is used to bring in heat from outside the house during the winter months and to remove heat from inside during the summer months. It may come as a surprise to learn that there is heat even in "cold" air. You saw an example of the cooling process in Fig. 20-14. This same process also gives off heat. With a heat pump, this heat is used instead of discarded.

Radiation is the transfer of energy in waves through space. If you hold your hand below a lighted electric bulb, you can feel the heat. Since your hand is below the lamp, it is obvious that the heat does not reach your hand by convection. Why not? Conduction will not explain the warmth either, since air is a poor conductor of heat. Your hand is warmed by radiation, a third method of energy transfer.

Radiant energy is best transferred in a vacuum. For example, the earth receives heat from the sun by radiation. The other forms of heat transfer, conduction and convection, cannot take place in a vacuum.

Some homes are heated by radiation. In a radiant-heating system, steam or hot water is pumped through pipes in the floor, walls, or ceilings. These warmed surfaces radiate heat into the room.

Dark-colored, rough surfaces are good absorbers of radiant heat. Light-colored, shiny surfaces reflect more radiant heat than they absorb. Can you see why light-colored clothes are cooler in hot weather than dark clothes? Thermos bottles have a mirror-like coating that reflects the radiant heat from hot liquids back into the bottle. This helps to keep the liquid hot.

Have you ever opened a car door to find that the inside of the car was very hot? Where did the heat come from? Short-wave energy from the sun goes through the car windows, warming the surfaces inside. The longer-wave heat reflected from the surfaces does not go through the glass and is trapped inside.

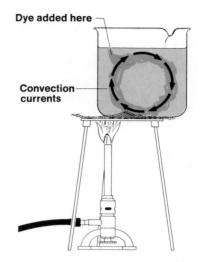

Dye added here

Convection currents

20-17 What would happen to the direction of the convection current if the source of heat were moved to the right side of the beaker?

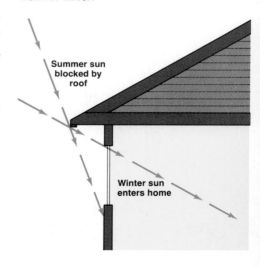

20-18 How does the greenhouse effect help to keep this home warm in winter?

Summer sun blocked by roof

Winter sun enters home

20-19 The cloud cover of Venus.

This process is sometimes used to help heat homes in winter. See Fig. 20-18 on the previous page.

Greenhouses are built with large amounts of glass to make use of the sun's heat. This heating effect is called the *greenhouse effect*. The greenhouse effect is a vital factor in warming the surface of the planet Venus. Some of the radiant energy from the sun goes through the thick cloud cover, warming the surface of the planet. The long-wave radiation from the surface, however, cannot get back out through the clouds. Therefore, Venus is very hot. A Soviet spacecraft found the temperature on Venus to be 420°C. See Fig. 20-19.

activity

Ice-keeping contest. Divide your class into three or more teams. Give each group a tin can with an ice cube in it and one of several kinds of insulating materials. Have each group find a way to keep its ice cube from melting, using only the insulating material. All teams should keep all melted water (and not absorb it in the insulation). Therefore, the winning team can be found at the end of the period by pouring the melted water into test tubes. What factors give the best results? How does this activity relate to keeping things cold in a refrigerator or hot in an oven? How does it relate to keeping your house cool in the summer and warm in the winter?

Benjamin Thompson (Count Rumford)

Benjamin Thompson (1753–1814) was born in the American colonies and as a teenager fought in the Revolutionary War on the British side. Thus, as the war came to an end, he thought it wise to move to Europe.

Thompson had a brilliant career in Europe and was granted the title of Count Rumford. Among his many tasks, he supervised the boring of cannon barrels.

The boring machine had a set of metal cutters attached by wheels and gears to a long boom. Several horses hitched to the boom plodded a circular path hour after hour as the cutters slowly bored a hole through the cannon barrel.

The cutting action produced a large amount of heat. The accepted theory to explain the heat was that it "flowed" from the iron barrel as it was being cut. That is, the metal itself contained the heat, but it was not released until the metal was cut.

Count Rumford found that this theory did not work. He noted that heat was produced even when the cutters were too dull to cut! Therefore, the heat could not have come from the metal itself. Where did the heat come from? Count Rumford concluded that the heat came not from the barrel, but from the horses! The horses did work to turn the cutters, which produced heat by friction against the metal.

summary

1. Temperature is the average kinetic energy of molecules.
2. Heat is the total kinetic energy of all the molecules of an object and is measured in calories or joules.
3. Each substance has its own specific heat.
4. Heat causes matter to expand in both length and volume.
5. When water is cooled from 4°C to 0°C, it expands.
6. Heat is absorbed and given off during melting and freezing.
7. Heat can be transferred by conduction, convection, and radiation.

review

Match each item in the left column with the best response in the right column. *Do not write in this book.*

1. Kelvin
2. temperature
3. heat
4. calorie
5. Celsius
6. heat of fusion
7. heat of vaporization
8. conduction
9. convection
10. radiation

a. energy of a substance owing to its molecular motion
b. transfer of heat energy through space
c. heat absorbed by a solid when it melts
d. heat capacity of a substance
e. heat absorbed by a liquid in changing to a gas
f. thermometer scale with a freezing point of 273°
g. thermometer scale with a freezing point of 0°
h. heat needed to warm 1 g of water 1 C°
i. increase in length owing to increase in warmth
j. transfer of heat by movement of fluids
k. measure of the average kinetic energy of molecules
l. transfer of heat from molecule to molecule

questions

Group A

1. How is heat energy related to the action of molecules?
2. What is the difference between temperature and heat?
3. Describe how you could check the accuracy of a Celsius thermometer.
4. Does a thermometer in a kettle of hot water measure the amount of heat contained in the water? Explain.
5. What is meant by absolute zero?
6. Does water become hotter if it is boiled more vigorously? Explain.
7. Why is the specific heat of a substance an important property?
8. Name two ways in which the expansion of gases differs from the expansion of liquids and solids.
9. When a railroad is built during cold weather, why is a space left at the joints between the steel rail sections?
10. What provision is made in the building of concrete highways to allow for expansion and contraction?
11. Why do solids expand when they become warmer?
12. As water is cooled from 8°C to 0°C, what happens to its density?
13. Why does a piece of ice float in water?
14. Why does a car standing in the sun get hotter inside than outside?

Group B

15. The specific heat of aluminum is 0.21. What does this statement mean?
16. Why do oceanic islands have a fairly even climate?
17. Which way will a bimetal bar of copper and steel bend if it is placed in a freezing mixture of water and ice? Explain.
18. Does the coefficient of linear expansion depend on the unit of length used? Explain.
19. A platinum wire could be tightly sealed into the end of a heat-softened glass rod. A copper wire, however, would form a loose seal with the glass. Can you suggest a reason why this is so?
20. Why does a cold, thick drinking glass often break when boiling water is poured into it?
21. The air above the ice in a pond is −10°C. What is the likely temperature of (a) the upper surface of the ice, (b) the water just beneath the ice, and (c) the water at the bottom of the pond?
22. Why does steam at 100°C cause more severe burns than hot water at 100°C?
23. Explain why a block of ice in a warm room takes a long time to melt.
24. Why would a hot drink in a metal cup be more likely to burn your mouth than one in a china cup?
25. Why is shiny aluminum foil sometimes used in the walls and ceilings of a building? Explain.

problems

1. How much heat could be obtained by burning 6 kg of coal if a test sample shows that the coal releases 8000 cal/g?
2. How much energy is needed to heat 1050 g of lead from 25°C to 285°C if the specific heat of lead is 0.03 cal/g C°?
3. Which takes more energy: to heat 40 g of aluminum from 40°C to 120°C, or to heat 60 g of zinc from 40°C to 140°C? How much more? What are the values in joules? in calories?
4. Check Table 20-1, Energy Values of Some Common Foods, and find the values in kilocalories of the following amounts of food: (a) 100 g milk; (b) 1 medium egg; (c) 100 g chocolate candy; and (d) 3/8 cup of peanuts.
5. A steel rod is exactly 39 cm long at 0°C. How much longer will it be at 25°C if the coefficient of linear expansion for steel is 0.000013/C°?
6. An oil tank holds 500 L at 30°C. What will be the volume of the oil when it cools to 10°C if the coefficient of volume expansion for oil is 0.001/C°?
7. The tallest building in the world is about 400 m high. How much taller is that building in the summer than in the winter? Assume a temperature difference of 50°C and base the answer on the expansion of steel.
8. Find the number of calories of heat needed to change 40 g of water at 20°C to steam at 100°C.
9. Find the number of calories of heat needed to change 10 g of ice at 0°C to water at 100°C.
10. A chip of ice weighs 90 g. How many calories will it absorb when it melts? If the water freezes again, how many calories of heat will be released?

further reading

Balestrino, P., *Hot As an Ice Cube.* New York: Thomas Y. Crowell, 1971.

Cobb, Vicki, *Heat.* New York: Franklin Watts, 1973.

Experiments: Properties of Gas and Heat Energy. Educational Services, American Gas Association, 1515 Wilson Blvd., Arlington, Va. 22209, 1980.

Froman, R., *Hot and Cold and in Between.* New York: Grosset & Dunlap, 1971.

Stone, A. Harris, and Siegel, Bertram, *The Heat's On.* Englewood Cliffs, N.J.: Prentice-Hall, 1970.

Wade, Harlan, *Heat.* Raintree Pubs. Ltd., 1977.

21

engines

The simple machines that you have already studied were used for hundreds of years. People or animals were needed to supply the input force in these machines. Later, water and wind supplied the input force for a limited number of uses. It was not until people found a way to use engines that it was possible to expand greatly the amount of useful work that could be done. In this chapter, you will study the major kinds of engines that run modern machines.

objectives:

☐ To describe the effects of engines on the world.

☐ To distinguish between rotary and reciprocal engines.

☐ To distinguish between internal and external combustion engines.

☐ To compare the four-stroke engine with the diesel.

☐ To describe how jet and rocket engines work.

power for a modern world

Engines help provide food, clothing, and shelter. How do engines affect your life in the modern world? Think of what you did today. It is likely that engines made possible the food you ate this morning, the clothes you are wearing, the home you live in, and the method of travel that you used to get to school.

How do engines improve the food supply? The answer becomes clear when you think of what it is like to grow crops *without* engines. What is needed just to plant a garden? First, the soil has to be turned with a shovel and the lumps broken up with a hoe. Then the soil has to be worked with a rake until it is free of lumps. How many people do all this work with a shovel, hoe, and rake? Not many. Most people find that the use of a tractor and plow or a power tiller makes this job much easier. With the help of engine power, the garden can be made much larger than if all this work were done by muscle power alone.

Farmers have an even greater need of engines. They use large tractors and power equipment to plow and till the soil. Engines are used to keep crops weed-free and to add fertilizer. Then engines are used to cut and harvest the crops, and finally to carry them to market. See Fig. 21-1.

Even at this point, most of the food is still not ready to be used in your home. The food must be processed into a form that is easier to use. For instance, most people eat bread rather than raw wheat or rye. Engines supply the power used to grind the grain into flour, to mix the dough, to bake the bread, and to wrap it. Then the loaves of bread are trucked to stores for consumers to buy. Without engines, there would be less food for everyone.

Now think of the clothes you are wearing. Did someone gather the raw material and weave the cloth by hand? Not likely. Very few people wear clothes made entirely by hand. Engines were used to gather raw materials, to process them into usable cloth, and to cut and sew the cloth into dresses, pants, sweaters, and shirts. Finally, engines were used again to transport the clothes to nearby stores.

How are engines used in your homes? Turn a handle and water comes from a faucet. What brings this water to your home? The water is pumped and treated using engine power. Flick a switch and the lights go on in your room. Here again, engine power is needed to make the electric current that is sent to most homes. What about the walls and framework of your home? The walls are most likely large sheets of "dry-wall"

21-1 Farmers use engines to help them grow and harvest crops. How are engines being used in this photo?

nailed to a framework of lumber. Both the dry-wall sheets and the lumber are made by using engine power.

Can you think of other ways in which engines affect your life?

The era of power began with the steam engine. As stated earlier, human and animal muscle power were used to do work hundreds of years ago. However, wind and water power also had some limited uses. But it was not until the steam engine was invented that chemical energy (from burning fuel) was used to provide the input forces needed to run large machines.

James Watt, an eighteenth century Scottish engineer, is sometimes given credit for the invention of the steam engine. Even though steam engines had been

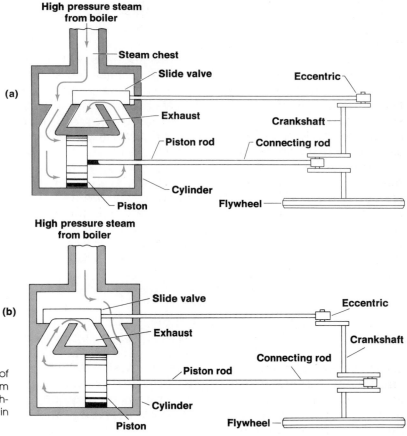

21-2 Two views of the inside of Watt's simple slide-valve steam engine. Can you see how the high-pressure steam pushes the piston in (a) and (b)?

unit 3 motion, forces, and energy

built earlier, Watt improved the early, wasteful models and made them more practical. His invention marked the beginning of the machine age.

Have you seen steam push up the cover on a tea kettle? The steam engine works on the same principle. Instead of pushing up against a cover, the steam pushes against a sliding piston within a cylinder. The piston moves first one way and then the other. See Fig. 21-2. The back-and-forth movement of the piston identifies it as a *reciprocating* (re-*sip*-row-kay-ting) engine.

In contrast, steam *turbines* rotate with a steady, constant motion. They work smoothly, with almost no vibration, and waste less energy than the reciprocating engine. Turbines are more efficient and more compact; they take up less space for a given amount of power. All large electric power plants and large ships use turbines rather than reciprocating engines.

The steam turbine works like a high-speed windmill, as shown in Fig. 21-3. Nozzles direct the steam at an angle against blades much as air moves over a pinwheel. Many electric generators are run by steam turbines that have a power output great enough to take care of the needs of a city of 50,000 people. Yet, such a turbine can fit into an average classroom. The steam for modern turbines may have a pressure as high as 1300 N/cm^2 and be as hot as 600°C.

Steam turbines work well only at high speeds. In making electric power, the turbine shaft is joined directly to the shaft of a high-speed electric generator. In a ship, however, where propeller speeds are much slower, the turbine is attached to the propeller by gears that reduce the speed. On some ships and trains, the turbine runs an electric generator. Electric motors are used to drive the ship or train.

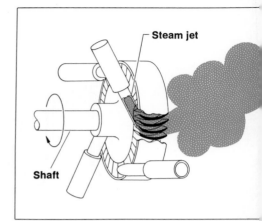

21-3 How is the steam turbine different from the reciprocating engine? Which one has fewer parts?

power for cars

Gasoline engines and diesels are internal combustion engines. Steam engines and steam turbines are classified as *external combustion engines*. That is, the fuel burns under a boiler located outside the engine itself. Gasoline and diesel engines are two examples of *internal*

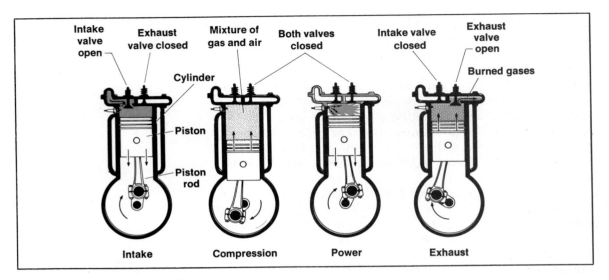

Intake valve open · Exhaust valve closed · Cylinder · Mixture of gas and air · Both valves closed · Intake valve closed · Exhaust valve open · Burned gases · Piston · Piston rod

Intake · Compression · Power · Exhaust

21-4 How many times does the cylinder go down and up in a four-stroke cycle?

combustion engines. In such engines, the fuel burns inside the cylinders.

The most common gasoline engine is the *four-stroke* engine. This term does not refer to the number of cylinders, but to the number of steps in the engine's operation needed to complete one full cycle. Engines of this type are used widely in cars.

Figure 21-4 shows the steps in the operation of the four-stroke engine. The first stroke is the *intake*. The piston slides down during the intake stroke, lowering the pressure in the upper part of the cylinder. A mixture of air and gasoline vapor is drawn into the cylinder from the carburetor through the open intake valve. The exhaust valve is closed at this time.

In some engines, a fuel-injection system is used instead of a carburetor to force a measured amount of fuel into the cylinder at the proper time.

The *compression stroke*, during which the piston travels upward, follows the intake stroke. At this stage, both the intake and the exhaust valves are closed. As the piston moves up, it squeezes or compresses the air-gasoline mixture, thus heating it up.

The *power stroke* takes place next. Just before the piston reaches the top of the stroke, an electric spark jumps across the gap of the spark plug to explode the air-gasoline mixture. The hot gases force the piston

unit 3 motion, forces, and energy

down in the cylinder. *This power stroke is the only stroke during which the engine does work.*

During the fourth stroke, the piston travels upward again, pushing the burned gases out through the open exhaust valve. Therefore, it is called the *exhaust stroke.*

Since power is provided during only one of the four strokes, there must be some way of keeping the piston moving during the other three strokes. This is done in several ways. One way is by connecting a heavy *flywheel* to the crankshaft of the engine. The flywheel's inertia keeps the piston moving during the other strokes. Another way is to connect several cylinders to the same crankshaft.

While the intake stroke takes place in one cylinder, the exhaust stroke is taking place in another, and the crankshaft is kept turning by the power stroke in still another cylinder. See Fig. 21-5.

If car engines had only one cylinder, the car would not ride smoothly. The earliest car engines had only

21-5 The strokes in a four-cylinder engine. Can you explain what is happening in each cylinder?

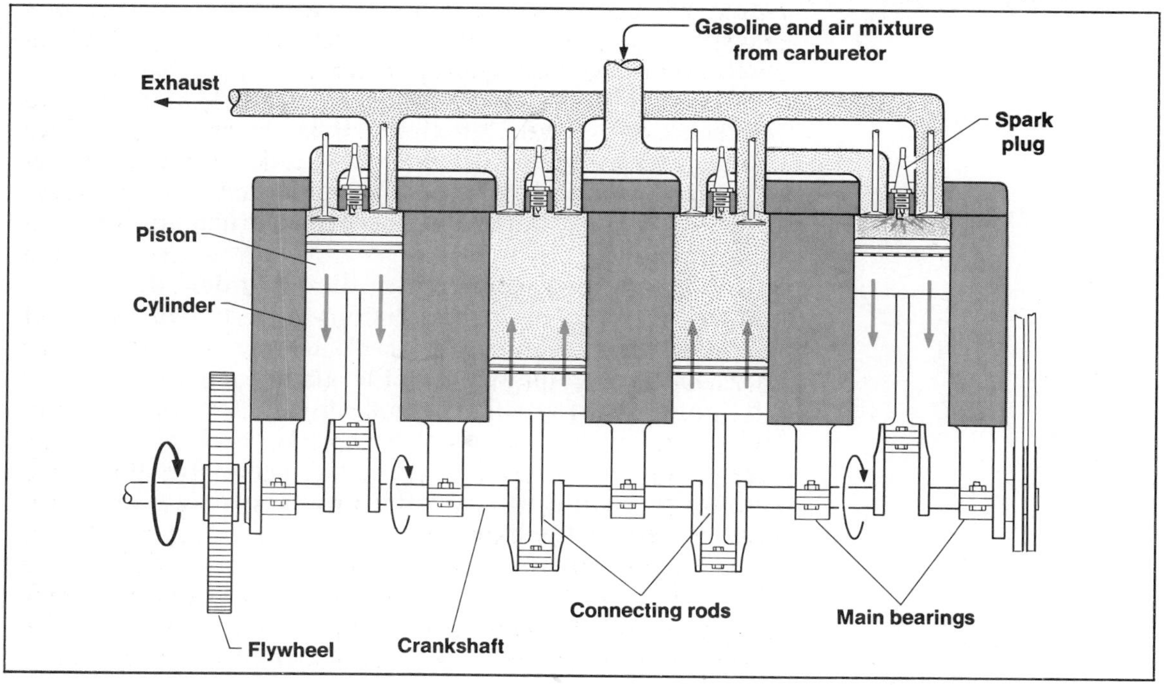

Gasoline and air mixture from carburetor

Exhaust

Spark plug

Piston

Cylinder

Flywheel

Crankshaft

Connecting rods

Main bearings

one cylinder, and each time that it fired, the whole car shook. The more cylinders an engine has, the smoother it runs. For a long time, most cars were made with eight cylinders to give the most power as well as a smooth ride. Now, however, most cars have four or six cylinders to provide better fuel economy.

Gas engines have strengths and weaknesses. The best gasoline engines change about 20–25% of the energy of their fuel into useful work. The cooling system, the exhaust gases, and friction use up the rest of the energy. By contrast, early steam engines used only 5–8% of their fuel's energy. Large high-speed steam turbines used in modern power plants are about 30–40% efficient.

A gas engine can be started or stopped on a moment's notice. A steam engine, however, cannot be started until the steam pressure has built up. Gasoline engines also develop a great deal of power in a fairly small space.

Gasoline engines also have some weaknesses. For instance, four-stroke engines can run only in one direction. They cannot be made to run backwards. This weakness is solved by having a *reverse gear*. Reciprocating steam engines can be run either forward or backwards simply by shifting a sliding valve.

Diesels are more efficient than gasoline engines. There are three main differences between diesels and gasoline engines. Diesels have much higher compression ratios, burn a kerosene-like fuel, and use no spark plugs.

In the diesel, air enters the cylinder during the intake stroke. The rising piston then squeezes the air to about one eighteenth of its former volume. When it is so highly compressed, the air becomes hot enough to ignite the fuel. Therefore, at the top of the compression stroke, an *injector* sprays fuel oil into the cylinder. The fuel explodes at once. The hot gases push the piston down in the power stroke. This stroke is followed by the exhaust stroke.

The **compression ratio** of a diesel engine may be as high as 18 to 1. In contrast, the compression ratio of the gasoline engine of a car is commonly about 8 to 1. *The compression ratio of an engine is the ratio of* *the volume within the cylinder when the piston is at the bottom of its stroke to the volume within the cylinder when the piston is at the top of the stroke.* See Fig. 21-6.

In recent years, car engines have been built with lower compression ratios because they tend to reduce air pollution. However, the lower ratios are not without fault. Lower ratios use more fuel, making the energy shortage worse.

Diesel fuel is much like kerosene. It provides somewhat more energy per unit of volume and is somewhat cheaper than gasoline. Finally, diesels are more efficient than gasoline engines, largely because of their higher compression ratios. The efficiency of diesel engines can be as high as 40%.

With all these assets, why are diesels not used more widely in cars? There are a number of reasons. Diesels are hard to start when cold, tend to be noisy, and need heavy engine castings to withstand the high pressures formed inside the engine. These problems are costly to overcome. As a result, diesels are used in large units like tractors or trucks, where the cost of the engine is only a small part of the total cost. In cars, however, the diesel engine is a greater part of the total cost. Nevertheless, as the cost of fuel goes higher and higher, the fuel-saving diesel becomes more and more attractive for use in cars.

Diesels are commonly used in submarines, tugboats, fishing vessels, and even small ocean liners. Large trucks, buses, and construction equipment use diesel power. Small plants that generate electricity also operate on diesels. On diesel-electric trains, the diesel engine runs an electric generator. The electric current, in turn, is used to power the train.

Power is sent to the car's wheels. Look back at Fig. 21-5 and notice that each piston is attached to the *crankshaft.* Thus, if the piston moves, the crankshaft must also move. However, there is an important change in the motion. The pistons move up and down, but the crankshaft turns in a smooth rotary motion. This change in motion is similar to the change in motion that takes place when you ride your bike. Your knees and upper legs move up and down, but the pedals go in a circle. In the engine, the pistons act like your upper legs, the connecting rods like your lower legs,

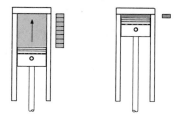

Automobile engine 8 to 1

Diesel engine 18 to 1

21-6 A diesel engine has a compression ratio of 18 to 1. What are some advantages of the diesel's high compression ratio?

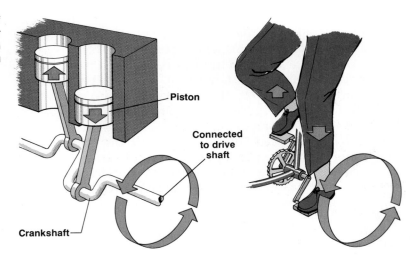

21-7 The up-and-down motion of the pistons is changed to circular motion. The same change in motion occurs when you ride a bicycle.

Piston

Connected to drive shaft

Crankshaft

and the crankshaft like the pedals. See Fig. 21-7.

The cylinders do not all fire at the same time. Instead, they are timed so that first one, then another, then a third fires, and so on. A steady series of power strokes is sent to the crankshaft. The crankshaft is connected through transmission gears to a drive shaft, which delivers power to the wheels through a set of differential gears.

Figure 21-8 shows a car with the engine in the front and the drive wheels in the back. Many cars have front-wheel drive, thus doing away with the long drive shaft down the middle of the car.

The transmission permits the driver to shift gears to get the best *gear ratio*, or mechanical advantage. In a car with manual transmission, the driver depresses a *clutch* to shift gears. With an automatic transmission, the driver does not need to shift gears; it is done automatically. In low gear, the drive shaft turns much more slowly than the engine. This supplies a large torque to the wheels for starting and for steep climbs.

The drive shaft is attached to the rear axles through a *differential* (dif-er-*en*-shal). The differential is a set of gears that permits the rear wheels to turn at different speeds. Without the differential, a car could not make a turn without "slipping," because when a car turns, the outside wheels must turn faster than the inner ones.

unit 3   motion, forces, and energy

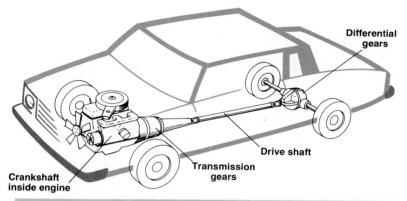

21-8 In a rear-wheel drive car, the power from the engine goes to the drive wheels through transmission gears, drive shaft, and differential gears.

Differential gears

Drive shaft

Transmission gears

Crankshaft inside engine

engines for air and space flight

Aircraft use jet engines. Although gasoline engines are used in small aircraft, almost all large aircraft are powered by jet engines. Jet engines all work on the action-reaction principle defined in Newton's third law of motion.

The force that blows the high-speed gases out of the rear of the engine represents the "action." The "reaction" is the force directed back on the engine by the escaping gas. This "reaction" is the thrust, or force, that drives the plane forward. The jet plane uses a large amount of fuel, but it produces a thrust that can push a plane faster than the speed of sound.

The *ramjet* is the simplest of all jet engines. See Fig. 21-9(a). It is used only after another engine has given the aircraft a high speed. It cannot function unless air is rammed into the intake. The ramjet has no moving parts and can deliver great amounts of power at high efficiency when moving at high speed.

The *turbojet* has a long shaft through its center. Air-compressor blades are mounted on the front end and turbine blades in the rear, as shown in Fig. 21-9(b). The purpose of the compressor is to pull in air and force it into the combustion chamber. Here a fuel, somewhat like kerosene, is sprayed in and ignited.

The hot exhaust gases striking the turbine blades make the turbine turn at high speed. Then the hot

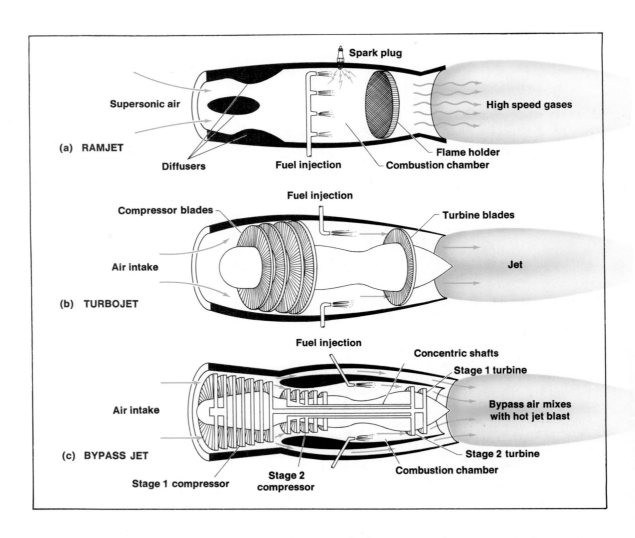

(a) RAMJET

Spark plug

Supersonic air

High speed gases

Diffusers

Fuel injection

Flame holder

Combustion chamber

Fuel injection

Compressor blades

Turbine blades

Air intake

Jet

(b) TURBOJET

Fuel injection

Concentric shafts

Stage 1 turbine

Air intake

Bypass air mixes
with hot jet blast

(c) BYPASS JET

Stage 2 turbine

Combustion chamber

Stage 1 compressor

Stage 2
compressor

21-9 In what way are the three types of jet engine alike?

gases rush out of the jet at the rear of the engine, giving the engine its forward thrust. Notice that the spinning turbine is connected by a shaft to the compressor at the front.

Another type of jet is the *bypass jet,* or *fanjet.* Note in Fig. 21-9(c) that a large stream of cool air bypasses the combustion chamber and goes directly into the exhaust gases.

The noise of jet engines comes mostly from the high-speed exhaust gases. In the fanjet, the noise is reduced by the bypass exhaust layer, which acts as a buffer between the exhaust gases and ordinary air.

　　　　　　　unit 3　motion, forces, and energy

Also, the fanjet is more efficient at low speeds than other types of jet engines.

Some jet engines, mostly on high-speed military aircraft, also have *afterburners*. The afterburner is a small, modified ramjet that is attached to the tail pipe of a turbojet. You may recall that the ramjet works only when high-speed gases enter the front of its combustion chamber. When a ramjet is attached to the tail pipe of a turbojet, the exhaust gases of the turbojet supply the high-speed gases that make the ramjet efficient. See Fig. 21-10.

Rockets are space engines. Rocket engines, like jets, operate on the action-reaction principle. They differ, however, in that the rocket engine carries its own *oxidizer*, while the jet does not. An oxidizer reacts with the fuel to supply energy. It may or may not contain oxygen. Therefore, unlike the jet, which needs oxygen from the air, the rocket engine can operate outside the earth's atmosphere.

Of all types of engines, rockets produce by far the greatest thrust for a given amount of weight. Rocket engines burn either solid or liquid fuels. See Fig. 21-11.

Solid-fuel rocket engines work much like Fourth-of-July rockets. The solid fuel and oxidizer are formed to act like a combustion chamber.

One of the major assets of solid fuel is that it can be stored in the rocket. The rocket is always ready for a quick launching. Liquid-fueled rockets, on the other hand, must have the fuel pumped in just before launch.

The size of some rocket engines is truly amazing. The first stage of the giant Saturn V rocket ship, which sent the first astronauts to the moon, was 11 m across and 45 m high. It carried 2,000,000 kg of liquid

21-10 What type of engine is used to power modern military and commercial aircraft?

21-11 In what way is the rocket engine an example of Newton's third law of motion? Why can a rocket engine operate in space?

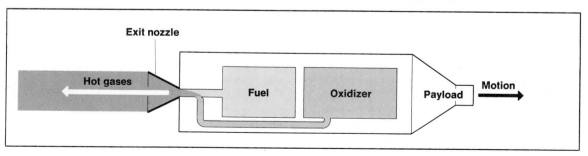

21-12 Spacecraft have no use for wings and other aerodynamic control surfaces. Why? What is the purpose of these surfaces on the space shuttle? Why is the shuttle different from previous spacecraft?

fuel. The fully assembled spacecraft stood more than 115 m high.

The engines produced enough thrust to lift a load equal to 4500 midsize cars off the ground. In December, 1968, the Saturn V sent a mass of 40,000 kg to the moon. In later flights, it carried astronauts to the moon and back.

In more recent years, a *space shuttle* was built to continue the exploration of space. The shuttle is similar to other space vehicles except that the main section of the shuttle can be recovered and used again. You can see by looking at Fig. 21-12 that the space shuttle has wings and tail surfaces designed to permit it to maneuver in the air. These control surfaces allow a pilot to land the shuttle like an airplane.

Jobs in Mechanics and Repair

Where do you take your car for a tune-up? Who do you call when your furnace or refrigerator does not work? Who installs a new phone in your home? These jobs and many others are done by people who are skilled in mechanics and repair.

Mechanics are people who work mostly on engines and machines. Engines are used in vehicles such as cars, buses, trucks, airplanes, farm tractors, heavy construction equipment, boats, and lawn mowers. All these machines and engines need mechanics who can service and repair them.

Some mechanics do light-duty repair such as tune-ups and adjustments. Others do complete engine overhauls. There are so many cars and other vehicles with engines that there seems to be an endless number of jobs available for people who are willing and able to do the work.

The repair field includes a wide range of additional jobs, including the repair of telephones, bicycles, industrial machinery, television sets, vending machines, locks, watches, and cameras.

Many jobs in mechanics and repair are also available in the heating and cooling industries. In Chapter 20, you learned some of the principles of heat and refrigeration. These principles are put to good use by people who install, service, and repair furnaces, freezers, and air conditioners.

How does one get such a job? Many industries have apprenticeship programs, which give on-the-job training. These programs provide the training that is needed to gain the skills.

Most apprenticeship programs are open to high school graduates who have had good grades. Many of the industries voluntarily register their programs with the Bureau of Apprenticeship and Training of the U.S. Department of Labor. The bureau sees to it that the apprentices in these programs receive a fair wage and good on-the-job training.

summary

1. Engines help provide food, clothing, and shelter.
2. The external combustion steam engine sparked the era of power.
3. Diesel and gasoline engines are internal combustion engines.
4. A typical gasoline engine is a four-stroke engine.
5. Diesel engines are more efficient than gasoline engines.
6. Jet engines are reaction engines used in aircraft.
7. Rocket engines can be used in space because they need no outside oxidizer.

review

Match each item in the left column with the best response in the right column. *Do not write in this book.*

1. diesel engine
2. steam engine
3. steam turbine
4. gasoline engine
5. ramjet
6. turbojet
7. fanjet
8. rocket engine

a. jet engine with no moving parts
b. internal combustion engine that uses spark plugs
c. internal combustion engine that uses no spark plugs
d. device that mixes air and gasoline
e. nonreciprocating external combustion engine
f. only engine that will operate in the vacuum of space
g. reciprocating external combustion engine
h. a rotary internal combustion engine
i. jet engine in which some air bypasses the combustion chamber
j. an engine with an air compressor used in most high-speed aircraft

questions

Group A
1. What is an engine? Does an engine create energy? Explain.
2. How does a steam turbine work?
3. How does the efficiency of a steam turbine compare with that of a reciprocating steam engine?
4. What is the difference between an internal and an external combustion engine?
5. Describe the operation of the four-stroke car engine.
6. List three advantages that a gasoline engine has over a steam engine.
7. Explain how a diesel engine operates.
8. State three differences between diesel and gasoline engines.
9. Explain the function of each of the following parts of the automobile: differential, transmission, and drive shaft.

Group B
10. How do engines help increase the supply of food, clothing, and shelter?
11. How do engines help reduce the cost of food, clothing, and shelter?
12. How is the back-and-forth motion of a piston changed to rotary motion?
13. Explain the difference between reciprocating engines and turbines.
14. Why are steam engines not used to power airplanes?
15. What is the main advantage of a six-cylinder over a one-cylinder engine of equal power?
16. What is rocket-fuel oxidizer? Does it always contain oxygen?
17. Why is the ramjet rarely used even though it is the simplest and most efficient of all jet engines?
18. In what major way are jets and rocket engines alike? How are they different?

further reading

Automobile Factory (Industry at Work). New York: Franklin Watts, 1975.

Hinkelbein, A., *Energy and Power*. New York: Franklin Watts, 1971.

Liebers, Arthur, *You Can Be a Mechanic*. New York: Lothrop, Lee & Shepard, 1975.

Meyer, Jerome, *Engine*. Cleveland: Collins-World, 1972.

Ross, F., *The Space Shuttle, Its Story and How to Make a Flying Paper Model*. New York: Lothrop, Lee & Shepard, 1979.

Urquhart, David I., *The Internal Combustion Engine and How It Works*. New York: Walck, 1973.

22

wave motion and sound

Have you ever dropped a stone into a still pond? Did you see the circular waves that moved outward from the point where the stone hit the water? These waves are a form of energy transfer.

You have already learned about other ways in which energy is transferred, such as heat moving from one point to another by conduction or convection. In this chapter, you will see how energy is transferred by means of wave motion, and especially how sounds are transmitted by waves.

objectives:

☐ To identify the properties of wave motion.

☐ To distinguish between two types of wave motion (longitudinal and transverse).

☐ To demonstrate that sound is caused by vibrations.

☐ To state a rule showing the relationship between frequency and pitch.

☐ To explain the concept of loudness as a measure of wave amplitude.

☐ To interpret the relationship between sound and music.

☐ To describe how sounds are made in musical instruments.

unit 4   wave motion and energy

transmitting energy by wave motion

All waves are alike in many ways. There are many kinds of waves, and most of them cannot be seen. You cannot see radio, heat, light, or sound waves. To help you learn about waves, you will first study a wave that *can* be seen: the water wave.

If you tap a pool of water with your finger, you will send out a regular series of waves, such as those shown in Fig. 22-1. The highest point of the wave is its *crest*. The lowest part is its *trough*. See Fig. 22-2. *The distance measured from the crest of one wave to the crest of the next wave (or from trough to trough) is called the* **wavelength**. *The number of waves that pass a given point in one second is the* **frequency**. The shorter the wavelength, the higher the frequency.

The material that carries the wave or transfers the energy is called the *medium*. In this case, the medium is water. The distance that a particle of the medium moves up or down from its rest position to the top of a crest or to the bottom of a trough is called the *amplitude*.

The energy of a wave depends upon its size or amplitude. If you tapped the water gently, the energy and amplitude of the wave would be very small. However, if you dropped a large boulder into the water, you would produce a wave of large amplitude.

As the water wave travels, it loses some of its energy through friction. The energy also spreads out over a wider area. As a result, the amplitude becomes smaller and smaller, until finally the wave dies out. The energy of the wave has been changed to heat.

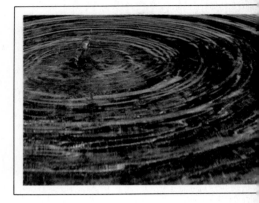

22-1 What form of energy transfer is shown in this photo?

22-2 A side view of water waves. What is the distance from one trough to another called?

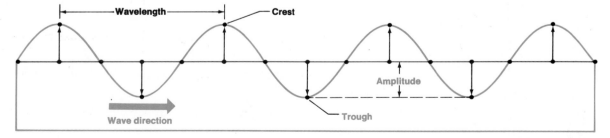

Wavelength Crest Amplitude Wave direction Trough

chapter 22   wave motion and sound

389

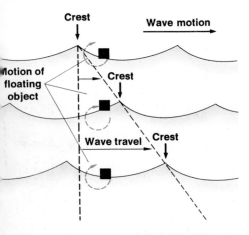

Crest

Wave motion

Motion of floating object

Crest

Wave travel   Crest

22-3 What is the difference between the motion of the water and the motion of the wave?

Again, suppose you tap the surface of a pool of water, sending out circular waves. A cork floating on the surface bobs up and down as the waves pass it. The cork is not carried along with the outward motion of the wave.

The cork's motion shows what happens to the water surface when a wave passes through it. As the wave passes, the water itself simply rises and falls a little, and returns to its starting point. Only the wave pattern of crests and troughs moves outward. The water itself does not move forward. See Fig. 22-3.

An exception to the fact that wave motion and the movement of water are unrelated occurs when waves reach a shoreline. At that point the waves "touch bottom" and turn into breakers. Then the water actually moves along with the wave up onto the beach.

There are two types of wave motion. The motion of a wave can be seen in another way. Tie one end of a rope to a door knob and shake the other end sharply, as shown in Fig. 22-4. You can see the waves moving along the rope. If you shake the rope hard enough, the waves will reflect back from the door before they die out.

The parts of the rope (like the particles of water) move up and down while the wave itself moves ahead. The type of wave produced by water and the rope is called a **transverse wave**. *A transverse wave is a wave in which the particles of the medium move at right angles to the direction of the motion of the wave.*

There is another type of wave motion. You can demonstrate it by using a long, coiled spring on a smooth table top. Attach the spring at one end and give the free end a quick push-pull motion. This type of wave is called a **longitudinal wave**. *A longitudinal*

22-4 How would you describe the motion of any given part of the rope? How does it compare with the movement of the wave?

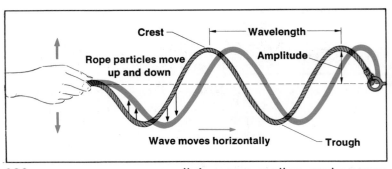

Crest   Wavelength

Rope particles move up and down   Amplitude

Wave moves horizontally   Trough

390   unit 4   wave motion and energy

wave is defined as one in which the particles of the medium move back and forth in the same direction as the wave itself moves. See Fig. 22-5.

Instead of crests and troughs, the coils are compressed and spread out. The compressed part of the wave is called a *compression,* and the spread-out part of the wave is called a *rarefaction* (rare-eh-*fak*-shun).

Observe that as the wave moves lengthwise along the spring, no turn of wire moves along with the wave. Any given turn of wire moves back and forth only a short distance, but the energy of the waves is passed along the entire length of the spring.

The wavelength of a longitudinal wave may be thought of as the distance from one compression to the next, or from one rarefaction to the next. The amplitude is the maximum distance forward or backward from its rest position that any particle (in this case a single turn of wire) of the medium moves. Sound waves are good examples of longitudinal waves.

Earthquakes produce both types of waves. The *longitudinal earthquake wave* is sometimes called a *primary wave.* This primary wave exerts a push-pull type of force, passing through all parts of the earth, including the central liquid core. A primary wave travels about 8 km/sec, passing through the earth's diameter in about 26 min.

The *transverse earthquake wave,* also called a *shear wave,* has a snake-like motion and travels at about 4 km/sec through the crust and mantle of the earth. The transverse wave is the major cause of damage by an earthquake.

A transverse earthquake wave, however, travels only through solids, not liquids. This fact supports the theory that the core of the earth is liquid. The transverse earthquake waves do not pass through the liquid core to reach those parts of the earth that lie opposite the earth's core. See Fig. 22-6.

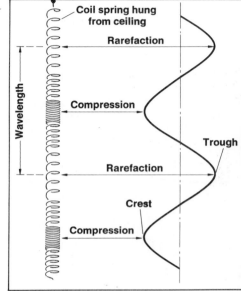

22-5 How does the longitudinal wave (left) compare with the transverse wave (right)?

22-6 How can geologists decide from a study of earthquake waves that the core of the earth is liquid?

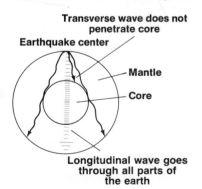

Transverse wave does not penetrate core

Earthquake center

Mantle

Core

Longitudinal wave goes through all parts of the earth

sound: a form of wave motion

Sound is caused by vibrations. Sound waves and water waves are alike in at least one way. Sound moves away in all directions from its source, just as waves move from the point where a stone is dropped into water.

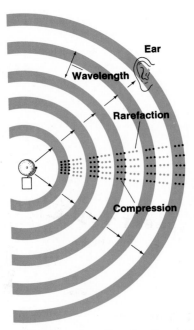

22-7 How does the vibrating bell affect the molecules of air? What kind of wave is formed?

However, sound waves are also different in some ways from water waves. A sound wave is a longitudinal wave, while a water wave is a transverse wave. Thus, sound waves can pass through all phases of matter—solids, liquids, and gases.

You can show how sound waves travel by using the coiled spring again. A quick shove against the free end of the coiled spring represents sound energy. The coils represent molecules of matter. The shove causes the coils to compress and spread apart just as sound causes the molecules in matter to compress and spread apart. Air is a very common medium for sound waves, and it can act just like the coils of a spring. Air molecules bunch together to form a compression and spread apart to form a rarefaction. See Fig. 22-7.

Sound waves are produced only by vibrating matter. If you place tiny paper riders on a piano string and strike the proper key, the paper dances as the string vibrates. Place some sand or paper clips on a drumhead and watch them move about as you beat the drum. The sound lasts only as long as the vibrations last.

Sound can also be produced by colliding objects. Dropping a book on the floor or letting a drop of water fall into an empty pan will produce sound. The collisions cause vibrations. **Sound** *is a form of energy produced by the vibration of matter.* Sound energy is transmitted by longitudinal waves.

activity

Sound from vibrating matter. Strike a tuning fork with a rubber mallet. If you do not have a mallet, tap the tuning fork on the heel of your shoe. (CAUTION: Do not strike a tuning fork against furniture or walls.) Dip the tips of the tuning fork into a container of water. What do you observe?

Now strike a tuning fork and hold one of the tips lightly against a hanging ping-pong ball. What happens?

Finally, hold a vibrating tuning fork lightly against a loose sheet of hanging paper. What happens?

Repeat the tests with several tuning forks. Do you see similar reactions?

Sound travels in solids, liquids, and gases. Some substances are good conductors of sound and others are not, but sound cannot be transmitted at all without a medium. A simple test can show that sound energy does not travel through a vacuum. See Fig. 22-8. As air is pumped from the bell jar, the sound of an alarm clock grows weaker and weaker and finally disappears.

Now think of what it is like on the moon or in space. Since there is no air, how can astronauts talk to each other? They use radios, because radio waves can move through a vacuum, whereas sound waves require a form of matter such as a solid, liquid, or gas to move from place to place.

You know that sounds travel through the air because you can hear other people speak. You also know that sounds travel through solids because you can hear noises through walls and ceilings in your home and school. If you have ever had a tooth filled, you know that the sound of the dentist's drill carries very well through the solid bones of your skull.

You can do a few simple activities to see how much better sound travels through some solids than through air. Listen to how loud the sound of a stretched rubber band is when you pluck it while holding one end in your teeth. Hold the *base* of a vibrating tuning fork firmly against one end of a meter stick. Press the other end of the stick against a table top or blackboard. Note how much better the sound travels through the solid stick than through air.

Liquids are also good conductors of sound. If you have ever swum underwater in a swimming pool, you may have noticed that you can hear the filter motor and other sounds. Certain animals, such as porpoises and whales, signal each other by making sounds. You will learn more about underwater sounds when you read about *sonar* later in this chapter.

What causes sound to travel so well through some substances but not through others? Could the density of the substance be a factor? Steel is quite dense and carries sounds very well. For instance, railroad workers will often listen to the rails to find out if a train is coming. They can hear the sound of the train while it is still a long distance away.

Yet density is not the answer. Concrete and brick are also dense substances, but they are poor sound

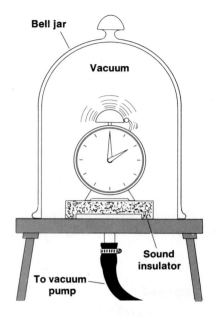

22-8 What happens to the sound of the alarm as air is pumped out of the bell jar?

Table 22-1:

Speed of Sound at 20°C	
Air	344 m/sec
Water	1450 m/sec
Steel	5000 m/sec

conductors. In fact, concrete blocks are often used in the walls of schools to block out noise. If all other factors are equal, the more dense a substance is, the worse it conducts sound.

Tests have shown that the best sound conductors are elastic substances. This means that the molecules are easily moved by sound energy. They transmit the sound energy from one molecule to the next throughout the length of the substance. The movement of the molecules is extremely small. For most sounds, the molecules move less than a millionth of a millimeter. The speed of sound varies *directly* as the *elasticity* of the medium, and *inversely* as its *density*. The greater the elasticity, the greater the speed; the greater the density, the slower the speed.

Careful measurements of the speed of sound show that it is greater on warm days than on cold days. Sound travels through air at about 332 m/sec when the temperature is 0°C. As the air gets warmer, sound travels about 0.6 m/sec faster for each rise of 1°C. This means that if the air temperature is 30°C, sound will travel 18 m/sec faster (0.6 × 30), or a speed of 350 m/sec (332 + 18).

We have stated that the speed of sound increases as the air gets warmer. Although the air is no more elastic when it is warmer, the density decreases. The net result is that sound travels somewhat faster. See Table 22-1.

You can estimate how far away a lightning flash is by noting how long it takes before you hear the thunder. Assume that sound travels about 350 m/sec. Therefore, if the sound takes 3 sec to reach you, the lightning is about 1 km away (350 m/sec × 3 sec = 1050 m).

sample problem

A hunter fires a rifle and 2.5 sec later hears an echo from a cliff. How far away is the cliff if the air temperature is 10°C?

solution:

Step 1. The speed of sound in air is 332 m/sec at 0°C. At 10°C, sound travels 6 m/sec faster (0.6 × 10) than at 0°. Therefore, the speed is 332 m/sec + 6 m/sec = 338 m/sec.

Step 2. In 2.5 sec, the sound travels 338 m/sec × 2.5 sec = 845 m.

Step 3. Therefore, the cliff is 845 ÷ 2, or about 422.5 m away.

Our ears detect sound waves. Follow the movement of sound in the ear. Sound waves enter the ear, as shown in Fig. 22-9. The *outer ear* is shaped to collect the waves and lead them down the *ear canal* to the *eardrum*. Sound waves striking the eardrum cause it to vibrate. As the eardrum vibrates, three small hinged bones in the *middle ear* pass the vibration along to the *inner ear*.

The inner ear is filled with liquid in which are located many *end-fibers* of the *auditory nerve*. As the sound waves spread through the liquid, they vibrate these nerve endings. The auditory nerve picks up the vibrations and sends electric pulses to the *auditory center* of the brain. The brain, in turn, produces the sensation you know as sound.

Pitch is one measure of a sound wave. You learned earlier that sounds are caused by vibrations. Now look more closely at how the speed of the vibrations, or **pitch**, affects the sound. The number of vibrations per second is referred to as the *frequency* of the sound.

Pitch refers to the effect of frequency on the ears. The greater the number of waves per second that strike the eardrums, the higher the pitch of the sound. The common unit for expressing frequency is *vps*, meaning vibrations per second. Vps is often also called Hertz (Hz). Humans are able

22-9 Can you trace the path of the sound from the telephone receiver to the auditory nerve? How many bones are located in the inner ear?

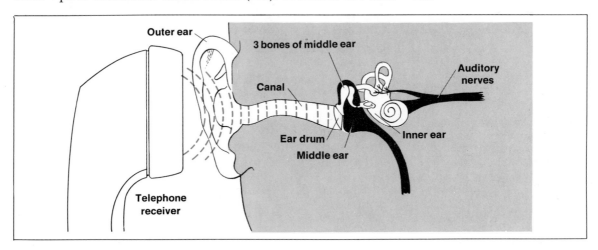

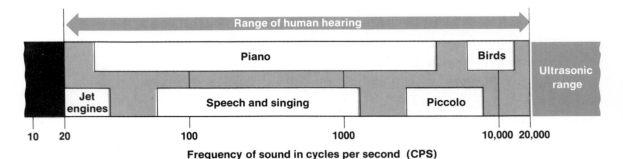

Range of human hearing

| Jet engines | Piano | Speech and singing | Piccolo | Birds | Ultrasonic range |

10 20   100   1000   10,000 20,000

Frequency of sound in cycles per second (CPS)

22-10 Compare the frequencies of several sounds. Can the human voice produce sounds beyond the range of human hearing?

22-11 Which of the two wave patterns shows a higher frequency? How do their amplitudes compare?

to hear sounds with a frequency between 20 vps and 20,000 vps. These sounds are called *audible* sounds. See Fig. 22-10. These limits vary greatly in different people and can change with age. The upper limit of the hearing range is reduced sharply as a person gets older. Many animals, such as dogs and bats, can hear sounds of much higher frequency than humans can hear.

Low-frequency waves, from 20 to 200 vps, produce deep bass (base) tones. High-frequency waves, above 8000 vps, produce high-pitched tones. See Fig. 22-11.

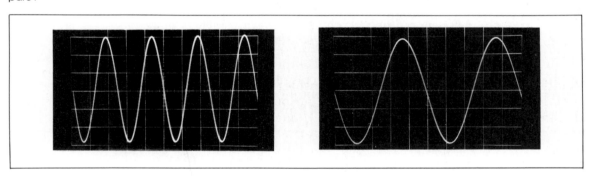

activity

Contradicting sounds. Fill two soda bottles with different amounts of water. Blow across the top of each bottle to determine which bottle produces the higher pitch. Now, tap the bottles with a pencil and again determine which bottle produces the higher pitch. How does the pitch produced by blowing compare with the pitch produced by striking? Explain.

High-frequency sounds are used in industry. Sounds with frequencies above 20,000 vps are called *ultrasonic.* These sounds cannot be heard by humans.

Industry makes wide use of ultrasonics for cleaning and testing various objects. Objects to be cleaned are dipped into a liquid. The ultrasonic vibrations are then directed into the liquid. The high-frequency vibrations cause a scrubbing action that cleans the objects.

To test for cracks or flaws, objects are subjected to ultrasonic waves. Any cracks in an object will scatter the waves and cause an echo. Ultrasonic waves are used by some dentists in cleaning teeth and in a variety of other medical uses.

Sonar is an electronic method of using ultrasonic waves. Sonar operates by means of a narrow beam of ultrasonic waves just as radar uses a beam of high-frequency radio waves. In sonar, a pulsed beam is sent through the water and reflected back to the receiver. The time it takes the reflected wave to return is a measure of how far away the reflecting surface is.

Sonar is used to survey the ocean floor, to detect objects, and to navigate. Submarine crews use sonar to locate other ships. Fishing boats carry sonar to help find schools of fish. Certain animals such as bats and porpoises have well-developed, natural systems similar to sonar.

Velocity and frequency are related. Consider what happens when a tuning fork vibrates, sending out waves in all directions. Suppose the fork has a frequency of 500 vps. In other words, the fork sends out 500 complete waves in 1 sec. The waves are each 0.7 m long and are placed end-to-end, as in the diagram in Fig. 22-12 (p. 398). The first wave is 350 m away at the end of 1 sec (500 waves, each 0.7 m long, have left the fork). Therefore, the wave must have traveled at a speed of 350 m/sec. Notice that this speed is obtained by multiplying the frequency (500) by the wavelength (0.7). This relationship can be expressed by the equation

$$V = F \times L$$
$$350 \text{ m/sec} = 500/\text{sec} \times 0.7 \text{ m}$$

where V is the velocity, F is the frequency of vibration, and L is the wavelength. This equation is very useful in

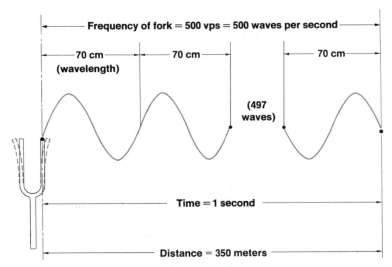

Frequency of fork = 500 vps = 500 waves per second

70 cm (wavelength) — 70 cm — 70 cm

(497 waves)

Time = 1 second

Distance = 350 meters

22-12 Five hundred waves per second, each 0.70 m long, placed end-to-end extend 350 m. How can you determine the speed of the sound from the figures given in the diagram?

science because it works for all kinds of waves, including sound waves.

For a wave in any one medium at a given temperature, the velocity *V* does not change. If the frequency is doubled to 1000 vps, the wavelength becomes half as much, or 0.35 m. If the frequency is cut to 250 vps, the wavelength becomes 1.5 m. High-pitched, high-frequency sounds have short wavelengths. The wavelengths of audible sounds range from 25 mm to about 17 m.

sample problem

The musical tone called middle A is produced by a sound wave with a frequency of 440 vps. What would be the wavelength of sound of this frequency if the temperature of the air were 33°C?

solution:

Step 1. The velocity of sound at 33°C is 332 m/sec + 0.6 m/sec × 33°C, or about 352 m/sec.

Step 2. $V = F \times L$, or
$L = V/F$

Step 3. Substituting,

$$L = \frac{352 \text{ m/sec}}{440 \text{ waves/sec}}$$

$$= 0.8 \text{ m/wave}$$

unit 4 wave motion and energy

The pitch of sound changes when the sound or listener moves. You have probably heard the pitch of the horn on an approaching car drop suddenly as the car passed you. Why should the pitch drop suddenly? The horn itself does not change its pitch. Yet, the pitch that strikes your ears does change.

The number of vibrations per second that strike your eardrum is changed if the source of the sound moves, or if you move. If the source is moving toward you, or you are moving toward it, the pitch is higher. More waves reach your ear per second than would be the case if you and the source were not moving.

On the other hand, the pitch is lowered if the vibrating object is moving away from you. *The rise or fall of pitch that is due to relative motion between the observer and the source of sound is called the* **Doppler effect.** The Doppler effect applies to all types of waves, not only to sound waves. Study Fig. 22-13.

Loudness is a measure of wave amplitude. The more energy a wave contains, the greater its size or amplitude. The effect of this energy on our ears is called **loudness.**

A sound becomes fainter as you move farther away because the sound spreads out to cover a larger area. When you are twice as far away, the sound is only one fourth as loud. When you are three times as far away from

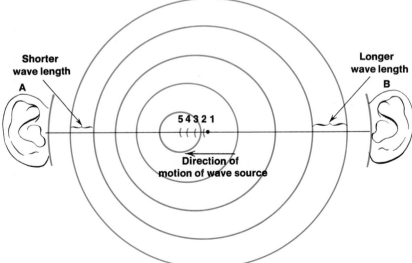

Shorter wave length
A

Longer wave length
B

5 4 3 2 1

Direction of motion of wave source

22-13 As the source of the sound moves from 1 to 5, what happens to the pitch heard by ear A? by ear B?

the source of the sound, it sounds one ninth as loud. *The loudness of sound varies inversely with the square of the distance.* This is the same inverse-square law that applies to gravity, heat, light, and radio waves.

The inverse-square law applies only in "free space," when the waves are not reflected or focused. It does not hold true in your classroom. Here sound is reflected from walls and ceiling, keeping the sound from weakening as fast as in free space.

You can measure the loudness of sound by using a *decibel meter.* The sound waves that strike a microphone in the meter produce an electric current. The strength of the current depends on the loudness of the sound. The current is then amplified and registered on a meter that is marked in *decibels,* the units of sound intensity.

A sound that can barely be heard by a person with good hearing is given a value of 0 decibels (db). This value is called the *threshold of hearing.* A sound is rated as 1 db louder than another sound if the slight difference in loudness can be detected. Decibel ratings of some common sounds are given in Table 22-2.

Table 22-2:

Decibel Ratings of Some Common Sounds	
Sound	Decibels
Threshold of hearing	0
Whisper	10–20
Quiet office	20–40
Automobile	40–50
Conversation	60–70
Heavy traffic	70–80
Air drills, riveters	90–100
Thunder	110
Threshold of pain	120
Jet airplane engine, 30 m away	140

Exposure to sounds that are too loud are likely to cause a hearing loss. The U.S. Department of Labor has set limits on noise levels for workers. See Table 22-3.

unit 4 wave motion and energy

Table 22-3:

Noise Exposure	
Sound level (decibels)	Time (per day)
90	8 hours
95	4 hours
100	2 hours
105	1 hour
115	15 minutes

People exposed to noise levels greater than the values in Table 22-3 must wear sound-muffling devices. Can you see why members of the ground crew who work around jet aircraft wear sound mufflers? Check the decibel rating of a rock band. How long can you listen to it safely?

Acoustics is the study of sound quality. The *acoustics* (uh-*koo*-stix) of a building means the "hearing qualities" it has. If the design of a theater or concert hall is such that the audience can hear well, it is said to have good acoustics. Acoustics also refers to the branch of science that deals with the study of sound. See Fig. 22-14.

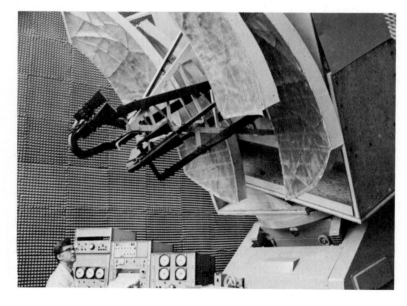

22-14 What do you think this special room is used for?

An *echo* has been referred to as a reflected sound. However, you cannot hear the echo if it is reflected back to you in less than about 0.10 sec. Since sound can travel a little over 30 m in this time, it is clear that the reflecting surface must be more than 15 m away if you are to hear the echo.

In some places, such as a concert hall, sound waves may be reflected back and forth several times. This gives rise to *reverberations* (ree-vur-ber-*ay*-shuns), or multiple echoes. If the walls of the concert hall strongly reflect the waves, the echoes remain audible for a long time, producing poor acoustics. To prevent this, the walls and ceilings may be covered with substances that absorb sounds.

the science of musical sounds

Some sounds are pleasing to the ear. If you strike your desk with a ruler, or let a drinking glass fall to the floor, a jumble of irregular vibrations is produced. The resulting sound, which is unpleasant to the ear, is termed *noise*. A noise has an irregular pattern of vibration.

Musical tones, on the other hand, have regular wave forms. Their wave patterns may be simple or they may be complex. However, a musical tone has a regular wave pattern that is repeated over and over again.

For the most part, musical tones have a pleasing effect on the ear, while noise has an unpleasant effect. The difference between noise and music, however, is not whether you like or dislike the sounds, but if the wave pattern is regular or not.

A musical tone has the properties of *pitch, loudness,* and *quality*. The first two of these, pitch and loudness, have already been discussed. You learned that pitch depends upon the frequency and that loudness depends upon the energy, or the amplitude, of the sound wave. But what is meant by quality?

Suppose someone is playing a piano and someone else a trumpet in the next room. If they each take turns playing a note of the same pitch and loudness, would you be able to tell the notes apart? Could you tell which instrument was being played even if you could not see them? The property called tone quality helps you to identify a musical sound. The quality of a musical tone depends on the shape of its wave pattern. Each kind of

22-15 How is the pitch changed on a horn? What is the source of the sound?

musical instrument has its own unique wave pattern. In an orchestra, these waves combine to produce music.

There are three groups of musical instruments. Music is made in many different ways. Musical instruments are played by such methods as plucking, blowing, striking, rattling, or rubbing. These methods are used on instruments that can be placed into three groups: the *wind instruments, stringed instruments,* and *percussion instruments.*

Each of these groups can be divided further. For instance, the wind instruments include one class in which a reed causes a column of air to vibrate. The reed vibrates because the player blows on it. This class includes the clarinet, saxophone, harmonica, and accordion.

A second class of wind instruments is the horn group. In this group, the vibration of an air column is started by the vibrating lips of the musician. To change the pitch, a horn may have a valve or a sliding section, which changes the length of the air column. The horns include the bugle, cornet, trumpet, tuba, and trombone. See Fig. 22-15.

A third class of wind instruments is one in which air is blown across an edge causing the air column to vibrate. The flute and piccolo are examples of such instruments.

The stringed instruments include those that are plucked, such as the banjo, ukulele, and guitar, and those

chapter 22   wave motion and sound   403

22-16 How can a player change the pitch and loudness of a cello?

that are rubbed with a bow, such as the violin, viola, string bass, and cello. See Fig. 22-16.

Finally, the percussion instruments are those that are played by striking an object with some kind of stick or mallet. The percussion instruments include vibrating membranes, such as drums, and vibrating solids, such as the xylophone and the bell.

Tones can be controlled in several ways. The pitch of wind instruments can be controlled by changing the length of air columns. When a musician moves the slide of a trombone in and out, the slide makes the air column shorter and longer. A short column is in resonance with the more rapid vibrations of the musician's lips and, as a result, a tone of higher frequency is produced. Resonance will be discussed in more detail later in this chapter.

In some instruments, such as a trumpet, clarinet, or saxophone, the player changes the length of the air column by using keys or valves to block off parts of the air column. A flute player uses the fingertips to open and close holes along the side of the flute.

In other wind instruments, such as the bugle, there is no way of changing the length of the air column. To produce different tones with the bugle, the tension of the lips and the force with which the player blows must be changed. This produces higher or lower tones. Tones of other instruments in the horn group can be changed in the same way.

Most wind instruments have air columns that are open at each end. Such air columns are known as *open pipes*. The length of the open pipe is related to the tones that the instrument produces. For the most part, the length of the open pipe is one half the wavelength of the main, or *fundamental*, frequency that it reinforces. You will learn more about fundamental wavelengths later.

If the pipe is closed at one end, it is referred to as a *closed pipe* and reinforces an entirely different sound. A closed pipe reinforces a wavelength four times as long as the pipe.

The pipe organ has many different kinds of pipes, some open and some closed. The xylophone is often provided with closed pipes hanging under the wooden bars. The effective length of each pipe is designed so that it will reinforce the frequency of the bar directly above it.

Three factors affect the pitch of stringed instruments.
The three factors that affect the pitch of a vibrating string
are its *length, tension,* and *mass.*

For a given mass and tension, the longer a string, the
slower it vibrates, and the lower its frequency. You know
that the deep bass notes of the piano or harp are produced
by the longest strings. A violinist can raise the frequency
of a string just by pressing down on it. This process
prevents the entire string from vibrating. The part that
still vibrates is shortened. Tests have shown that the
frequency of a vibrating string is inversely proportional to
its length. For example, if you reduce a string to half its
original length, its frequency doubles.

Increasing the tension of a string also makes it vibrate
more rapidly and raises the frequency. To tune a string
that sounds flat, or too low in pitch, the violinist will turn
a peg to tighten it, thus increasing its tension and frequency. The frequency of a string is directly proportional
to the square root of its tension. That is, the greater the
tension, the higher the frequency.

You have probably noticed that the bass strings of a
piano are not only longer and under less tension, but they
are also heavier. These wires are often wrapped with a
tight coil of silver or bronze wire to increase their mass.
The heavy wire has more inertia and, therefore, vibrates
more slowly at a lower frequency.

Instruments produce fundamental tones and overtones.
If you pluck a stretched string lightly in the middle, it
vibrates as a whole, as shown in Fig. 22-17**A.** The tone you
hear when this is done is called the *fundamental.* It is the
lowest note that can be made by a vibrating string.

However, if you pluck the string one quarter of the way
from the end, it not only vibrates as a whole but also in
segments, as shown in Fig. 22-17**B.** The tone you hear has
a different quality. Besides the fundamental, the *first
overtone* is now present. Overtones are higher-pitched
tones produced by an object as it vibrates in segments.
Overtones are tones whose frequencies are whole-num-
ber multiples of the fundamental frequency. For example,
if the frequency of the fundamental is 100 vps, the frequency of the first overtone is 200 vps. The frequency of
the second overtone is 300 vps.

The number and strength of the overtones determine
the quality of a musical tone. When overtones are formed,

22-17 If the frequency of the fundamental tone (**a**) is 100 vps, what
are the frequencies of the first, second, and third overtones?

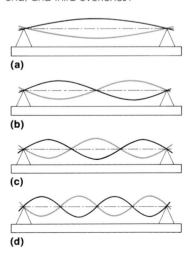

(a)

(b)

(c)

(d)

the vibrating patterns of instruments become very complex. A string can vibrate not only as a whole but also in segments to produce overtones. Most of these overtones blend well with each other and with the fundamental, producing a pleasing effect. Overtones are often referred to as *harmonics*.

activity

Overtones. Set up two identical strings and place them under equal tension as shown in Fig. 22-18. Pluck the first string to produce a fundamental tone. Use a block of wood to slowly shorten the second string while at the same time plucking both strings. How short must the second string be to produce a note at the first overtone? Shorten the second string further and see if there are other places where the two strings vibrate together to produce a pleasant sound. Compare the lengths of the two strings at those points where overtones are produced. What did you find? What happened to the sound when one string was shortened very slightly by the movable block?

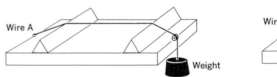
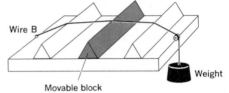

22-18 How does the movable block affect the frequency of the string? How does the size of the weight affect the frequency?

Rhythm, melody, and harmony are factors in musical composition. To the composer, the performer, and the listener, music is an art. But music is also a science.

The principal factors of a musical composition are rhythm, melody, and harmony. Rhythm is the timing pattern of music. Melody is the effect of single tones following each other in succession. Melody is often called the *tune*. Harmony is the effect of two or more tones sounded together. Musical tones sounded together in harmony form a *chord*.

From experience, you know that certain tones harmonize well. Other tones produce an unpleasant effect, or *discord*, when you hear them sounded together. Whether there is harmony or discord depends almost entirely upon how one frequency blends with another.

unit 4 wave motion and energy

Pleasing chords have specific ratios. If the frequency ratios of two tones are small whole numbers, they form a pleasing chord. These ratios may be expressed as 1:2, 1:3, 2:3, 3:4, and so on. If the frequency ratio is 1:2, the chord is called an *octave*. For example, if middle C on the piano has a frequency of 262 vps, then the frequency of the first octave, the eighth note above middle C, is twice 262, or 524 vps.

You have probably heard of the *major chord*, which is made up of the three musical tones called *do, mi, sol*. For at least 2500 years, the major chord has been known as the "sweetest" of the three-toned chords. The frequency ratios of these tones are 4:5:6. If the *do* has a frequency of 400 vps, *mi* will have 500 vps, and *sol* 600 vps.

The major scale was first made from three of these 4:5:6 chords, properly placed. An octave on this scale includes the notes *do, re, mi, fa, sol, la, ti, do*. See Fig. 22-19.

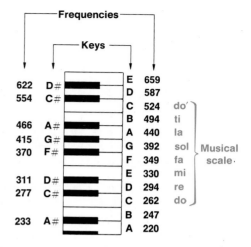

22-19 What is the ratio between the tones do mi sol in the musical scale?

Sounds are made louder by forced vibrations and by resonance. *Sounds made by instruments are made louder in one of two ways: (1) forced vibrations or (2) resonance.*

Stringed instruments rely on forced vibrations to increase the loudness. That is, the string is attached to a surface that is forced to vibrate when the string vibrates. Therefore, since the vibrating surface is larger than the string alone, the sound becomes louder.

The strings of a piano create louder tones than those of a zither or harp for this reason. Piano strings are mounted on a *sounding board,* in which they produce forced vibrations. In much the same way, forced vibrations in the wooden body of a violin increase the sounds produced by the strings.

activity

Forced vibrations. Strike a tuning fork and press its *stem* against the top of the table. What do you hear? Try tuning forks of different frequencies. Also try placing the stems against different surfaces such as windows, walls, cabinet doors, and similar surfaces. What happens to the sound? Can you explain why?

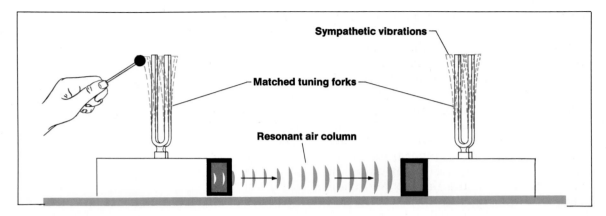

Sympathetic vibrations

Matched tuning forks

Resonant air column

22-20 Why does one tuning fork begin to vibrate when the other tuning fork is struck?

22-21 What is the relationship between a resonating air column and a vibrating tuning fork?

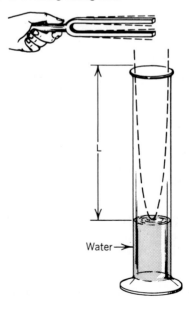

L

Water →

Another way in which sounds are increased is by resonance. Resonance occurs when an object vibrates at its own *natural frequency*. The natural frequency is the one fundamental frequency at which an object will vibrate when disturbed. Moreover, an object will normally absorb energy only at its own natural frequency.

Try the following experiment. Press down the loud pedal of a piano to free the strings. Then sing a note, such as "a-a-h-h," very loudly. The piano will respond by sounding the same tone. The sound waves of your voice strike all of the strings. However, only those strings having the same natural frequency as your voice are able to pick up the energy and start vibrating. The strings that are vibrating are in resonance with the tone of your voice.

You can show how resonance works in the lab by placing two mounted tuning forks of the same natural frequency close together, as shown in Fig. 22-20. If one is struck, the other tuning fork will also begin to vibrate. However, wrapping a heavy rubber band around the end of a tine of one tuning fork changes the natural frequency. The tuning fork will no longer resonate.

Columns of air also have natural frequencies. This can be shown by holding a vibrating tuning fork over a cylinder, as shown in Fig. 22-21. As the tines of the fork vibrate over the air column, they send down waves that reflect back up from the surface of the water.

As water is slowly added, the air column inside the cylinder is shortened. At a certain point, the tone of the tuning fork suddenly sounds quite loud. At this point the

unit 4 wave motion and energy

air column has the same natural frequency as the tuning fork.

The sound grows louder because the sound waves of the tuning fork are in step with the sound waves inside the cylinder. How does this happen? First, a sound wave from the tuning fork goes down the cylinder. The wave is reflected by the water and goes back up the cylinder. At this point, it meets the next vibration of the tuning fork and is reflected back down the cylinder. Now the sound wave has not only the energy of the tuning fork but also the energy of the reflected wave. As long as the new waves from the tuning fork are in step with previous waves in the air column, the amplitude of the sound increases sharply. In other words, the column of air in the cylinder resonates with the vibrating fork.

There is a relationship between the wavelength of the sound and the length of the resonating air column. *The length of a resonant air column that is closed at one end is one fourth the fundamental wavelength of the sound it reinforces.*

Resonance is of great importance in many other branches of science. For example, when you tune a radio, you are adjusting its frequency to match that of the radio waves you want to receive. Energy of this frequency is then amplified while other radio frequencies are rejected.

sample problem _____

If the length of the resonant air column in Fig. 22-21 is 50 cm, what is the frequency of the tuning fork? Assume the temperature of the air to be 0°C.

solution:

Step 1. At resonance, the fundamental wavelength is four times the length of the resonant air column, or $L = 4 \times 50$ cm $= 2$ m. The speed of sound at 0°C $= 332$ m/sec.

Step 2. $V = F \times L$, or

$$F = V/L$$

Step 3. Substituting given and known values,

$$F = (332 \text{ m/sec})/2 \text{ m}$$
$$= 166 \text{ vps}$$

Robert A. Moog
Electronic Music

Robert A. Moog is a new kind of instrument maker. Instead of working with wood, string, and valves as the old artists did, Moog works with electronic filters, transistors, and wave generators.

Moog is an amateur musician with a PhD in engineering physics. He combined these talents to build the Moog Electronic Music Synthesizer, an instrument that looks something like a computer. The synthesizer can reproduce the sounds usually expected of more common wood and metal instruments, as well as an odd collection of beeps, thumps, buzzes, and squawks. The sounds are formed into music by a system of electronic controls.

An early version of a synthesizer was built by RCA in 1955. It was able to imitate the sounds and styles of a harpsichord, an organ, a hillbilly band, and a dance band. However, the effects were mostly for demonstration.

The Moog Synthesizer, however, is far more skillful. It has been used to play everything from rock to the classics.

ʃummary

1. All waves have amplitude, frequency, wavelength, and shape.
2. Particles in a transverse wave move up and down; particles in a longitudinal wave move back and forth.
3. Sound is caused by vibrations.
4. Sound travels in solids, liquids, and gases, but not in a vacuum.
5. Pitch refers to the effect of frequency on our ears.
6. Velocity of a sound equals frequency times wavelength ($V = F \times L$).
7. Acoustics is the study of sound quality.
8. Noise has an irregular wave form while music has regular wave forms.
9. Musical instruments can be divided into wind, stringed, and percussion.
10. The pitch of a stringed instrument depends on the length, tension, and mass of the string.
11. Sounds are made louder by increasing the vibrating surface or by causing objects to vibrate in resonance.
12. The factors that describe a musical composition are its rhythm, melody, and harmony.

review

Match each item in the left column with the best response in the right column. *Do not write in this book.*

1. wavelength
2. transverse wave
3. longitudinal wave
4. decibel
5. reverberation
6. noise
7. musical tone
8. pitch
9. fundamental
10. resonance

a. effect of frequency of sound wave on our ears
b. distance from one part of a wave to corresponding part of the next wave
c. condition in which vibrations of two objects are in tune with each other
d. has a frequency ratio of 4:5:6
e. lowest tone produced by a vibrating body
f. multiple echoes
g. back-and-forth motion
h. up-and-down motion
i. sound frequency above 20,000 vps
j. sounds having a regular wave pattern
k. sounds having irregular wave patterns
l. unit of sound intensity

questions

Group A
1. What is sound? What causes sound?
2. What is the amplitude of a sound wave?
3. What are compressions and rarefactions? Do these terms refer to transverse or longitudinal waves?
4. Describe the difference between transverse and longitudinal waves.
5. Through what phase of matter does sound travel fastest? slowest?
6. Explain the meaning of ultrasonics. List several of its uses.
7. How many times greater than the speed of sound in air is the speed of sound in steel?
8. How much faster does sound travel through air at 30°C than at 0°C?
9. What is the Doppler effect?
10. What is meant by the acoustics of an auditorium?
11. How does noise differ from a musical tone?
12. How long must a closed pipe be to reinforce a particular fundamental tone?
13. How are overtones produced in vibrating strings? How are they related to the fundamental?
14. What are the three principal factors of a musical composition?
15. Give two examples of forced vibrations.
16. How is the pitch of a violin string raised?

Group B

17. Does a transverse wave travel through a solid? a liquid? Explain.
18. How would you demonstrate that sound does not travel through a vacuum?
19. If you shouted at a wall 10 m away, could you hear an echo? Explain.
20. If all gases are equally elastic, do you think sound waves travel more rapidly through air or through hydrogen? through air or carbon dioxide? Explain.
21. Why does sound travel more rapidly through steel than through air?
22. Why does sound travel more rapidly through warm air than through cold air?
23. One noise is rated at 10 db and another at 11 db; would they sound different?
24. Explain three properties of a musical tone.
25. What factors determine the frequency of a stretched string?
26. What is meant by the fundamental frequency of a string? How is it produced?
27. The vibrating lips of a trumpet player start the sound, but what produces the musical tones that you hear?
28. If two tones form a pleasing chord, what can be said of their frequencies?
29. In what two ways can a trombonist raise the pitch?
30. What is the relationship between pitch and the size of musical instruments?

problems

1. What is the wavelength of the lowest pitched sound you are able to hear? Assume a frequency of 20 vps and a temperature of 20°C.
2. An ultrasonic generator sends 500,000 waves per second through the ocean. If each wave is 0.3 cm in length, what is the speed of sound in ocean water?
3. Find the wavelength of water waves that have a frequency of 0.75 wave per second and a speed of 6.75 m per second.
4. Assume that you first hear a police siren 240 m away. By the time it is only 60 m away, how much louder is the sound?
5. You hear an echo 2 sec after firing a shot. How far away is the reflecting surface if the temperature is 20°C?
6. A piano wire is vibrating as a whole, in one segment, at 440 vps. (a) What is the frequency of its fundamental tone? (b) What is the frequency of its first overtone? (c) What is the frequency of its second overtone? (d) What is the frequency of the tone three octaves above the fundamental? (e) What is the frequency of the tone one octave below the fundamental?
7. If the *do* of a chord has a frequency of 880 vps, what is the frequency of the *mi* above? the *sol* above?
8. A violin string with a frequency of 262 vps is shortened to two thirds its former length. (a) What is the new frequency? (b) If its first tone is called *do*, what is the new tone called?
9. (a) What is the wavelength of a tone that resonates with a closed pipe 0.4 m long (temperature, 0°C)? (b) What is the wavelength of a sound that resonates with an open pipe 0.4 m long? (c) Which pipe sounds a note of the higher frequency?

further reading

Chedd, Graham, *Sound*. Garden City, N. Y.: Doubleday, 1970.

DeSautter, D. M., *Your Book of Sound*. Levittown, N.Y.: Transatlantic, 1971.

Fisher, R., *Heros of Music*. New York: Fleet, 1971.

Fox, L. M., *Instruments of the Orchestra*. New York: Roy, 1971.

Freeman, Ira M., *Science of Sound and Ultrasonics*. New York: Random House, 1968.

Kentzer, Michael, *Waves*. Cleveland: Collins-World, 1977.

Tannenbaum, B., and Stillman, M., *Understanding Sound*. New York: McGraw-Hill, 1971.

Wade, Harlan, *Sound*. Raintree Pubs. Ltd., 1977.

Wilson, Robina B., *The Voice of Music*. New York: Atheneum, 1977.

23

the nature of light

Early humans, watching the flickering flames of a campfire, probably thought the yellow light was a magical, unknown spirit. Today, however, scientists know that light is a form of energy that travels in waves. You will recall that in the last chapter you learned about the wave nature of sound. In this chapter, you will learn about the properties of light.

objectives:

☐ To describe how light is produced and transmitted.

☐ To demonstrate how light energy can be measured.

☐ To describe the properties and uses of lasers.

☐ To demonstrate the law of reflection.

☐ To identify the types of images formed by plane, convex, and concave mirrors.

☐ To understand that light refracts because its speed changes.

☐ To identify the types of images formed by concave and convex lenses.

☐ To describe the operation of optical systems such as magnifiers, projectors, telescopes, and the human eye.

the properties of light

Light has the properties of both waves and particles. The first major studies on the nature of light were carried out late in the seventeenth century by Isaac Newton and Christian Huygens (*high*-gens). Newton believed that a beam of light was made up of a stream of particles. Huygens formed a different theory. He felt that light traveled in waves.

Both scientists tried to find evidence about light by observing how it formed shadows and how images were formed by mirrors and lenses. Newton thought that if light traveled in waves, then the wave action would cause the edges of shadows and images to be blurred. Huygens, however, felt that the blurred edges were too small to be seen by the human eye.

In the centuries that followed, the debate continued. More evidence was found by both sides to support their theories. It wasn't until the early part of this century that Max Planck and Albert Einstein discovered that light had the properties of *both* particles and waves. Planck and Einstein found that light travels as waves but is emitted and absorbed as particles.

Prior to these findings, it was believed that the wave and particle theories were opposites that could not exist together. However, this dual nature of light is the theory that scientists accept today.

Planck and Einstein found that electrons were knocked from a metal surface when hit by a beam of light. This *photoelectric* effect is caused by particles of light.

A particle of light is called a **photon.** See Fig. 23-1. Each photon consists of a tiny packet of energy. The energy of a photon depends upon the frequency of the light wave. For example, a photon of violet light, which has twice the frequency of red light, has twice the energy. The photons of ultraviolet light and X rays have even more energy. The photoelectric effect can only be explained by the particle theory of light.

Light can be polarized. The discovery that light can be *polarized* seemed to support the wave theory of light. Figure 23-2 on p. 416 shows how light can be polarized.

When light passes through the transparent sheet, nothing changes. However, as it passes through sheet B,

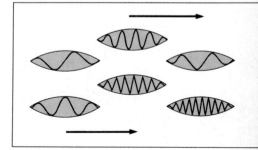

23-1 What does this highly symbolic drawing of photons tell you about the wavelengths and frequencies of light?

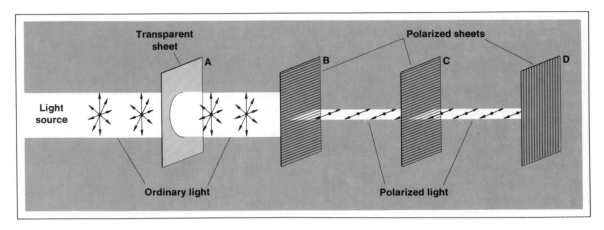

Transparent sheet

Polarized sheets

A   B   C   D

Light source

Ordinary light

Polarized light

23-2 What happens to ordinary light as it passes through the transparent sheet?

which is polarized, only light waves vibrating in one direction páss through. The rest of the vibrations are blocked. It is as if the light were "combed" through parallel slots.

The light that remains after going through sheet B also goes through sheet C, which is polarized the same way as sheet B. Polarizing sheet D, however, is turned at right angles to sheet C. Therefore, the light waves that pass through sheet C are not lined up with the slots of sheet D and all light is blocked.

Certain crystals act like polarizing sheets because they block certain light waves. When the atomic planes of the crystals are lined up in a given plane (crystal lattice), they permit the passage of light waves vibrating only in that same plane. In other words, only those waves that vibrate in line with this given set of atomic planes can pass through. This screening process is called *polarization*, and the light that has been screened is said to be polarized.

The fact that light can be polarized shows that light travels in transverse waves. If light traveled in longitudinal waves, such barriers would have no effect on the light. Why?

Polarized light can be useful. Have you ever used Polaroid sunglasses? Much of the annoying glare you may experience in driving comes from sunlight being reflected off various surfaces. This glare is largely blocked out by polarized glasses. See Fig. 23-3.

Polarized light is a help to engineers in designing objects. A transparent plastic model of the object is stressed by twisting, stretching, and compressing. When

unit 4 wave motion and energy

viewed through polarized light, colored patterns in the plastic models show where the strains in the object are greatest and therefore, where the object is most likely to break. See Fig. 23-4.

Light is not the only type of wave that can be polarized. All transverse waves can be polarized.

Light is produced by excited atoms. Light can be produced by any of the following methods:

1. by *heating* something until it becomes so hot that it gives off light. The term used to describe a glowing hot object is *incandescent.* An electric light bulb, a burning match, or a candle are good examples of incandescence. The light is given off by particles of carbon heated to the point where they glow.

23-3 What is the effect of using a polarized lens (right) in this scene?

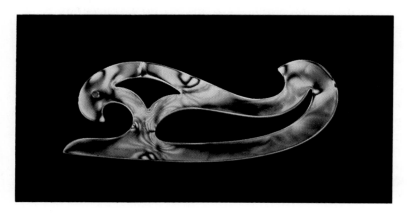

23-4 Where are the points of greatest strain revealed by polarized light?

2. by passing an *electric current* through a gas, as in neon lighting.
3. by the process called *fluorescence*. Certain substances glow when hit by strong rays, as in fluorescent bulbs.

Whatever method is used, *light is produced whenever electrons fall from one energy level in an atom to a lower level.* To understand this, compare electrons in orbit around the nucleus of an atom with satellites in orbit around the earth.

Suppose you want to lift a satellite to an orbit 600 km above the earth. A great deal of energy (fuel) is needed to do this. Kinetic energy of the rocket is transferred to the satellite as potential energy. While the satellite is in orbit, the energy it has absorbed stays with it. But when the satellite falls back to earth, it releases this energy in the form of heat.

If you were to raise the satellite to an orbit only 300 km above the earth, the energy needed would be less. As a result, the amount of energy released when the satellite falls to earth would be less than if it fell from a higher orbit.

Electrons in atoms behave in a similar way. The nucleus of the atom can be pictured as having "satellites," called electrons, in orbit at certain energy levels. *An electron can absorb energy and move to a higher energy level in an orbit farther away from the nucleus.* When this electron falls back to the starting energy level, it gives off this extra energy in the form of radiation, just as the earth satellite gives up its extra energy in the form of heat. See Fig. 23-5.

The higher the energy level to which an electron is lifted, the greater the energy it gives off when it falls back to a lower level. In the next chapter, you will see how the amount of energy given off is related to the color you see.

This comparison between electrons and satellites cannot be carried too far. First of all, electrons do not move in fixed orbits as earth satellites do. Also, satellites can be placed in an almost endless number of energy levels while electrons can exist only in certain distinct energy levels.

The methods of making light are all just ways of *exciting* atoms by raising their electrons to higher energy levels. In each case, some other form of energy, such as heat or electric current, is used to excite the atoms.

23-5 Which electron gives off the greatest amount of energy when it drops to a lower level?

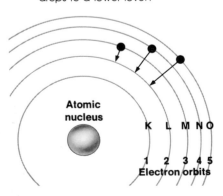

unit 4 wave motion and energy

Natural light on earth comes from stars. Daylight is caused by the sun, the nearest star. The intense light and heat of the sun and other stars is due to nuclear reactions.

Since light travels from distant stars through space, you know that light can travel through a vacuum. Light needs no solid, liquid, or gas to carry its waves. In fact, matter only hinders the passage of light. Therefore, light waves must be unlike sound waves, which cannot pass through a vacuum at all.

Light makes objects visible. You can see only when light strikes your eyes. The light may come directly from a source or it may be reflected off some object. You see most things by reflected light. In either case, if you are to see an object, light must come from the object and then enter your eyes.

When light waves strike a substance, they may be *reflected,* *absorbed,* or *transmitted,* as shown in Fig. 23-6. If the substance transmits none of the light waves, it is described as *opaque* to light. You cannot see through opaque objects.

If light waves pass through a substance so that objects can be seen clearly, that substance is *transparent* to light waves. Glass, water, and air are common examples of transparent matter.

Many substances are neither opaque nor transparent, but are somewhere in between. These substances, such as frosted glass and waxed paper, are said to be *translucent.* Light passes through translucent matter, but things cannot be seen clearly through them.

Any light that is absorbed is changed into heat, warming up the object that absorbed it.

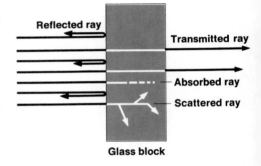

23-6 What can happen to light when it strikes an object?

Light can be measured. The brightness of a light source can be measured by comparing it with the brightness of some standard source of light. One such source is the *standard candle,* a light made to be an exact model.

The light from a candle (standard candle) was actually not an exact model because the brightness varied from one candle to another. In recent years, the standard of light brightness has been revised by the National Bureau of Standards. Now, the standard used is the light given off when platinum is heated to 1772°C and viewed through a long sighting tube with an opening 0.016 cm² in area. In practice, it is common to use light bulbs that have been compared with the standard. The unit of brightness given off by a standard candle at its source is referred to as one *candela* or one candle (cd).

The rate at which light energy is given off from a source is measured in *lumens* (lm). One candle is equal to about 12.5 lumens. When new, a modern 100-watt light bulb (incandescent type) has a brightness of about 1600 lumens.

People are often more interested in the amount of light falling on a surface than in the brightness of the source itself. When you are reading, for instance, you are more concerned with the amount of light that strikes the page than with the brightness of the bulb.

The light on the page depends not only upon the brightness of the bulb, but also upon how far away the page is from the bulb. Light brightness follows the inverse square law. For instance, suppose you measure the light from a lamp at a given distance from you. If you put the lamp twice as far away, you get only one fourth as much light.

The brightness of light can be measured by a *light meter*. A light meter is often used on (or in) a camera. Such a meter measures the light with a *photocell*. The photocell contains a substance that produces a small electric current when light hits it. The greater the light the stronger the electric current. The current is an example of the photoelectric effect discovered by Einstein.

Sometimes light meters are used to measure the light brightness in homes. It is very common to find homes with light levels too low for reading, sewing, or other close detail work. On the other hand, the lights in hallways do not need to be as bright as are often found in homes. Lower light levels often serve just as well in hallways and save energy as well.

23-7 What happens to the grease spot when the front and back of the paper receive the same amount of light?

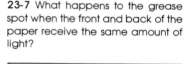

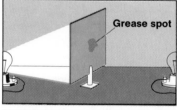

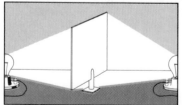

activity

Measuring light brightness. Rub a drop of oil, grease, butter, or fat into a sheet of paper and hold the sheet in front of a light as shown in Fig. 23-7. What do you see? Now remove the light source. What does the spot look like now? Place one light in front of the paper and another light in back of it. Move the paper back and forth between them.

Questions:
1. What does the grease spot look like with a light source behind it?
2. What does the spot look like when a light is placed in front of it?
3. What happens to the spot if the paper is moved back and forth between the two light sources? Can you find a point between the two lights where the spot disappears?

Another factor that is important in the brightness of a room is the color of the walls. Lamps that might give enough light in a room with white walls could be too dim in a room with red or blue walls. For example, white walls reflect as much as 85% of the light. Yellow reflects about 60%, but dark blue reflects only about 10% of the light.

Other factors that should be considered in home lighting are shadows and glare from uncovered bulbs. The softer and more *diffuse* (spread out) the light, the better.

Some light sources are more efficient than others. Not all sources give the same amount of light for a given amount of input. One common source, the incandescent light bulb, has a low efficiency. A 25-watt bulb uses 25 watts of electrical power but changes only about 1.5% of this power into visible light. The rest is lost as heat.

A far more efficient light source is a white fluorescent bulb. Light from such a bulb is often called "cool" light because it gives off so little heat. See Table 23-1.

You can see from the table that the efficiency of incandescent bulbs varies with size. How many 25-watt bulbs would be needed to equal the total lumens emitted by one 1000-watt bulb? How many watts would be needed to light all of these 25-watt bulbs?

Table 23-1:

Efficiencies of Light Sources			
Type Bulb	Watts Input	Total Lumens	Lumens Per Watt
Incandescent	25	250	10.0
Incandescent	100	1,600	16.0
Incandescent	1000	20,500	20.5
Fluorescent	40	2,320	58.0
Ideal	1	680	680.0

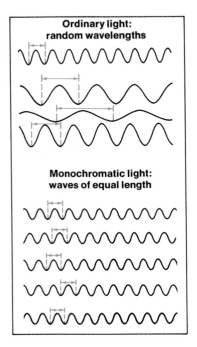

Ordinary light: random wavelengths

Monochromatic light: waves of equal length

23-8 What is the major difference in the wavelengths of common light and laser light?

23-9 What is meant by the statement that laser light is coherent?

Coherent light: waves in step

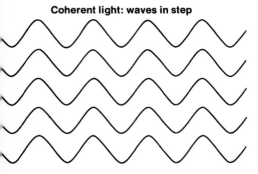

The laser produces coherent light. In 1960, scientists developed the first *laser*. The laser produces a kind of light that never before existed in nature. "Laser" is a word taken from the first letters of *L*ight *A*mplification by *S*timulated *E*mission of *R*adiation. The light from a laser has three main properties:

1. It is *monochromatic*. White light consists of a wide range of wavelengths. Even a light of one color contains a wide range of wavelengths. The light from a laser, however, is all of the same wavelength or frequency. See Fig. 23-8.

2. It is *coherent*. Normally, light waves travel in random phase; that is, the crests and troughs of one wave are not lined up with the crests and troughs of another. In a laser, however, the crests and troughs of waves are in step, as shown in Fig. 23-9.

3. It travels in a *plane wave front*. Light normally tends to spread out in all directions in a curved wave front, the way water waves travel when you drop a pebble in a quiet pond. At distances over 1 km even searchlight beams spread out. Light from a laser, however, travels in almost parallel lines with very little spreading, even at great distances. A laser beam may spread as little as 1 cm in 2 km. An early laser was aimed at the moon 384,000 km away, yet its beam spread to a diameter of only 3 km. See Fig. 23-10.

The frequency and wavelength of light produced when electrons drop from one energy level to another depend upon the difference in these energy levels. Refer back to Fig. 23-5 (p. 418). In a laser, only a single change in energy level is produced. As a result, light of only one frequency is emitted.

Before an atom can emit a photon of light, one of its electrons must absorb energy and move to a higher energy level. In the laser, the needed energy is "pumped in." Sometimes this energy comes from an outside light source. An electric discharge in the gas within the laser tube can also provide the needed energy. At this point, the atoms are said to be excited.

Since excited atoms are unstable, they easily release their energy. Normally, the energy is released at random. In a laser, however, the emissions are *stimulated*. The first photon in the highly excited substance to be emitted will be reflected back and forth within the laser. In the process, this photon will strike billions of other excited

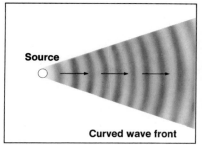

Curved wave front

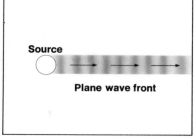

Plane wave front

23-10 What is the effect of the curved wave front and the plane wave front on the light's energy?

atoms in a fraction of a second. This, in turn, triggers (or causes) the emission of more photons. These photons are all in phase, or coherent. A chain reaction is set up, and a great deal of energy stored in the atoms is released at the same time.

The triggering action is referred to as *amplification*. Because one end of the laser is only partially mirrored, a certain small fraction of the radiated energy leaks through. However, this small fraction still contains a great amount of energy.

The first laser devices were able to produce only pulses of light. Now, however, *continuous wave* lasers emit a constant beam of light. The substances used in lasers include gases, liquids, and solids.

Lasers have many uses. There is a wide range of uses for the laser. For example, it can be used as a range finder. The time interval between the sending of the laser pulse and receiving the reflection can be recorded and the distance computed. Using the laser reflector placed on the moon by astronauts Armstrong and Aldrin, scientists were able to measure the distance to the moon to within a few centimeters and were able to locate the exact spot of each moon-landing site.

The laser can amplify heat as well as light. Laser heat has been used to drill tiny holes in diamonds. Drilling, which formerly took three days, was done in 2 min by the laser beam.

The laser is also used in medicine. The highly focused beam is used for delicate surgery. For example, a laser can be used to "weld" a detached retina to the back of the eye. The pulse of energy acts within a thousandth of a second. In fact, laser surgery takes place so fast that normal eye movements do not cause a blurring of the selected

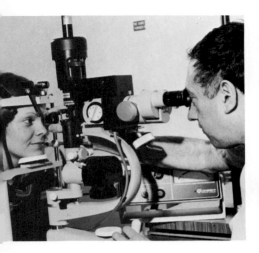

23-11 How is the laser used in eye surgery?

23-12 Which surface reflects the light evenly? unevenly?

"welding" site. See Fig. 23-11. Lasers have also been used in brain surgery and in the treatment of skin cancers.

The laser also has great promise as a communications device. A single beam from a laser may one day be able to carry all the TV programs of the world at once.

Lasers are used to make highly life-like photos called *holograms*. A hologram gives a view that changes depending on the angle at which it is viewed.

the reflection of light

Images are formed by even reflection. Light that strikes an object can be reflected in two basic ways. The reflected light is either *even* or *uneven*. See Fig. 23-12.

If light strikes a smooth surface the light is reflected in an even pattern. If the surface is smooth enough, as in a mirror or highly polished metal plate, you can see an image in that surface. If the surface is rough, as in a plastered wall or a sheet of paper, the reflected light is scattered or *diffused*. The reflected light is uneven and cannot form an image. Instead, diffusion makes the surface itself visible.

A perfect plate-glass wall mirror would not be visible. You could only see images of other things reflected in it, not the glass itself. On the other hand, when you look at a plastered wall, you see no images, only the wall itself.

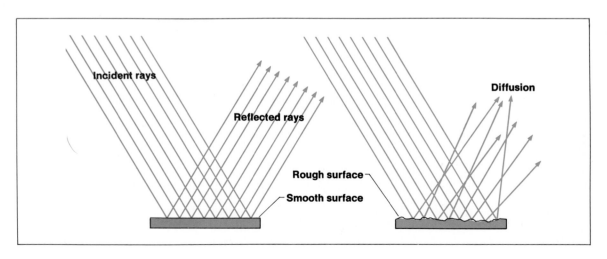

Incident rays

Reflected rays

Diffusion

Rough surface

Smooth surface

Plane mirrors obey the law of reflection. A flat, polished surface that reflects light without diffusing it is called a *plane mirror*. As shown in Fig. 23-13, the angle at which light is reflected from a mirror is the same as the angle at which it strikes the mirror. In other words, *the angle of reflection equals the angle of incidence.* A ball bouncing on the sidewalk obeys the same **law of reflection.**

When light is reflected from a plane mirror, the pattern of the light rays is even or undisturbed. In a sense, the rays are simply "folded over." This is why you can see an *image* by looking into a mirror.

Figure 23-14 shows how an image is formed by a plane mirror. Note the path of the rays as they travel from the object P, to the mirror, and then to the eyes. Your eye has no way of telling that the light rays have been reflected. Instead, the light rays seem to come from a point behind the mirror. You "see" the object at that point. In Fig. 23-14, that point is marked P' and is called the image.

You have seen that images formed by plane mirrors are life-sized, equal in size to the object. The image is located as far behind the mirror as the object is in front of it. Images in plane mirrors are said to be "left-handed," or mirror images. Stand in front of a mirror and reach out with your right hand to shake hands with your image. Which hand does the image extend?

Images formed by plane mirrors are called *virtual images*. They cannot be projected on screens but can be seen only by looking into the mirror itself.

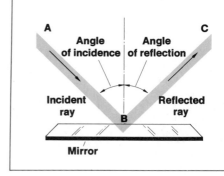

23-13 Can you tell what the law of reflection is by studying this diagram?

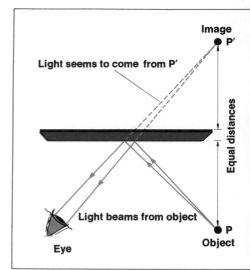

23-14 Compare the distance from the object to the mirror and from the image to the mirror. How do they compare? Where does the object appear to be?

activity

Reflections from a plane mirror. Place a clock in front of a plane mirror. Observe the image of the clock face in the mirror. Do the numbers appear different?

Now hinge two plane square mirrors together with masking tape and stand them on a table, one at a 90° angle from the other. Place the clock in front of these mirrors and again observe the images. What do you observe? How do you account for the difference between the images?

Curved mirrors also obey the law of reflection. There are two basic kinds of curved mirrors: *concave* or *convex*. The inside of an orange peel has a concave shape, while the outside is convex.

Curved mirrors are often spherical, or shaped like a round section cut from a sphere. However, mirrors may have any kind of curve whatsoever.

Mirrors all follow the law of reflection discussed earlier. At whatever point on the mirror a ray of light falls, it will be reflected in such a way that the angle of reflection equals the angle of incidence. Because the surface is curved, however, the images that result are quite different from those formed by a plane mirror.

You may have seen the odd images in the curved mirrors in an amusement park "fun house." You may also have seen the images in a silver bowl or spoon, in soap bubbles, or in the curved chrome on cars. These images are reflected from surfaces that are, or act like, curved mirrors.

However, curved mirrors have more important uses. In the world's largest telescopes, curved mirrors focus the light from distant stars. Curved mirrors are also used in searchlights, auto headlights, and flashlights to direct light to a certain spot.

23-15 Is the image formed by this curved mirror real or virtual? Is the image right side up or upside down?

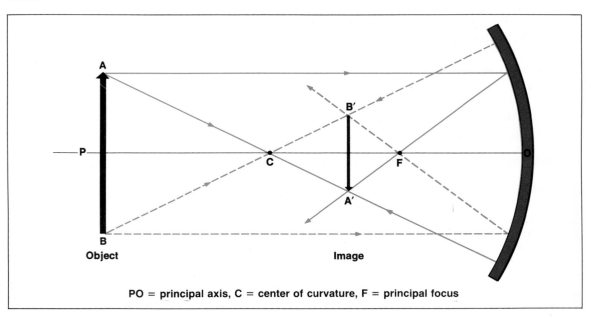

PO = principal axis, C = center of curvature, F = principal focus

unit 4 **wave motion and energy**

Concave mirrors can form real, inverted images. In Fig. 23-15, a spherical concave mirror is shown. The center of the sphere, C, of which the mirror is a section, is called the *center of curvature*. Line PO is called the principal axis and point F is the principal focus. All light rays that reach the mirror parallel to the principal axis will be reflected through the principal focus. By tracing the path of several rays from the object, you can locate its image.

An object, such as the arrowhead AB, is placed in front of the mirror at a point beyond C. A light ray from the tip A, parallel to the principal axis, hits the mirror and then reflects back through F. Another ray from A through C hits the mirror "head on" and reflects back upon itself along the same path. The image of the object at point A is located where these two reflected rays cross, at A'. Points B and B' are located in the same way.

Concave mirrors are used in reflecting telescopes. One of the largest telescopes in the world, at Mt. Palomar, California, has a mirror more than 5 m in diameter. The reflecting surface is commonly coated with aluminum. See Fig. 23-16.

Concave mirrors make good magnifiers. If you put an object inside the focus, F, of a concave mirror, the virtual image formed in the mirror is larger than the object. This is the reason you see the magnified image of your face in a shaving or makeup mirror.

The reflecting surface of a convex mirror bulges outward. The silvered balls used to decorate Christmas trees are good examples of convex mirrors. Convex mirrors are sometimes used as rear-view mirrors for cars and trucks. Although the image formed by a convex mirror is smaller than the object, it does provide a wide field of view. The image formed is a virtual image and cannot be projected. Virtual images are always erect.

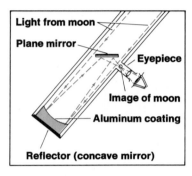

23-16 What kind of mirrors are used in this reflecting telescope? How many mirrors are there?

activity

Images in a concave mirror. The fact that a concave mirror forms an image (Fig. 23-15) can be shown in the following test. In a dark room, place a lighted candle several feet in front of a concave mirror. Move a small card back and forth between the candle and the mirror until an image of the flame can be seen. Is the image larger or smaller than the candle? Is it erect or inverted?

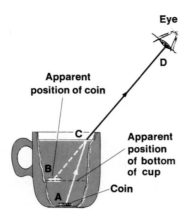

Apparent position of coin

Apparent position of bottom of cup

Coin

23-17 Why does the coin become visible when water is added to the cup?

the refraction of light

Seeing is not always believing. Have you ever ridden in a car on a hot, dry afternoon and seen that at times the highway ahead looked wet? The road looked as if there were reflections in it, as in a pool of water. But when you reached the spot, the water was gone.

This illusion is an example of a **mirage.** *A mirage is an optical illusion caused by the bending of light rays as they pass through air layers of different densities.*

This bending can also be viewed by putting a penny in the bottom of an empty cup and standing back so that the cup's rim just barely hides the penny from view. See Fig. 23-17. Without changing your position, pour water into the cup being careful not to disturb the penny. What happens? The penny seems to rise in the cup until you can see it clearly.

If you dip a straight stick at an angle into water, the stick appears to be sharply bent at the water line. This effect is also due to the bending of light rays.

Light travels in a straight line only so long as it passes through a single medium, such as air, water, or glass. When light goes at an angle from one medium into another of different density, it is bent or curved.

Light refracts because it changes speed. When light goes from air into glass, it slows down from 300,000 km/sec to 200,000 km/sec. If it enters glass at an angle, light is bent into a new angle.

The bending of light can be compared to a column of soldiers marching from firm ground into a soggy swamp. As the first row of soldiers steps into the swamp, they slow down. The second and following rows keep coming and bunch up as they too enter the swamp. The soldiers continue through the swamp, traveling more slowly and closer together than they did on the firm ground. See Fig. 23-18**A**.

Much the same thing happens when the column enters the swamp at an angle. However, in addition, the column is swung around slightly into a new angle of travel. Can you see how this might happen?

The soldiers in the front row do not all step into the swamp at the same time. Instead, they enter it one at a

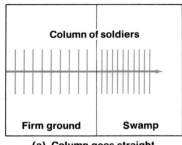

(a) **Column goes straight into swamp**

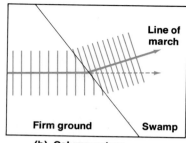

(b) **Column enters swamp at an angle**

23-18 What happens when the soldiers go directly into the swamp? when they enter the swamp at an angle? How does the column compare to light beams?

time. As they enter, each soldier falls slightly behind the soldier to the right. Thus, the line of soldiers who were formerly all side-by-side is now turned to a new angle. The following rows of soldiers bunch up behind. See Fig. 23-18**B**.

Which way is light bent when it goes from one medium to another at an angle? If light passes at an angle from a substance like air into a denser substance like water or glass, it bends into the denser substance. This is shown in Fig. 23-19. When light passes at an angle from a dense substance into one that is less dense, it again bends toward the denser substance.

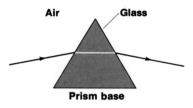

23-19 In which direction is light bent when it enters a denser medium? a less dense medium?

Light travels at different speeds in different substances. As you can see in the example of the marching soldiers, the more they slow down, the more the lines bend. So the degree to which a substance is able to bend light is related to the speed of light in that substance. This refracting or bending ability of a substance is related to its *index of refraction.*

You can find the index of refraction of a substance by dividing the speed of light in a vacuum by the speed of light in that substance. In most cases, the speed of light in air, rather than in a vacuum, is used because there is very little difference between the two speeds.

The index of refraction is a property of a substance and can be used to identify that substance. For instance, diamond has an index of 2.42, while glass has an index of about 1.5. A jeweler can easily tell a fake from a real diamond by measuring its index of refraction.

Not all glass has an index of refraction of 1.5. The exact value depends on the kind of glass. Several indexes of refraction are shown in Table 23-2.

chapter 23 the nature of light

429

Table 23-2:

Indexes of Refraction	
Vacuum	1.0000
Air	1.0003
Ice	1.31
Water	1.33
Glass (crown)	1.52
Glass (flint)	1.61
Diamond	2.42

sample problem

Find the index of refraction (IR) of glass in which light travels at 200,000 km/sec.

solution:

Step 1. Index of refraction = speed of light in air ÷ speed of light in glass, or

$$IR = \frac{V \text{ air}}{V \text{ glass}}$$

Step 2.

$$IR = \frac{300,000 \text{ km/sec}}{200,000 \text{ km/sec}}$$

$$= 1.5$$

Air can also cause refraction. Most often you can disregard the difference between the speed of light in air and its speed in a vacuum. However, this difference does result in a slight bending of light as it enters the atmosphere. This bending allows you to see the sun before it rises and after it sets. Figure 23-20 shows how this happens.

Suppose that you are at C, looking toward the setting sun. As the ray AB enters the air, it bends as shown. This refraction occurs because the ray is traveling into denser air and, as a result, slows down. As you can see, the effect of this refraction is to "lift" the sun above the horizon. Refraction lengthens the hours of daylight because the earth receives the sun's rays before the sun actually rises and after it sets.

You can see other ways in which air affects light. Look at a distant object using a line of sight that is directly above a heat source such as a candle, Bunsen burner, or hot plate. The dancing of "heat rays" on a hot metal roof, the shimmering of a hot roadway in the distance, and the twinkling of stars are all examples of distortions caused by the changing refraction of the light as it passes through air of varying densities. In working out star positions, astronomers must correct their readings to allow for the refraction of air.

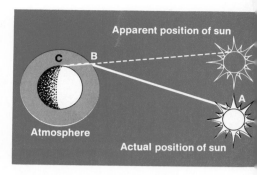

23-20 Why is the sun visible when it is still below the horizon?

Lenses refract light. A *lens* is a piece of transparent substance having at least one curved surface. Although lenses differ in shape, size, and substance, the chief purpose of all lenses is to refract light. The amount of refraction produced by a lens depends both upon its shape and upon the index of refraction of the substance of which it is made.

Lenses are used in telescopes, cameras, eyeglasses, microscopes, spectroscopes, movie projectors, and many other optical devices. To understand how lenses affect light, first look at the path of light through a simple prism.

Refer to Fig. 23-19 (p. 429) showing the path of a light ray through a glass prism. The ray bends into the prism as it enters the glass and away from the prism as it leaves. The effect is to bend the light beam toward the base, or thick part, of the prism.

If two prisms are set base-to-base, as in Fig. 23-21**A** on p. 432, light rays come together, or *converge*, after passing through the two prisms. The light is refracted toward the thicker part of each prism. The double convex lens in Fig. 23-21**B** is very much like two prisms joined together with their side points smoothed down. Note that a convex lens is thicker at the center than at the edges. Since it always brings parallel light rays together, it is called a *converging lens*.

If two prisms are placed together point-to-point, refraction causes the light rays to spread apart, or *diverge.* As shown in Fig. 23-21**C,** a double concave lens is like two prisms joined together, point-to-point. A concave lens is

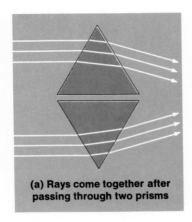

(a) Rays come together after passing through two prisms

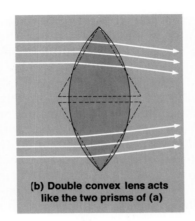

(b) Double convex lens acts like the two prisms of (a)

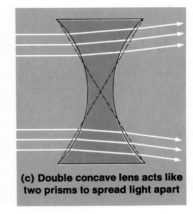

(c) Double concave lens acts like two prisms to spread light apart

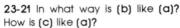

23-21 In what way is (b) like (a)? How is (c) like (a)?

thinner at the center than at the edges and tends to spread light rays apart. Therefore, it is called a *diverging lens*.

A lens has a focal length. A common magnifying glass is a good example of a convex lens. With it, you can do some tests that will help you see how lenses affect light. First, focus the light from the sun on a sheet of paper by moving the lens back and forth until a dazzling spot appears on the paper. The spot appears where the light from the sun has converged after passing through the lens. This spot is so hot that the paper may char or even burn. See Fig. 23-22. CAUTION: *Do not look directly at the sun.*

The light of the sun comes together at what is called the *principal focus* (or *focal point*) of the lens. This is the point at which the rays parallel to the principal axis converge after passing through the lens. The distance from the center of the lens to the principal focus is called the *focal length* of the lens. To find the focal length of a lens, simply measure the distance between the magnifying glass and the bright spot on the paper.

23-22 How can you find the focal length of a convex lens?

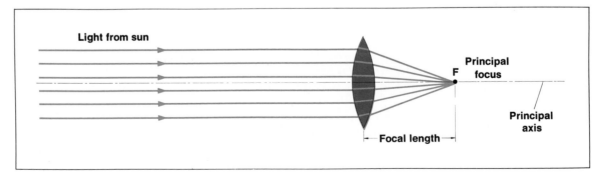

Light from sun

Principal focus

F

Focal length

Principal axis

Using a convex lens in a darkened room, focus the image of a bright, distant object on a sheet of paper. This kind of image is called a *real image*. Notice that it is in full color, inverted, and in this case, smaller than the object. The image is located very close to the principal focus of the lens. See Fig. 23-23.

The image is smaller than the object because the object is so far away. The closer the object is to the lens, the larger the image will be. The following formula shows how the distances and sizes of objects and their images are related to each other:

$$\frac{\text{size of image}}{\text{size of object}} = \frac{\text{distance of image}}{\text{distance of object}}$$

$$\frac{s_i}{s_o} = \frac{d_i}{d_o}$$

For example, if the image were 10 times as far away from the lens as the object, the image would be 10 times larger than the object. The magnification of the lens is 10, or 10**X**.

To see how lenses form images, use a luminous object, such as a lighted candle, in a darkened room. Set up the equipment as shown in Fig. 23-23. Move the lens slowly along the stick until you get a clear, inverted, real image of the lighted candle on the screen.

23-23 What is the relationship between object size and image size? object distance and image distance? Why is the image "real"?

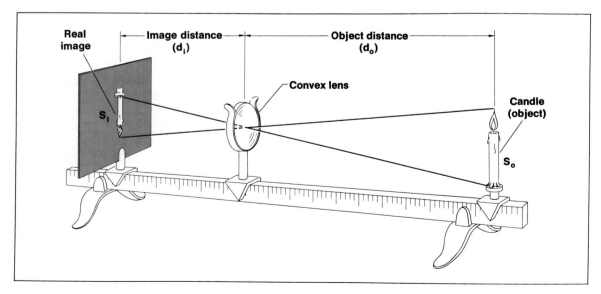

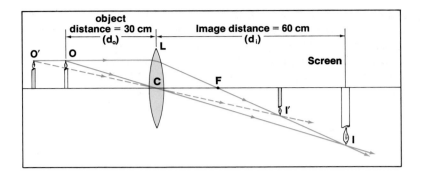

23-24 What happens to the image size and distance if the candle is moved from point O' to O?

Suppose you moved the candle from one point to another farther away. What happens to the image? From Fig. 23-24 you can see that the image moves closer to the lens and becomes smaller.

If the lens had a focal length of 20 cm and the candle was 30 cm from the center of the lens, where would the image be? To find the image of point O, for example (the top of the candle flame), you must find where rays of light from that point come together again after passing through the lens. You know that the ray parallel to the principal axis OL will refract through F, the principal focus. Another ray, OC, drawn through the center of the lens, will pass straight through without refracting. At the point I where both rays intersect is the image of point O.

Try following the same procedure for the base of the candle. A ray of light from the base, parallel to OL, goes straight through C and will not be refracted.

If you repeat this process for other points on the object candle, you will be able to draw the whole image of the candle. The image will be at the point shown in the diagram that is 60 cm from the lens. Since the image is twice as far away as the object, it is also twice as large. Like all real images, it is inverted.

Images can be located. As you move the candle nearer and farther away from the lens, you will find that you will have to move the screen in order to focus the image. The closer the candle is to the focal point, the farther away is the image. This relationship between object and image locations is shown by the following formula:

$$\frac{1}{f} = \frac{1}{d_o} + \frac{1}{d_i}$$

As shown in Fig. 23-24, d_o is the object distance, d_i is the image distance, and f is the focal length. Instead of locating the image by drawing a diagram, you can find its location by using the formula. If the object is 30 cm from the lens and the focal length is 20 cm, you can solve for d_i:

$$\frac{1}{20} = \frac{1}{30} + \frac{1}{d_i}$$

$$d_i = 60 \text{ cm}$$

sample problem

A 6-cm stick stands upright 12 cm from a convex lens that has a focal length of 3 cm. (a) How far away from the lens is the image located? (b) How large is the image?

solution:

(a) Given: $d_o = 12$ cm
$\qquad\qquad f = 3$ cm
$\qquad\qquad d_i = ?$

Use the following formula:

$$\frac{1}{f} = \frac{1}{d_o} + \frac{1}{d_i}$$

Substituting, $\quad \dfrac{1}{3} = \dfrac{1}{12} + \dfrac{1}{d_i}$

$$d_i = 4 \text{ cm}$$

(b) Given: $d_i = 4$ cm
$\qquad\qquad d_o = 12$ cm
$\qquad\qquad s_o = 6$ cm

Use the following formula:

$$\frac{s_i}{s_o} = \frac{d_i}{d_o}$$

Substituting, $\quad \dfrac{s_i}{6} = \dfrac{4}{12}$

$$s_i = 2 \text{ cm}$$

Lenses have many uses. In Fig. 23-24, you saw that as the object is moved closer to the lens, the image moves farther away, becoming larger. If the object is placed just outside the focal point, the image is very large and far away. This is how slides and movie films are projected on a screen. The object is the lighted slide or film. It is placed just outside the focal length of the projector lens. Thus, a large, inverted image is formed on the distant screen. Why is the slide placed into the projector upside down?

If the object is moved closer to the lens, until it is at the focal point, no clear image appears on the screen. Instead, the rays from the object are parallel after passing through the lens. If a point source of light is placed at the focal point, a parallel beam of light can be projected.

If the object is moved still closer to the lens, inside the focal point, the rays will spread apart after they go through the lens. See Fig. 23-25. Remember, real images are formed only when rays coming from the object converge after passing through the lens. When the object is located inside the focal point, no real image can be formed.

However, a virtual image can be seen by looking through the lens. As shown in the diagram, the image is enlarged, and right-side-up. To read with a magnifier, you hold it closer to the page than the focal length of the lens and keep your eye close to the lens.

Have you ever wondered how a camera works? A convex lens focuses an image on the film. In taking pictures of distant objects, the distance between the lens and the film is almost equal to the focal length of the lens. Can you see why?

To get sharp pictures of nearby objects, the distance between the film and the lens must be increased. Most cameras are equipped with an adjustment to change the film-lens distance. A good lens, with a large opening, can

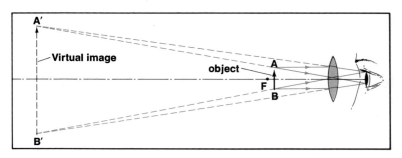

23-25 Why is the image upright in this diagram of a microscope?

unit 4 wave motion and energy

give sharp pictures with a short exposure time. Why is a short exposure time needed to get a clear picture of a moving object?

The eye is like a camera. In many ways, your eye acts like a camera. However, it does not record an image on film. Instead, your eye is linked to your brain somewhat like a television camera is connected to its transmitter.

Light enters your eye through the tough transparent skin called the *cornea*, where it is refracted. It then goes through a flexible convex lens, where it is bent even more. This refraction causes an image to form on the *retina* (ret-ih-nuh). Like the film of the camera, the retina is sensitive to light. Light striking the retina produces nerve impulses that travel through the *optic nerve* to the *visual center* of the brain. Here in the brain the sensation of "seeing" occurs.

The amount of light that can enter your eye is controlled by the *iris,* a colored ring that opens up in the dark and closes down to a tiny opening in bright light. The iris, which is just outside the lens, acts like the *diaphragm,* an adjustable opening of the camera. Your *eyelids* correspond to the shutter of the camera. They both open and close to expose the film. See Fig. 23-26.

The lens of your eye is flexible. It can bulge out or flatten to focus on objects that are near or far away. To focus on nearby objects, muscles in your eye cause the lens to bulge out, shortening the focal length of the lens.

When focusing on objects over 6 m away, the lens is almost flat and the muscles are relaxed. When relaxed, the focal length of the lens is almost equal to the distance from the lens to the retina.

Sometimes, however, images do not form clearly on the retina. A *nearsighted* person has eyeballs that are too long, or lenses that are too thick. In either case, the image is brought to focus in front of the retina instead of on it. For this reason, a nearsighted person sees a fuzzy image of distant objects. See Fig. 23-27.

Nearsightedness can be corrected by using concave lenses in front of the eye. These lenses cause light rays to diverge or spread out. As a result, the converging effect of the person's own eye lens is reduced.

The *farsighted* eyeball is too short or has a lens that is too flat. This produces a fuzzy image of nearby objects. As

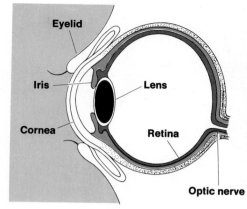

23-26 How does the iris of the eye act like the diaphragm of a camera?

23-27 How does a concave lens correct nearsightedness?

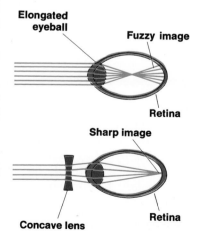

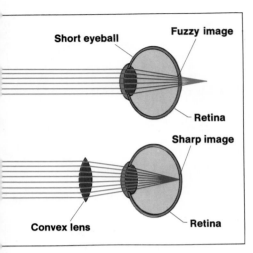

Short eyeball — **Fuzzy image** — **Retina**

Sharp image — **Convex lens** — **Retina**

23-28 How does a convex lens correct farsightedness?

shown in Fig. 23-28, a convex lens can correct this fault. Middle-aged persons lose the ability to focus on nearby objects because their lenses become stiff. Sometimes they need to wear *bifocals* (lenses with two different focal lengths) or *trifocals* (lenses with three different focal lengths) to be able to focus on objects at different distances.

Movies are illusions. You are familiar with how a movie projector throws an enlarged image of the film on a screen. Movies are an illusion formed by a rapid succession of such projected images. Why do they seem to move?

When you look at a lighted lamp, a real image of the lamp forms on your retina, producing the sensation of vision in your brain. When the lamp is off, the image disappears, but you still "see" it for about 0.10 sec. This effect is called *persistence of vision*.

Persistence of vision makes movies possible. The movie projector throws a still picture on the screen for just a fraction of a second. Then, the picture is replaced by a slightly different one. This process is repeated 24 times per second in commercial movies, 16 times per second in home movies, and 30 times per second on a television.

Because of persistence of vision, you retain the image of one picture while the next one is shown, and the two pictures blend together. Since each picture differs only slightly from the previous one and merges smoothly with it, you see what appears to be constant motion on the screen.

activity

Flip movies. You can get an illusion of motion by making some "flip-pictures." Using a small note pad, start with the *last* page and make a simple line drawing, such as a human "stick" figure who is walking. Then turn to the *preceding* page and draw the same figure in the same place on the page, with very slight changes in stride. Keep on doing this until you have enough pages to flip the pictures rapidly with your thumb. Can you explain the result?

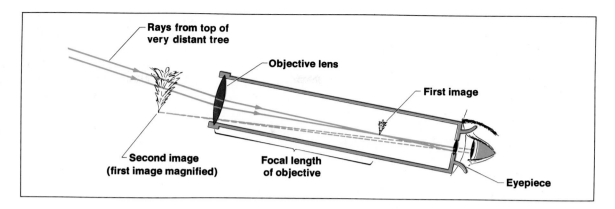

Rays from top of very distant tree

Objective lens

First image

Second image (first image magnified)

Focal length of objective

Eyepiece

Some optical devices use more than one lens. The eye and the optical devices discussed up to now all had only a single lens to refract light. To correct for a vision problem, one lens is added in front of the eye's own lens. Now, you will see how two lenses are used together in an optical device.

One such device is a *refracting telescope.* You will recall that you learned about a *reflecting* telescope earlier in the chapter, when a mirror was used to focus the light. In a refracting telescope, however, a lens instead of a mirror is used to focus the light. See Fig. 23-29.

A large convex lens, called the *objective,* is placed at the outer end of the long telescope tube. The objective lens of a telescope has a long focal length. A real inverted image of a distant object is formed by the lens near the lower end of the tube, very close to the principal focus of the lens. The image is viewed through an *eyepiece,* a small convex lens that magnifies the image.

The *magnifying power* of this kind of telescope is found by dividing the focal length of the objective *(F)* by the focal length of the eyepiece *(f):*

$$\text{magnification} = \frac{F}{f}$$

Although the image is inverted, this is not a problem when viewing the moon or stars. Why? When the refracting telescope is used to view objects on the earth, however, another lens is added to turn the image right-side-up.

While the telescope permits people to see objects that are far away, the *microscope* lets them view objects that are very near and very small. See Fig. 23-30.

23-29 How does the formation of an image in a refracting telescope differ from that in a reflecting telescope?

23-30 Compare this diagram of a compound microscope with that of a simple microscope in Fig. 23-25. What are the major differences?

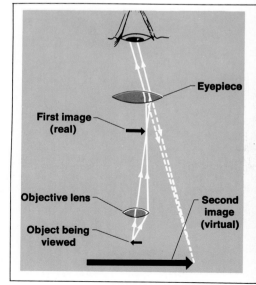

Eyepiece

First image (real)

Objective lens

Object being viewed

Second image (virtual)

The microscope has two convex lenses with the objective lens at the lower end of the tube. The objective lens has a very short focal length. The object to be viewed is placed just beyond the principal focus of this lens. A real, inverted, and enlarged image of the object is formed near the upper end of the tube. The eyepiece is used to magnify the image still further.

The overall magnifying power of a microscope is found by multiplying the separate powers of each lens. For example, if a microscope has an objective lens that enlarges 44X and an eyepiece that enlarges 10X, the final enlargement is 440X.

The highest practical power possible with the *optical* microscope is about 2000X. The *electron* and *field-ion* microscopes give magnifications that are much higher than this, but they do not depend on the refraction of light for their operation.

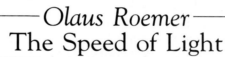

Olaus Roemer
The Speed of Light

How can you measure the speed of something as fast as light? Early scientists believed that it took no time at all for light to go from one place to another. In the seventeenth century, however, a Danish astronomer, Olaus Roemer, found a means of measuring the speed of light.

The discovery came through Roemer's study of the moons of Jupiter. Roemer timed the eclipses of the inner moon. He noted that it went behind the planet every 42 hr 28 min. Most people would have been content to record the periods of time for perhaps several dozen times, but Roemer kept at the job for several months.

Then he noticed a slight drift in the time from one eclipse to another. After six months, the total drift was over 16 min. To Roemer's surprise, the 16 min were gained back during the next six months. What would account for this drift? Roemer even built a special clock to be sure that his timings were accurate. He kept on sighting and recording for four years and each year 16 min were lost and then gained back.

Roemer finally reasoned that distance was the only variable. Therefore, it must take some time for the light to travel the changing distance. When the earth is on the same side of the sun as Jupiter, the two planets are

300,000,000 km closer together than when they are on opposite sides of the sun. Roemer found that by dividing the time (about 1000 sec) by the distance (300,000,000 km), he obtained the speed of light. His value of 300,000 km/sec has proved to be surprisingly accurate.

summary

1. Light has properties of both waves and particles.
2. Light is produced by excited atoms, that is, electrons moving from one energy level to a lower level.
3. Light striking your eyes allows you to see.
4. Light energy can be measured.
5. The laser beam produces monochromatic coherent light.
6. Images are formed by even reflections.
7. The law of reflection states that the angle of incidence is equal to the angle of reflection.
8. Plane and curved mirrors obey the law of reflection.
9. Light is bent when it passes at an angle from one medium to another.
10. The index of refraction is a ratio of the speed of light in air to that in another medium.
11. Convex lenses converge or bring light rays together.
12. Concave lenses diverge or spread light rays apart.
13. The eye acts like a camera.

review

Match the items in the left column with the best response in the right column. *Do not write in this book.*

1. photon
2. opaque
3. concave
4. convex
5. refraction
6. nearsightedness
7. farsightedness
8. reflection
9. real image
10. virtual image

a. bending of light rays
b. emitting its own light
c. image that can be projected
d. image that cannot be projected
e. lens of even thickness
f. light bouncing off a surface
g. not able to transmit light
h. shaped like the inside of an orange peel
i. shaped like the outside of an orange peel
j. single "packet" of energy
k. weakness in distance vision
l. weakness in near vision

questions

Group A
1. How do you know that light can pass through a vacuum?
2. When does an atom radiate light energy?
3. List four different ways of producing light.
4. Why is a candle flame visible? a tree?
5. What can happen to light when it strikes an object?
6. Describe some uses of lasers.
7. What is the unit of lamp brightness?
8. What is the law of reflection?
9. How fast does light travel in a vacuum?
10. Why does a stick appear bent when it is dipped into water at an angle?
11. How do convex and concave lenses affect parallel light rays?
12. Where do rays from a distant object converge after passing through a convex lens?
13. If the object is a great distance away from a convex lens, where is the image located?
14. Where would you hold a magnifier to look at an object?
15. What is persistence of vision?
16. What causes nearsightedness? How can it be corrected?
17. What causes farsightedness? How can it be corrected?
18. Name three parts of a camera that correspond to parts of a human eye.
19. In what part of a refracting telescope is the objective lens found?

Group B
20. What evidence is there to show that light is a stream of particles? that light travels in waves?
21. Does polarization show that light waves are transverse or longitudinal? Explain.
22. Distinguish between opaque, transparent, and translucent.
23. How do incandescent and fluorescent bulbs compare in efficiency?
24. What are coherent light waves?
25. How is normal light different from monochromatic light?
26. What type of surface produces an uneven reflection? an even reflection?
27. What is the relationship between the distance of an object from a plane mirror and the distance of its virtual image from the mirror?
28. Why are images formed in plane mirrors sometimes referred to as "left-handed" images?
29. How is the index of refraction found?
30. When refraction occurs, in which direction is the light ray bent?
31. What is the relationship between the distances and sizes of objects and their images?
32. Where would you put a convex lens to throw an enlarged image of a lighted bulb on a screen?

unit 4 wave motion and energy

problems

1. What happens to its apparent brightness as a torch moves from a point 10 m away from you to a point 40 m away?
2. How many 100-watt bulbs are needed to equal the total lumens of one 1000-watt bulb? (See Table 23-1 on p. 421.)
3. How many watts of input are needed to light all the 100-watt bulbs needed in problem 2?
4. When you stand 3 m in front of a plane mirror, exactly where is your image?
5. Find the speed of light in water if the index of refraction for water is 1.33.
6. If the image of a 5-cm candle is located three times as far from the lens as the object, what is the size of the image?
7. If an object is 60 cm away from a convex lens with a focal length of 20 cm, where is the image located? How large is the image in relation to the size of the object?
8. Find the magnifying power of a refracting telescope if the focal length of the object lens is 10 m and of the eyepiece is 10 cm.
9. What is the total magnification of a microscope if the objective lens magnifies the object 36X and the eyepiece magnifies the image 10X?
10. If the light from the sun takes 500 sec to reach the earth, how far away is the sun?

further reading

Adler, Irving, *Story of Light*. New York: Harvey, 1971.
Brown, Sam, *Homebuilt Telescopes*. Edmund Scientific.
Froman, Robert, *Science, Art, and Visual Illusions*. New York: Simon & Schuster, 1970.
Hawken, William R., *You and Your Lens*. Garden City, N.Y.: Amphoto, 1975.
Headstrom, Richard, *Adventures with a Hand Lens*. New York: Dover, 1976.
Hoffman, Banesh, and Dukas, Helen, *Albert Einstein: Creator and Rebel*. New York: Viking, 1972.
Jacobs, Lou, Jr., *You and Your Camera*. New York: Lothrop, Lee & Shepard, 1971.
Kentzer, Michael, *Waves*. Cleveland: Collins-World, 1977.
Meyer, Jerome S., *Prisms and Lenses: How They Work*. Cleveland: Collins-World, 1972.
Mims, Forrest M., *Lasers: The Incredible Light Machines*. New York: David McKay, 1977.
Schwalberg, Carol, *Light and Shadow*. New York: Parents', 1972.

color

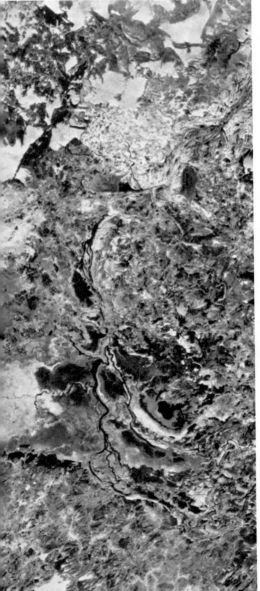

In Chapter 23, you learned that light is a form of energy that behaves both as a particle and as a wave. In this chapter, you will see that the frequency of light causes color. You will also see that visible light is only a small part of a wide range of radiations.

objectives:

☐ To identify the place of light in the electromagnetic spectrum.

☐ To demonstrate how a prism is used to separate white light into colors.

☐ To identify the primary colors of light.

☐ To identify the primary colors of pigments.

☐ To describe how the spectroscope is used in research.

☐ To describe the Doppler effect.

the electromagnetic spectrum

Light is part of the electromagnetic spectrum. You learned from your study of sound waves that the shorter the wavelength, the higher the frequency of the sound. The same principle applies in studying the wavelength of light. High-frequency light has a short wavelength.

The wavelength of violet light is the shortest wavelength that can be seen by the human eye. The frequency of violet light is 2500 waves per millimeter. Red light, however, has a frequency of only about 1300 waves per millimeter. You can easily tell from these figures that violet light has a much shorter wavelength than red.

There are many waves with frequencies higher and lower than those of visible light. These waves are all alike except for frequency and wavelength. Frequency and wavelength account for all the different effects observed. The entire range of waves is known as the *electromagnetic spectrum.* See Fig. 24-1.

Frequencies below those of red light are called *infrared* rays. These rays, given off by hot objects, are commonly referred to as *radiant heat.* Infrared rays give a feeling of warmth when they strike the skin.

The frequencies just beyond the violet are called *ultraviolet light.* Since it has such a high frequency, ultraviolet light has more energy per photon than visible light.

You may already know about some of the effects of ultraviolet light. Camera film is very sensitive to ultraviolet *(black light),* but your eyes are not. This does not mean that ultraviolet light has no effect on your eyes. Quite the opposite is true, in fact. Because of its high energy per

24-1 The electromagnetic spectrum.

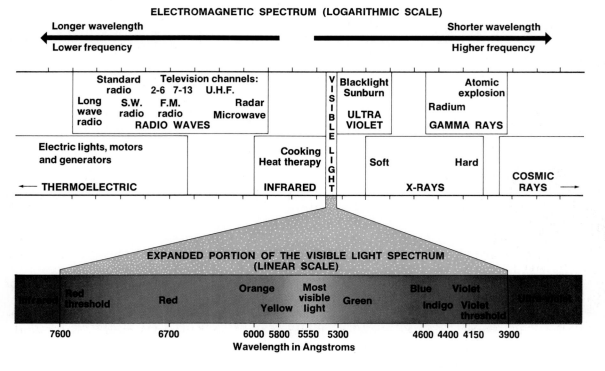

ELECTROMAGNETIC SPECTRUM (LOGARITHMIC SCALE)

Longer wavelength   Shorter wavelength

Lower frequency   Higher frequency

EXPANDED PORTION OF THE VISIBLE LIGHT SPECTRUM
(LINEAR SCALE)

Wavelength in Angstroms

photon, ultraviolet light can harm your eyes. This is why you should not look at an ultraviolet sunlamp when it is on. Ultraviolet rays tan and burn your skin, kill bacteria, and cause some substances to *fluoresce,* that is, to give off visible light.

Beyond ultraviolet are the *X rays,* with still higher frequencies and also greater photon energy. You may already know some uses of X rays and will study more about them later.

Still higher in frequency are the *gamma rays,* which are produced in the nucleus of the atom. Most radioactive elements, such as radium, emit gamma rays.

You will notice that there is no sharp dividing line between the different sections of the spectrum but that they overlap. For example, the low infrared can be used as radio waves; some electronic tubes produce X rays with the same wavelength as gamma rays.

You should also notice that in Fig. 24-1, the scale of frequencies increases ten times for each move of one unit to the right. Likewise, the wavelengths along the bottom of the spectrum decrease by ten times for each move of one unit to the right.

If this kind of scale were not used, there would not be enough room on a page to show the entire electromagnetic spectrum. Even a small section of this scale would cover kilometers of paper when completed as an even scale. The visible part of the spectrum, however, is such a small part of the entire spectrum that it can fit on the page as an even scale.

You will also see that in the expanded part of the visible light spectrum, the wavelengths are given in *angstrom* units (Å). Because wavelengths of light are so short, scientists have adopted a smaller unit of length than the centimeter or millimeter. The angstrom is so short that ten million angstroms are equal to 1 mm.

The eye cannot see all the colors of the visible spectrum equally well. Yellow is the easiest color to see, and the values drop off as the frequencies go higher or lower. The effect of the spectrum of visible light on the average eye is shown in Fig. 24-2.

The colors of light can be separated. Sir Isaac Newton was the first to make a careful study of color. Newton allowed a beam of sunlight to pass through a glass prism. When he projected the beam onto a screen, Newton

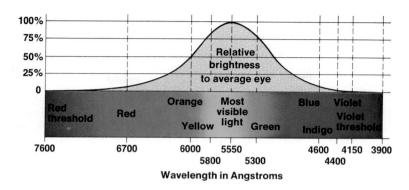

found that the light spread out into a band, or *spectrum,* of colors. See Fig. 24-3. Blending these colors together again by using a second prism, Newton regained the beam of white light.

The prism sorts out the colors, letting each one produce its own effect. How does the prism do this? You will recall that the speed of light changes as it passes from one

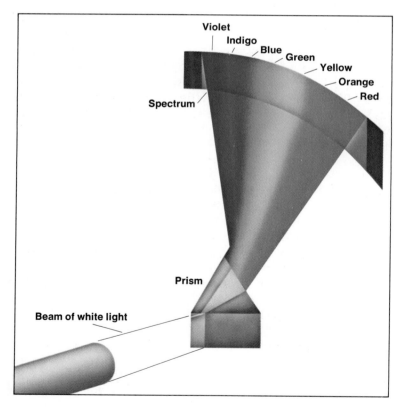

24-3 When white light is separated into a spectrum by a prism, why is violet bent most and red bent least?

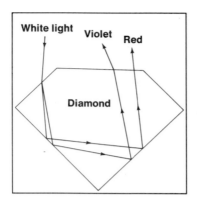

24-4 Why is it possible for a diamond or other clear gem to emit flashes of color, even in white light?

medium to another. In the last chapter, you learned that this change in speed causes light to bend or refract. Each color bends at a different angle. Violet light is slowed down the most when it passes into a glass prism, while red is slowed down the least.

Since the waves of violet light are slowed down more than the waves of the other colors in the glass prism, violet light is bent the most. Red light is slowed down the least, so it is also bent the least. The other colors, orange, yellow, green, blue, and indigo, are found between the red and violet. A gemstone, such as a diamond, acts like a prism to separate white light into colors. That is why flashes of color can sometimes be seen in a clear diamond. See Fig. 24-4.

You learned in the last chapter that a lens acts like a prism. Therefore, as light passes through a lens, it is not only refracted but separated into a spectrum of colors as well. How, then, does a camera lens focus a sharp image on film? Why aren't photos blurred?

The blurring is overcome by doing something similar to what Isaac Newton did with prisms. Newton used one prism to separate light and another prism to bring the light back together again. In much the same way, two or more lenses are used in cameras to produce clear pictures. The result is that light spread apart by one lens is combined again by another lens. In a good camera, seven or eight lenses are combined into two or three units to form a clear lens system. Such a system is a *color-corrected* lens and will produce a sharp image on color film.

Water, dust, and air separate colors. Raindrops sometimes act like tiny prisms. They can refract sunlight to separate out the colors. Figure 24-5 shows the path of two light rays through a raindrop. Note that the light is refracted, reflected, and then refracted again.

The combined effect of thousands of raindrops, all located at about the same angle from the observer, makes a complete rainbow. To see a rainbow in the sky, stand so that the sun is at your back and low in the sky. Try standing on top of a stepladder directly above a sprinkler when the sun is high in the sky. You should be able to see a full-circle rainbow in the spray below you.

Why is the sky blue during the day and red at sunset? The light of the sun must travel through several hundred kilometers of air on its path toward the surface of the

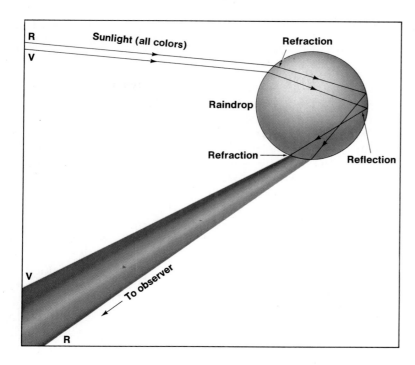

24-5 A rainbow is the result of refraction and reflection of light by raindrops. How could you make a rainbow?

earth. As sunlight passes through air, some of the light is scattered by tiny bits of air, water, and dust. Blue light, which has a shorter wavelength, is scattered most easily, and seems to come to your eyes from all directions. Green, yellow, and red light, on the other hand, have longer wavelengths, and can pass through the air with less scattering. This scattering of blue light gives the sky a blue color, and white objects viewed in direct sunlight have a slightly yellowish color.

Ultraviolet light, which you cannot see but which can give you a sunburn, is scattered even more easily than visible blue light. Therefore, scattered ultraviolet light can give you a sunburn even in the shade.

Sometimes the sky turns a bright reddish-orange at sunset. This effect is also explained by light scattering. In the evening, sunlight passes through the atmosphere at a sharp angle. Therefore, the light must travel farther through the atmosphere to reach the earth. During this long travel, blue light is scattered more than the other colors of the spectrum. Finally, only long-wavelength red light is able to get through the atmosphere to light the sky as the sun is setting.

chapter 24 color   449

the colors of objects

The color you see is reflected color. Why do some objects look white, others blue, and still others yellow when they are all lit by the same white light? The color you see depends on how these objects *absorb, reflect,* or *transmit* the light that falls on them.

You see an object as white if it reflects light of all wavelengths equally well. If an object does not reflect light at all, it is black. Black is the absence of light. A piece of black velvet absorbs light of all wavelengths and reflects none.

The amount of light reflected from different colored paper is shown in Table 24-1.

Table 24-1:

Reflecting Ability of Paper	
white paper	85%
ivory	67%
bright yellow	50–70%
dark red	14%
dark green	9%
dark blue	8%
flat black	2–4%

24-6 What effect do the filters have on the white light? What is reflected by the white paper? the red paper?

So far, only the color of objects seen in white light has been discussed. What happens when an object is viewed in filtered light? Hold a sheet of white paper in sunlight or in the beam of light from a projector. Now cover the source of light with a piece of red glass or plastic. The paper will look red.

The red filter absorbs light of all wavelengths except red. Since only red passes through, only red can be reflected from the white paper. See Fig. 24-6. Now use green glass or plastic instead of red. What color is the white paper now? Can you explain why?

Suppose you go one step further. Keep the red filter but use red paper instead of white. The red paper receives only red light. Red paper can reflect only red light, so, of course, the paper looks red. If, instead, you hold a green

unit 4 wave motion and energy

filter in the beam, only green light will strike the red paper. What color will the paper reflect? Since the red paper absorbs green light, it can reflect no light at all. As a result, the red paper looks black when green light shines on it.

Remember, an object's color depends on (1) the kind of light shining on it, (2) which wavelengths of this light it is able to transmit, if transparent, or (3) which wavelengths it reflects, if opaque.

activity

Changing colors. Darken the room and turn on a projector or other source of light. Cover the light with a colored filter. Hold up a series of colored objects in the light for the class to see. Use a variety of textures and colors such as cloths, foil wrapping paper, dishes, and papers. Ask the class to name the color of each object. Then view all objects again in white light. Do the objects look different?

The primary colors of light are red, green, and blue. One way to study the effect of mixing colors is to look at a spinning disc painted in various colors. Because of persistence of vision, which was studied in the last chapter, you would see the same effect as if all the colors were entering your eye at once.

A more practical way of mixing colors is to project beams of colored light on the same spot on a white screen. When all the colors of the spectrum are projected in this way, white is obtained. This result is like the second part of Newton's experiment, in which he used a second prism to join the colors that had been separated by the first prism.

White light can also be obtained by projecting the right amounts of only three colors: *red, green,* and *blue.* These colors are called the *primary colors* of light. (Note: These colors differ from the primary pigments, which will be discussed later in this chapter.) By changing the relative amounts of the primary colors, all of the colors of the spectrum can be produced. See Fig. 24-7.

Any two colored lights that produce white when mixed together are called *complementary colors.* Blue and yellow are such a pair of complementary colors. If blue is

24-7 What colors are produced when red and green lights are mixed? when blue is added?

24-8 Stare at the center of the flag for 20 sec. Then stare at a sheet of white paper. What colors appear on the white paper? Why?

24-9 The letter H is the same shade in each case. The disks are also exactly alike. But do they look alike? How do the background shades affect what you see?

removed from white light, the colors that remain will produce yellow. It would appear, then, that blue must contain all the colors except yellow. Therefore, if you mix blue with yellow, you will get white. Can you explain why laundry bluing helps whiten clothes that have become tinged with yellow? Red and blue-green are also complementary. When mixed together, they produce white.

There is a simple way to find the complement of any color. Stare steadily at a given color for about 20 sec. Then stare steadily at a sheet of white paper and you will see the complementary color appear before your eyes. For example, stare at a sheet of yellow paper. If you shift your gaze to a sheet of white paper, you will see a blue image appear. Why?

This effect is caused by *retinal fatigue*. That is, the cells on the retina of the eye become "tired" of a given color (yellow in this example), if it is viewed too long. Then, when you look at white paper, even though all colors enter the eye, the yellow does not "register." As a result, you see blue, the complement of yellow.

To see this effect with several colors at the same time, stare at Fig. 24-8. Now look at a sheet of white paper. Be sure that you hold your gaze steady both when looking at the colored object and again when staring at the white paper.

One theory of how the eye perceives color is that the retina of your eye has three types of cells. Each cell type is sensitive to one of the primary colors. If all three are equally stimulated, you see white. If only the cells

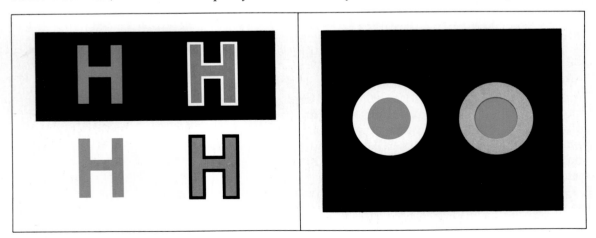

unit 4 wave motion and energy

sensitive to red or green are excited, you see yellow, which is between red and green in the spectrum. Other mixes of the three primary colors make it possible to detect all the colors.

Other factors also seem to affect the colors you see. For several examples of how background shades affect the brightness of colors, see Fig. 24-9.

Some people are *color blind*. That is, they have defective color vision. The very few people who are totally color blind see all objects in shades of gray. More often, however, a person is only partially color blind. Such a person cannot tell one color from another within a related color group. For example, red, green, and yellow might all look like shades of yellow, while green, blue, and violet might all look like shades of blue. See Fig. 24-10.

The primary pigments are red, blue, and yellow. Did you think mixing yellow light with blue light would give you green, not white? In fact, when you mix *pigments* instead of lights, yellow and blue do produce green. Until now, the discussion has been about mixing light. Mixing pigments produces a different effect. For example, mixing all the pigment colors produces black.

Grind some yellow chalk and blue chalk together with a little water. The mixture should be green. Yellow chalk appears yellow because it absorbs blue, indigo, and violet light. Blue chalk absorbs red, orange, and yellow light. Neither yellow nor blue chalk absorbs green light. Therefore, only green can be reflected from the mixture and enter your eyes.

The *four-color-printing process* is a good example of mixing pigments. A color picture is built up by printing the same piece of paper on four plates, one after the other. Each plate is inked with a different pigment. One plate is red, one blue, one yellow, and one black. Red, blue, and yellow are called the three *primary pigments*. See Fig. 24-11.

Color plays an important role in research. A prism is used in a device called a *spectroscope* to measure the wavelength of light. The prism refracts the light and a system of lenses is used to study the resulting spectrum. The spectroscope also has a scale on which the wavelength of light can be read directly.

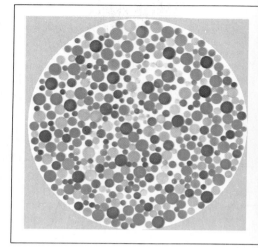

24-10 A typical test for red-green color blindness. What will a color-blind person person fail to see?

24-11 What color is formed when blue and yellow pigments are mixed? What happens if red is mixed with the blue and yellow?

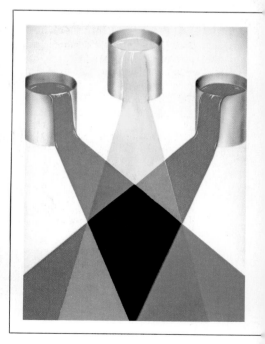

Some spectroscopes use a *diffraction grating* instead of a prism. The diffraction grating is a glass plate containing thousands of parallel lines per centimeter. As the light passes through or is reflected from these very narrow lines, the light spreads out. The spreading bands of light interfere with each other, strengthening some waves and weakening others. The result is a spectrum of light.

The grooves on a phonograph record act like a diffraction grating. Look at the reflection of a bare light bulb from the surface of a record to see the effect of diffraction.

The most common spectrum is the kind called a *continuous spectrum.* As the name implies, all the colors of the rainbow are present in this type of spectrum. Refer to Fig. 24-3 (p. 447). This kind of spectrum is produced when a solid, liquid, or very dense gas is heated until it glows.

Often, certain colors will stand out more brightly than others. This indicates how hot the light source is. For example, the brightest part of the sun's spectrum is in the greenish-yellow region. That color is brightest at about 6000°C. Therefore, the temperature of the sun's surface is about 6000°C.

A second type of spectrum is produced by a hot gas under low pressure. When this gas is excited by heat or electric current, it gives off light with only a narrow range of wavelengths. These different wavelengths form brightly colored lines on a dark background. This type of spectrum is called a *bright-line* or *emission spectrum.* Each chemical element produces its own unique bright-line spectrum. See Fig. 24-12.

A third type of spectrum is similar to the continuous spectrum, except that it has dark bands cutting through it. This type of spectrum is produced when light is passed through a gas at low pressure. The gas filters out, or absorbs, specific colors. Thus, the spectrum appears continuous except for the dark bands where colors are filtered out. This type of spectrum is called a *dark-line,* or

24-12 A bright-line spectrum of hydrogen. How is this spectrum formed?

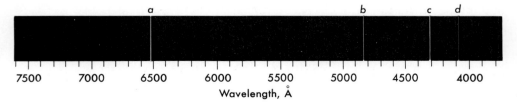

Wavelength, Å

unit 4 wave motion and energy

absorption spectrum. A cool gas absorbs those rays that it, itself, would give off if heated.

Most stars are dense masses of glowing gases. In viewing stars through a spectroscope, you would expect to see a continuous spectrum. Instead, a dark-line spectrum is seen. The dark-line spectrum is caused when some frequencies of the star's light are absorbed by the cooler gases above the star's surface. These dark lines represent the elements that the star is made of.

How do absorption and emission spectra give scientists information about an element? When atoms of an element are excited, they emit light of certain specific wavelengths. The wavelengths of light produced by one element are different from those produced by any other element. Thus, the light from each element is the element's "fingerprint."

For instance, if you looked through a spectroscope at the light from a neon sign, you would see a pattern of bright lines that are produced only by the element neon. If mercury gas is present, it will form its own special pattern. Scientists have used this method to show that neon and mercury, as well as over 50 other elements, exist on the sun. The elements are "fingerprinted" by the spectra they produce. With the aid of the spectroscope, a chemist can detect an element even if only a tiny trace of it is present.

A shift in frequency causes the Doppler effect. As you saw in Chapter 22, the Doppler effect is the shift in frequency that occurs whenever a wave source moves toward or away from you. In sound, the Doppler effect is heard as a higher or lower pitch. In light, the Doppler effect is seen as a shift in color towards either the blue or red end of the spectrum.

How does the Doppler effect cause a shift in color? See Fig. 24-13 on p. 456. Suppose that a light source, such as a star, is located at point 1. Light waves spread out in all directions, with point 1 as the center. Each wave is separated from the next by a distance, or wavelength, labeled a and b.

Now imagine the star moving rapidly toward the right. Each new wave starts from a new center, marked 2, 3, 4, 5, 6. What effect does this motion have on the wavelength of the light? The light received at A has a shorter

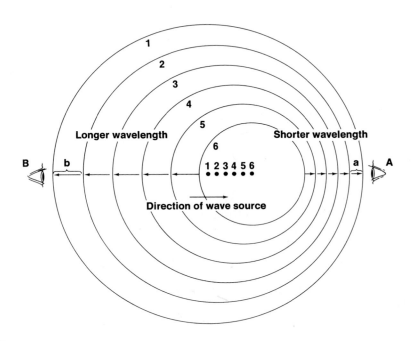

Longer wavelength

Shorter wavelength

B   b

1 2 3 4 5 6

a   A

Direction of wave source

24-13 A diagram of the Doppler effect. Toward which end of the spectrum would the color shift when viewed from A? from B?

wavelength (higher apparent frequency) than that at B.

If the star is green, its light would appear bluish to an observer at A, since blue light has a higher frequency, or shorter wavelength, than green. What about the person at B? To that person, the light would appear yellowish, since yellow has a lower frequency than green.

In studying the light from a star with a spectroscope, scientists can tell when the spectral lines of an element are shifted. If the lines are shifted toward the red end of the spectrum, scientists know that the star is moving away from earth.

Recent sightings have shown that the most distant galaxies within the range of the largest telescopes are moving away from earth at a speed of about 145,000 km/sec. This speed is almost half the speed of light. Other galaxies are moving away at speeds related to their distances from the earth. This discovery of the so-called *red shift* has led to the startling theory that the universe is expanding rapidly, perhaps even exploding.

The Doppler effect can be observed all along the electromagnetic spectrum. For example, radar uses the shift in frequency of radio waves to measure the speed of a car or of an aircraft.

Jobs That May Require Skills With Colors

There are many jobs available for people with special interest or skills in working with colors. The following jobs may make use of those skills.

Job	Duties
Architect	Designs homes and other structures.
Auto painter	Designs and applies color schemes for the interior or exterior of cars and vans.
Commercial artist	Needed in advertising and communications.
Display artist	Designs and sets up window and shelf displays in stores; needed in advertising.

(continued)

Job	Duties
Floral designer	Prepares flower arrangements and wreaths; needed in florist shops.
Furniture finisher	Hand-finishes furniture, including wood and upholstery.
Interior designer	Helps people decorate their homes, including wall and floor coverings and selection and placement of furniture.
Landscape architect	Designs outdoor landscapes; selects and places flowers, shrubs, trees, and grasses.
Letterer	Hand letters or paints signs and advertising displays; needed in publishing and advertising.
Painter and decorator	Needed in building trades.
Photographer	Takes pictures of people and events for newspaper and television news; studio photos of people.
Stained glass painter	Stains, paints, designs, and assembles stained glass windows.

ſummary

1. Visible light is only a small part of the electromagnetic spectrum.
2. Color is related to the frequency and wavelength of light waves.
3. White light can be separated into its colors by a prism.
4. The color of an object depends on the light reflected from that object.
5. The primary colors of light are red, green, and blue.
6. The primary pigments are red, yellow, and blue.
7. A spectroscope can be used to study the elements.
8. The Doppler effect is caused by a shift in the frequency of light to either the red or blue end of the spectrum.

review

Match each item in the left column with the best response in the right column. *Do not write in this book.*

1. Doppler effect
2. primary light colors
3. continuous spectrum
4. spectroscope
5. ultraviolet
6. infrared
7. gamma rays
8. white
9. black
10. yellow

a. band of rainbow-like colors formed when white light passes through a prism
b. device that separates light into its wavelengths
c. device used to make color-corrected lenses
d. produced by mixing lights of primary colors
e. produced by mixing pigments of primary colors
f. most visible color
g. radiant heat
h. "black light"
i. radiation produced by short-wave radio
j. radiation produced by certain atomic nuclei
k. red, green, and blue-violet
l. shift of frequency owing to motion

questions

Group A
1. Name six kinds of rays found in the electromagnetic spectrum.
2. What determines the color of visible light?
3. How are X rays and infrared rays alike? How are they different?
4. What happens to light energy when it is absorbed?
5. Name the colors of the visible spectrum in order of decreasing frequency.
6. What two factors determine the color of an opaque object?
7. Explain why one object looks black and another looks white.
8. Name a pair of complementary colors.
9. What happens to a beam of pure red light when it passes through a prism?
10. How does red light differ from blue light?
11. What color results when the three primary colors of light are mixed in proper proportions?
12. What color results when the three primary pigments are mixed in proper proportions?

Group B
13. What is the relationship between frequency and wavelength?
14. Explain how a prism spreads a beam of white light.
15. What causes a rainbow?
16. Why does a white object look red when it is viewed through a red filter?

17. Why does a red object appear black when viewed through a green filter?
18. What is retinal fatigue? How can it affect your color vision?
19. What color do you get when you mix blue and yellow light? blue and yellow pigment? Explain.
20. Explain how a spectroscope operates.
21. What produces a continuous spectrum? What produces a bright-line spectrum?
22. Explain how a dark-line spectrum is formed.
23. Describe the Doppler effect and give two applications.

further reading

Adler, I., *Story of Light*. New York: Harvey, 1971.

Hagen, Catherine, *Color*. Wolck, 1976.

Hutchings, Donald (Ed.), *Late Seventeenth Century Scientists*. Elmsford, N. Y.: Pergamon Press, 1969.

Kentzer, Michael, *Waves*. Cleveland: Collins-World, 1977.

Turner, Rufus P., *Frequency and Its Measurement*. Indianapolis: Howard W. Sams, 1975.

unit 5
electromagnetic
nature of matter

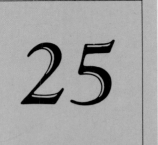

25

electrostatics

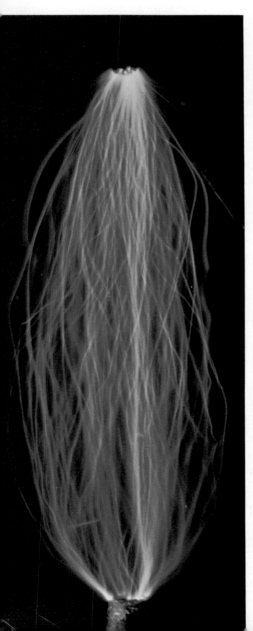

Electrostatics, or *static electricity,* was known very early in history. Our ancestors heard the crackle of sparks when they stroked a cat's fur. They may have felt small shocks after combing their hair. They may even have seen the glowing discharge from pointed objects on nights when the air was highly charged. Most obvious of all, they saw flashes of lightning during a storm. See Fig. 25-1.

By 600 B.C. the Greeks had learned that when they rubbed amber with wool, the amber would attract bits of lint and paper. The Greeks referred to this attraction as *electron,* the Greek word for amber. During the next 2000 years, not much was added to the knowledge of static charges. By the eighteenth century, however, Benjamin Franklin and others began a careful study of static charges and rapid advances were made.

objectives:

☐ To describe how static charges are formed and how they interact.

☐ To demonstrate how bodies can be charged by contact and by induction.

☐ To discover where charges concentrate on objects.

☐ To explain how some static charges can be dangerous.

☐ To describe some of the uses of electrostatic devices.

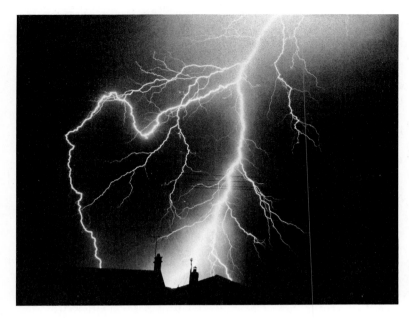

25-1 Lightning is an example of static electricity. Can you see why it is difficult to control static charges?

electric charges are all around you

Objects can receive a static charge. To understand how objects can be charged, you must first review the structure of the atom. Normally, an atom contains equal numbers of positively charged protons (+) and negatively charged electrons (−). In such a state, the atom is neutral. When an atom loses an electron, however, the charges are no longer equal. Since one electron is missing, the atom has a positive charge. The free electron has a negative charge.

When many atoms of an object gain or lose electrons, the object becomes charged. An object that receives extra electrons has a negative charge and one that loses electrons has a positive charge.

If you rub two objects together, the contact causes electrons to be transferred from one object to the other. For example, if you rub hard rubber (or almost any plastic) with wool or fur, the rubber picks up electrons from the wool. This is how your comb gets a negative charge when you run it through your hair. On the other hand, if you rub a dry glass rod with silk, the silk removes electrons from the glass, leaving the glass with a positive

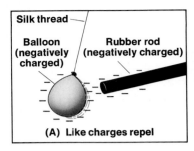

Silk thread

Balloon
(negatively
charged)

Rubber rod
(negatively charged)

(A) Like charges repel

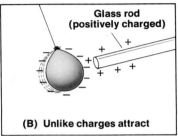

Glass rod
(positively charged)

(B) Unlike charges attract

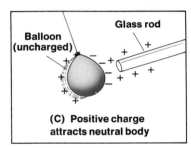

Balloon
(uncharged)

Glass rod

**(C) Positive charge
attracts neutral body**

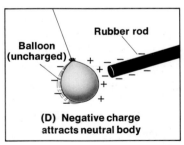

Balloon
(uncharged)

Rubber rod

**(D) Negative charge
attracts neutral body**

25-2 In which cases are objects attracted to each other? repelled?

charge. Notice that the part of the atom that moves in this process is always the electron.

When you walk across a rug, the contact between the rug and your shoes builds up a negative charge on your body. If you then touch a metal doorknob, the electrons travel from your hand to the doorknob. You experience a *shock*. Likewise, when you pet a cat, electrons travel from the cat's fur to your hand.

Charged and uncharged objects interact. To find out how static charges affect each other, hang a toy balloon from a thread, as shown in Fig. 25-2. Give the balloon a negative charge by rubbing it with wool or fur. Now charge a rubber rod in the same way and hold it near the balloon. What happens? You can see that the balloon is repelled by the rod. If two objects both have a positive charge, they repel each other. Tests like this show that *like charges repel each other*.

Now give a glass rod a positive charge by rubbing it with silk. Then hold it near the balloon. Notice that the positive rod attracts the negative balloon. Therefore, *unlike charges attract each other*.

Further tests show that if the balloon is uncharged, it is attracted to both the positive and negative charges. Thus, *charged objects attract uncharged (or neutral) objects*. (You will see why later in this chapter.)

Static charges attract and repel by means of a *field of force* they create around them. A field of force is the region around a charged object within which the force can be detected. When an electron or other charged object is brought into this field, it is affected by a force. A charged object will move either toward or away from another charged object, depending on the kind of charge each object has.

What affects the strength of the force between two charged objects? First, the larger the charge, the greater the force. Second, the closer the objects, the greater the force.

Suppose you measure the force between two charged objects a certain distance apart. When the objects are moved twice as far apart, the force between them becomes one fourth as much. If the distance between them is made three times as great, the force will be only one ninth as much. In other words, the force follows the inverse-square rule.

unit 5   electromagnetic nature of matter

activity

Attraction and repulsion. From a thread, suspend a tiny ball such as a kernel of puffed wheat or rice, or a bit of styrofoam. Rub a plastic comb with a woolen cloth and bring it near the ball.

Questions:
1. What happens to the ball?
2. Explain what happens to the ball after you touch it with the comb.
3. What happens if you bring your finger near the ball before the ball is touched with the comb? after? Why?
4. What happens if you wrap a layer of tinfoil around the ball and repeat the activity?

Electroscopes detect static charges. *Any device used to detect an electric charge is called an* **electroscope.** The balloon shown in Fig. 25-2 is one example. However, the *gold-leaf electroscope* is much more sensitive than a rubber balloon. In addition, this electroscope can also be used to tell how much charge is present.

In Fig. 25-3, notice that the metal rod holding the gold leaf passes through an insulating ring at the top of the round metal box. This ring keeps the charge from leaking to the box. Instead, the charge travels down the rod to the leaves of the electroscope.

Although the leaves can be made of aluminum or any thin metal foil, gold is the most sensitive metal because it can be made very thin and light. Gold leaves can detect even trace amounts of a charge. The leaves are mounted in a box or bottle to protect them from air currents.

When the metal knob of the electroscope is touched by a negatively charged rod, some of the electrons move from the rod to the knob. The electrons travel down the metal knob and spread over it. Some of the electrons reach the gold leaves, giving each leaf a negative charge. Therefore, the leaves repel each other. The greater the charge, the farther the leaves are spread apart.

What happens when an electroscope is touched by a rod with a positive charge? (Remember that a positive charge is due to a *lack* of electrons.) When a positive rod touches the knob of the electroscope, electrons leave the electroscope, giving the leaves a positive charge. Again, the

25-3 Why do the leaves of an electroscope spread apart when the top is touched by a charged rod?

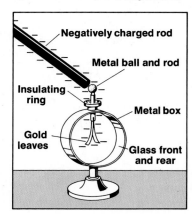

Negatively charged rod

Metal ball and rod

Insulating ring

Metal box

Gold leaves

Glass front and rear

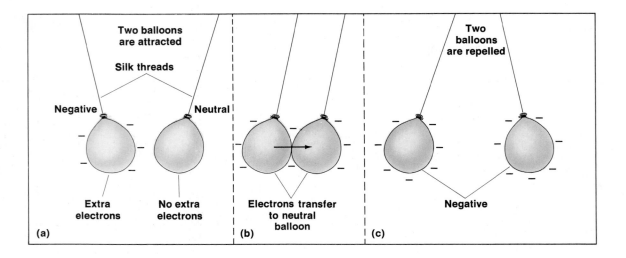

Two balloons
are attracted

Silk threads

Negative   Neutral

Extra   No extra
electrons   electrons

(a)

Electrons transfer
to neutral
balloon

(b)

Two
balloons
are repelled

Negative

(c)

25-4 Charging by contact. Why are the balloons attracted to each other in (a)? Why are they repelled in (c)?

25-5 Why are charges shown on the balloon if it is neutral?

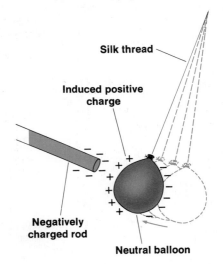

Silk thread

Induced positive
charge

Negatively
charged rod

Neutral balloon

leaves repel each other. In summary, an electroscope will detect either a positive or a negative charge.

Objects can be charged by contact or by induction. So far, you have seen how objects can be charged by touching them with other charged objects. This process is called charging by *contact*. In other words, a charged object transfers its charge to a neutral object when they touch. See Fig. 25-4.

To find out how to charge an object by *induction*, remember what happens to a neutral object when a charged object is brought near it (but does not touch it). You saw earlier that an uncharged object, such as the balloon in Fig. 25-5, is attracted to a charged object. This attraction occurs because some electrons on the balloon are repelled by the negative charge on the rod. These electrons move to the far side of the balloon, leaving the positive charge on the near side. The unlike charges on the balloon and the rod will then attract each other.

When the charged rod is removed, the displaced electrons return to their normal positions, and the balloon again becomes neutral. This brief shifting of electrons results in *induced charges*. An induced charge is one that appears when a nearby charge briefly upsets the electron balance.

Now suppose you "grounded" the balloon while the rod was close to it, as shown in Fig. 25-6. Grounding means absorbing the induced charge of an object by connecting it with a much larger object.

Most often, grounding means connecting with the earth. But in this case, just touching the balloon with your finger will ground the balloon because your body is large enough to drain off the excess electrons. The rod's negative charge repels the electrons from the balloon into your body.

If you then remove the ground (your finger), the electrons cannot get back to the balloon. The balloon is left with a shortage of electrons and thus has a positive charge. This process is called charging by induction.

Notice that when an object is charged by induction, it always receives a charge *opposite* to that of the inducing charge. For example, a negative charge on the rod will induce a positive charge on the balloon.

A charge can be placed on a conductor. You may have noticed that most of the charged objects in the above examples are nonmetals. Nonmetals are poor conductors of static charges. Charges do not move very well or very far on a poor conductor.

What happens when you put a charge on a conductor? As with the electroscope, the charge moves throughout the entire conductor very quickly. A conductor is a good charge carrier.

However, the charge does not spread evenly throughout all parts of a conductor. The charge will collect on outer surfaces and concentrate on sharp or pointed surfaces.

If a charge is placed on a metal cup, for instance, as shown in Fig. 25-7, most of the charge collects on the rim. The charge on the bottom and sides of the cup is much weaker.

Curiously enough, no charge is found on the inside of the cup. This is true even though the charge is put on the inside in the first place. The charge moves to the outside surface at once. The reason for this movement is the fact that like charges repel. Therefore, electrons move away from each other to the outermost surfaces of a conductor.

If a sharp needle is placed across the top of the cup, any charge placed on the cup will quickly disappear. The charge moves to the sharp point of the needle and leaks off into the air. This process is important in the operation

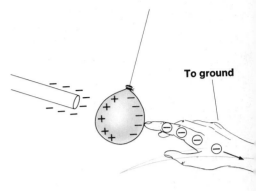

To ground

25-6 Charging by induction. Why is it important that the balloon is touched by the ground but not by the rod?

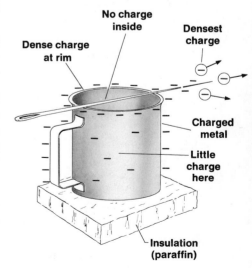

25-7 Where do charges build up on a conductor? Where is the charge most likely to leak off?

No charge inside

Dense charge at rim

Densest charge

Charged metal

Little charge here

Insulation (paraffin)

of many electrostatic devices, including lightning rods. Lightning rods will be explained later in this chapter.

Remember that there is no charge on the inside of the cup described in Fig. 25-7. Any hollow conductor acts in the same way. If the conductor is a hollow ball, the ball can collect and hold a huge charge without losing it.

Because charges move to the outside surface of a conductor, a space can be shielded from the effects of outside charges. This means that you are probably safe from lightning inside a metal car or in a building with a steel frame.

the hazards of static electricity

Static charges can produce sparks and shocks. Static charges can cause fires and explosions wherever there is flammable material. Static electricity has caused many disasters in the petroleum and dry-cleaning industries. Sparks from static charges have set fire to hydrogen balloons and to anesthetic gases used in surgery. Today, care is taken to avoid most of these hazards.

Did you ever get a shock when you touched the door handle of a car? The shock is caused by a static charge. A car can build up a static charge in a number of ways. Friction between the tires and the road or between the car and the air may cause a transfer of electrons.

You may have noticed upright, grounded wires just before a highway toll station. These wires neutralize the static charge built up on the car to prevent both you and the station attendant from getting shocked as you exchange coins. To neutralize a charge, the normal electron-proton balance must be restored. There are two common ways of restoring this balance to an object: (1) The charged object can be given an equal and opposite charge. (2) The charged object can be grounded. The earth acts as a storehouse of free electrons, which can neutralize positive charges. The earth also acts as a "sink" into which excess electrons can flow.

Charged objects also lose their charge simply by being exposed to air. The charge leaks off into the air. The amount of moisture in the air greatly affects the rate of discharge. On a humid summer day, the discharge is so fast that trying to do tests with static charges is hopeless.

On the other hand, very good results are achieved on a cold, dry winter day.

Lightning is caused by static charges. In the days of Benjamin Franklin, lightning was a mystery that frightened most people. Franklin, however, believed that lightning was a form of static electricity, similar to sparks, but on a larger scale. In 1752, with his famous "kite and key" experiment, Franklin proved his theory. See Fig. 25-8. The experiment showed quite clearly that lightning and static charges are the same.

How do moving electrons build up charges during a storm? Winds in a storm push around masses of rain, air, and clouds. The effect is somewhat like pushing a piece of fur on a plastic rod. These moving masses tend to gain or lose electrons, and build up positive or negative charges.

When a negatively charged mass of rain or cloud forms near the earth's surface, an opposite charge is induced on the objects directly below it. When the attraction between the two charges becomes great enough, the electrons jump from the cloud to earth. This discharge of electrons is lightning. The lightning heats the air and causes the air to expand rapidly. This expansion produces thunder.

You are not completely safe from lightning in a house. Lightning will often run along metal pipes, electric wires, and metal chimneys right through a house. Benjamin Franklin noticed that a pointed object loses a charge quickly. He reasoned that perhaps a building could be protected by attaching some grounded points "to draw off the electrical fluid."

How does a lightning rod work? Suppose a house, as in Fig. 25-9 (p. 470), has a number of pointed metal lightning rods connected to the ground. The charge built up in the air by a storm is slowly and quietly discharged by the lightning rods.

If a cloud has a negative charge, the lightning rod will drain extra electrons from the air to the ground. If the cloud charge is positive, the earth supplies the electrons needed to help neutralize the cloud. Sometimes, so many electrons stream from (or to) the metal points of the rods that they glow on a dark night.

If the charge is not neutralized fast enough, lightning will strike. In that event, lightning rods act as good conductors, which carry the charge safely into the

25-8 How did Benjamin Franklin show that lightning is the same as static electricity?

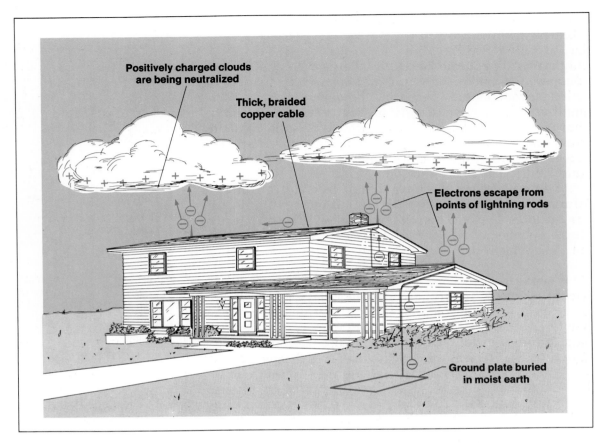

Positively charged clouds
are being neutralized

Thick, braided
copper cable

Electrons escape from
points of lightning rods

Ground plate buried
in moist earth

25-9 In what way is the lightning rod like the needle on the cup in Fig. 25-7?

ground. However, lightning rods can be dangerous if they are not properly grounded. Why do you suppose a steel skyscraper has no need of lightning rods?

You may think that a television antenna can serve the same purpose as a lightning rod, but this is not so. If the antenna is not grounded properly, lightning can come into the house, ruining the TV set and perhaps causing a fire.

using static electricity

Large static charges can be produced. By the seventeenth century, scientists learned that they could produce charges by rubbing various substances together. In 1672, Otto von Guericke, a German inventor, built a machine to do the rubbing for him. His machine looked something like a grindstone except that instead of the stone it used a ball of sulfur. See Fig. 25-10. From that day on, scientists have built bigger and better machines to produce static charges.

Huge charges can be formed in the Van de Graaff generator. To understand how a Van de Graaff generator works, refer to Fig. 25-11. The rubber belt becomes charged as it passes over a fleece-lined pulley wheel. In large generators, the electric charges are first produced in another device and then "sprayed" on the belt.

At the top, the charge is removed from the belt by a set of points attached to the inside of a large metal ball. The charge moves to the *outside* of this ball. (Remember the experiment with the cup?) The ball soon collects a large charge.

The capacitor can store a static charge. While Franklin was working in America, two scientists from Europe built a device in which static charges could be stored. They were astonished at the violent shocks they got from glass jars coated inside and out with metal foil, as shown in Fig. 25-12. These devices were called *condensers* because they

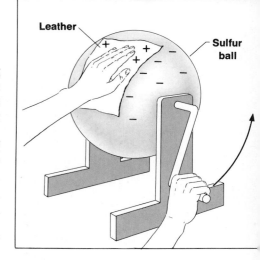

25-10 How is a static charge made by this early generator?

25-11 (left) Where is the static charge produced in this Van de Graaff generator? Where is the charge stored?

25-12 (right) How does the Leyden jar store a static charge?

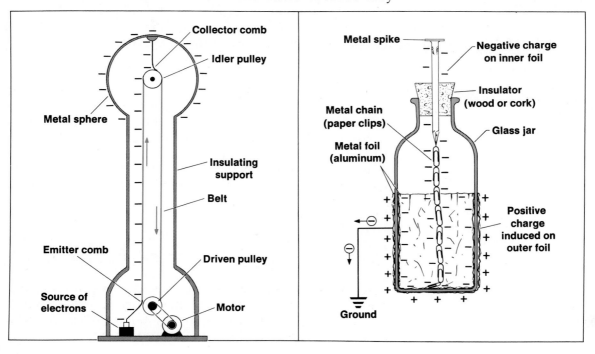

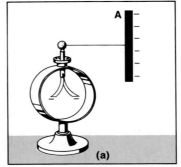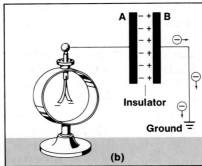

25-13 Why can plate A hold a larger charge if plate B is moved next to it?

(a)   (b)

were able to "condense" the so-called "electrical fluid." Today they are more correctly called *capacitors* and come in many shapes and sizes. Capacitors consist of metal plates separated by an insulator.

Suppose a small negative charge is put on metal plate A, shown in Fig. 25-13. The electroscope leaves will spread apart. Now if a grounded metal plate B is brought close to A, the electroscope leaves will come together. However, if a much larger charge is placed on A, the leaves will again spread apart.

As the diagram shows, the negative charge on A induces a positive charge on B. (Can you see how this happens?) This induced positive charge attracts the electrons on A to the side nearest B, making room for more electrons on A. In other words, a larger negative charge can now be placed on A. Without the capacitor effect, you could not put a larger charge on A. The charge would leak off as fast as you added it. The closer A is to B, the greater the charge the plates can hold. Also, the larger the plates, the more charge they can hold.

The capacitor is one of the most widely used devices in electronics. Sometimes capacitors are used to store large charges. More often, however, they are used in radios because of their natural frequencies.

You tune your radio by turning a *variable capacitor.* Such a capacitor has two sets of plates that mesh together without touching each other. As the plates open up, the capacity decreases because the effective area of the plates is reduced. When this happens, the frequency rises. If the plates are meshed together, the capacity increases because the effective area of the plates is increased. In this case, the frequency goes down. By this method, you can tune your radio to match the frequency of the station you wish to receive.

There are many uses of static charges. In 1938, Chester Carlson invented a process of copying printed matter without the use of liquid inks or chemicals. Instead, Carlson's process used tiny static charges and dry inks. The name he gave it, *xerography* (ze-*rog*-graf-fy), comes from the Greek words for "dry-writing."

In the copy machine, a beam of light transfers the image from the printed page onto a charged drum. The drum is coated with a photoconductor (a substance that conducts a charge only when exposed to light). When a strong light shines through the paper, it leaves a shadow of the printing on the drum. Where the light strikes the charged drum, the photoconductor carries away the positive charge. Where the light does not strike, however, the positive charge remains.

Now a dry ink with a negative charge is dusted onto the drum. The negative ink is attracted only to the positively charged area of the drum. The drum rotates and is pressed against a white sheet of paper, which attracts the ink. The ink in turn, is heated and fused into the paper, producing a lasting image. This use of static charges makes it possible to copy printed matter in seconds.

Among the many other uses of static charges are filters to clean the air. These filters range from small units used in home heating systems to remove dust and pollen to huge devices used to clean smoke from smokestacks. See Fig. 25-14.

These filters use static charges to ionize, or charge, the particles in the air. The air is then passed over plates or wires that are also highly charged. Some of the plates are positive and some are negative. The ionized particles are attracted to the charged plates and removed from the air.

High vacuums can be produced by ionizing air or other gases, and then removing the ions by electrostatic forces.

Some loudspeakers produce sound waves by using static charges. However, these speakers do not reproduce all frequencies equally well. Most often, they reproduce high tones better than low tones.

Static charges are also vital in the operation of many devices used in radar, television, and electronics.

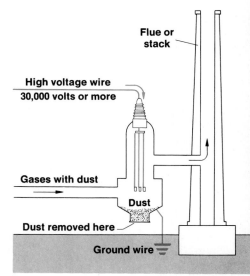

25-14 The Cottrell device removes dust and dirt from the flue gases. Can you describe how this device works?

Flue or stack

High voltage wire
30,000 volts or more

Gases with dust

Dust

Dust removed here

Ground wire

Benjamin Franklin
Jack of All Trades, Master of Many

Benjamin Franklin (1706–1790) was a gifted American scientist, scholar, and statesman. He excelled in such different fields as writing, printing, government service, science, oceanography, farming, optics, and architecture. Franklin even advised a change to daylight-saving time in the summer, saying, "It is foolish to sleep by daylight and work by candlelight."

Franklin's most well-known scientific work was in electricity. In 1752, with his famous kite experiment, Franklin proved that lightning is a static charge. First, Franklin tied a metal key to the lower end of a thin wire that was attached to a kite. He did not hold onto the wire or key, however. Franklin knew that a charge of lightning could be carried down the wire. Instead, he attached a short piece of string to his end of the wire.

When charges from the clouds flowed down the wire, a bright spark jumped from the key. The spark was like the kind you get when you touch a doorknob after scuffing your shoes across a rug on a cold winter night. Franklin showed that lightning is a form of static charge just like the charges that scientists produce in labs.

Franklin approached the study of science as he did all his other work. He felt that to be successful, a person has to work just a little harder than anyone else.

summary

1. When an object gains electrons, it has a negative charge.
2. When an object loses electrons, it has a positive charge.
3. Like charges repel each other; unlike charges attract each other.
4. The force between two charged bodies varies directly as the charge on each body and inversely as the square of the distance between them.
5. Charged objects attract uncharged objects.
6. Electroscopes detect static charges.
7. Objects can be charged by contact and by induction.
8. Static charges move to the outer surfaces of a conductor.

9. Lightning is caused by static charges.
10. Lightning rods reduce the build-up of charges and provide an easy path to the ground if lightning strikes.
11. Static charges can be stored in capacitors.
12. Xerography and electronic air filters are some uses of static charges.

review

Match each item in the left column with the best response in the right column. *Do not write in this book.*

1. static electricity
2. electrostatic filter
3. xerography
4. electroscope
5. lightning rod
6. neutral
7. capacitor
8. positive charge

a. device to draw off charges from buildings
b. device used to detect electric charges
c. device to reduce smoke pollution
d. device to store an electrical charge
e. "dry writing"
f. electricity at rest
g. excess electrons
h. process of charging by contact
i. shortage of electrons
j. uncharged

questions

Group A
1. What kind of charge does an electron have?
2. What charge does an object have if it loses some of its electrons? if it gains electrons?
3. How do like charges behave toward each other? unlike charges?
4. Where did the word "electron" come from?
5. What is the purpose of the electroscope?
6. When you rub hard rubber with wool, what kind of charge does each substance pick up? Explain.
7. What happens when a conductor with a positive charge is grounded?
8. In what two ways can a charge be neutralized?
9. What kind of charge is induced by a positive charge?
10. What are some dangers caused by a buildup of static charges?
11. What is lightning?
12. What is a capacitor used for? How is it made?

Group B
13. Explain why a neutral object is attracted by an object with either a positive or negative charge.

chapter 25 electrostatics   475

14. Why do you sometimes see a spark when you shuffle your shoes across a nylon rug and then touch a metal doorknob?
15. Describe two ways of putting a positive charge on an electroscope.
16. What happens to the leaves of a charged electroscope when you touch the knob with your finger? Explain.
17. Describe how a negative charge can be put on an insulated conductor using a rod with a positive charge.
18. A gold-leaf electroscope has a positive charge. As you bring a charged rod toward it, the leaves spread farther apart. What kind of charge is on the rod? Explain.
19. Explain how a space can be shielded from static charges. Give an example where such shielding is used.
20. Explain how an insulated metal ball becomes charged when you touch it with a negatively charged rod.
21. In what two ways can lightning rods protect buildings from damage?
22. Which has more need of lightning protection, a brick smokestack or a steel skyscraper. Why?
23. Explain how a capacitor stores up electric charges.
24. Explain how the image to be copied is transferred from the printed matter to the drum in an electrostatic copier.
25. Explain how the image to be copied is transferred from the drum to a plain sheet of paper in an electrostatic copier.

further reading

Asimov, Isaac, *How Did We Find Out About Electricity?* New York: Walker, 1973.

Buban, Peter, and Schmitt, Marshall L., *Understanding Electricity and Electronics*. New York: McGraw-Hill, 1975.

Hockey, S. W., *Fundamentals of Electrostatics*. New York: Barnes & Noble Books, 1972.

current and circuits

In the last chapter, you learned that static charges had been known for thousands of years. The energy in static electricity was in a form that could do little if any useful work. Scientists thought that electricity was some kind of "invisible fluid." (Even today people use terms like "current," "flow," and "juice" when talking about electricity.) However, scientists could only build up a static charge and release it again in a single flash of energy. How could this energy be controlled? How could the energy be released in a "current" or "flow" so that it could do useful work?

Today, currents are controlled and affect our daily lives. Flick a switch and the lights go on. Turn a handle and the water flows. Push a button and the TV set comes on. Cars, buses, and planes depend on electric currents. So do telephones, telegraphs, and radio and TV stations. Electric currents clearly affect us every day.

In this chapter, you will see how an electric current is different from static charges. You will also see how currents flow in some substances and not in others, and how currents can be controlled to do useful work.

objectives:

☐ To understand the nature of an electric current.

☐ To define the properties of conductors and insulators.

☐ To state the relationships among current, potential difference, and resistance.

☐ To compare the functions of series and parallel circuits.

□ To describe the safety features of fuses and circuit breakers.

□ To describe how an electric current can make heat and light.

□ To identify some of the practical uses of electric currents in the home.

electrons on the move

An electric current is a stream of moving electrons. In an earlier chapter, you learned that the electron is an atomic particle carrying a negative electric charge. Electrons exist in all atoms and can be removed from the atoms of certain elements quite easily. These free electrons can then be pushed along through substances called *conductors.* Such a movement of electrons in a conductor is called an *electric current.* The path along which the movement takes place is called an *electric circuit.*

Suppose that a copper wire is the pathway or circuit carrying a current. If the wire is cut and the loose ends attached to a *switch,* the current can be turned on or off. Opening the switch stops the flow of electrons just as an open drawbridge stops the flow of cars across a river. Thus, an open switch produces an *open circuit.* When the switch is closed, the pathway is again complete and the current starts again. This is called a *closed circuit.* See Fig. 26-1.

A flow of electrons can be compared with a flow of water. Just as a current of water can be measured in liters per second, so an electric current can be measured in electrons per second. However, the electron has such a tiny charge that vast numbers of them are needed to make even a small current.

A more useful unit is the *coulomb.* A coulomb consists of 6.25 billion billion electrons. One coulomb of electrons flowing through a wire per second is an *ampere* of current. A current of a little less than 1 ampere travels through a common 100-watt light bulb. The ampere is measured with an instrument called an *ammeter.*

A potential difference pushes electrons. To keep electrons flowing in a circuit, some sort of push or electric

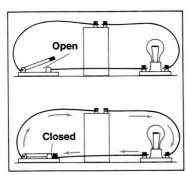

26-1 In which diagram is the path of electrons broken? Where is it complete? What is the function of a switch in an electric circuit?

Open

Closed

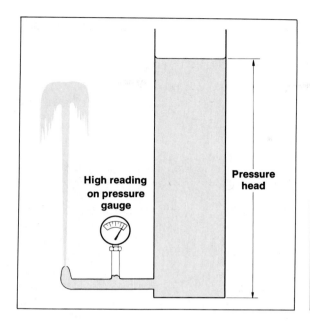
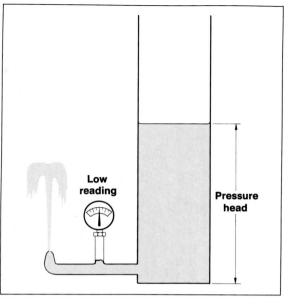

26-2 Compare water pressure with voltage. Which diagram represents a higher voltage?

"pressure" is needed. This push can be better understood if you again compare an electric circuit to a water fountain. See Fig. 26-2. In order to maintain a flow of water in the pipes, a push is needed. The push comes from the weight of the water that is at a level higher than the fountain.

Recall that anything lifted to a height above its starting point has potential energy that can be released when the object falls. In a water system, the water is raised by a pump, and the stored energy is released when the water falls. There is a *potential difference* between the starting and ending levels of the water.

Another way in which water can receive a push is shown in Fig. 26-3. The water in the pipe is pushed by a turning pump. In an electric circuit, a battery or generator acts as a "pump," taking electrons from atoms. These electrons create a difference in the "level" of the charge on the two *binding posts* of a cell or battery. Therefore, there is a potential difference between the two posts.

Electrons will flow from an object of high electron potential (the negative post on the cell) to one of low potential. This potential difference is measured in *volts* by an instrument called a *voltmeter*. An increase in voltage produces a larger current. A common flashlight cell gives

chapter 26 current and circuits

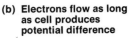
(b) Electrons flow as long as cell produces potential difference

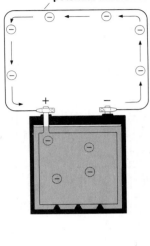

(a) Water flows as long as pump operates

26-3 What is the source of potential difference in the electric circuit? How does it compare to the source of pressure in the water system?

a potential difference, or voltage, of 1.5 volts. If you use two cells, the voltage is doubled, and the current produced is twice as large.

When electrons have a steady flow in one direction only, the flow is called a *direct current* (dc). A dry cell is a source of direct current. The current travels in the same direction for the life of the cell.

When the direction of the current keeps jumping back and forth in a circuit, it is called an *alternating current* (ac). Electric generators called *dynamos* are sources of alternating current. Almost all power plants supply ac to your homes. A later chapter will describe how these generators work.

Substances differ in how well they conduct currents. Given a source of electric current, how are the electrons carried from one place to another? The word *conductor* was used in the last chapter in discussing static charges. A conductor is any substance that allows electrons to move through it easily. Therefore, conductors are used to carry electric currents from place to place. The best conductors are metals such as silver, copper, gold, and aluminum. Because of the high cost of silver and gold, only copper and aluminum are commonly used.

Some materials will not carry a current at all, even if the potential difference or voltage is very high. Such materials are called *nonconductors* or *insulators*. These are mostly nonmetals such as plastics, rubber, porcelain, and glass.

Under normal conditions, air is a nonconductor, but when it contains moisture, air becomes a conductor. Certain elements, such as germanium and silicon, are *semiconductors*.

You know that water flows more freely through a clean water pipe than through one that is partly clogged. The clogged pipe offers **resistance** to the flow of water. In somewhat the same way, the current is greater if the resistance of the circuit is reduced. *The electrical resistance of a conductor depends upon its thickness, composition, length, and temperature.*

Just as the same water pressure will send a greater current through a large pipe than a small one, so the same voltage sends a greater current through a thick wire than a thin one. See Fig. 26-4. Assuming that all other factors

26-4 In what way are the carrying capacities of the pipes like those of the wires?

(a) Short, thick pipe: small resistance, large current

(b) Long, thin pipe: large resistance, small current

(c) Short, thick wire: low resistance, large current

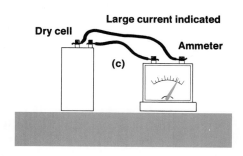

(d) Long, thin wire: high resistance, small current

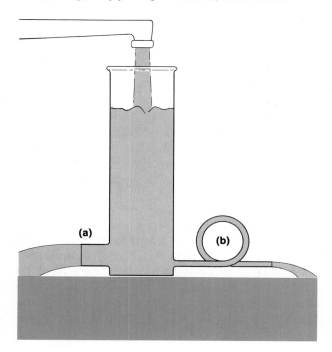

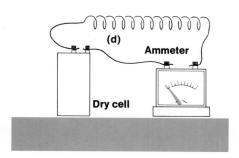

are the same, a thick wire will always have less resistance than a thin wire.

The unit of electric resistance is the *ohm*. The symbol for ohm is the Greek letter omega, Ω. To give you an idea of what the ohm is, 300 m of No. 10 copper wire (2.5 mm thick) has a resistance of 1 ohm. If it were twice as thick, this wire would have four times the cross-section area. The resistance would be only 0.25 ohm, and the wire could carry four times as much current given the same voltage. Therefore, *the resistance of a wire conductor varies inversely as the square of its diameter.*

The resistance of a substance also depends on its composition. Silver, copper, gold, and aluminum are the best conductors. A poor metallic conductor, called *nichrome,* is made from an alloy of nickel and chromium. Nichrome offers about 50 times as much resistance as copper. However, even nichrome is a better conductor than most nonmetals. The resistance of a semiconductor is between that of a metal and a nonmetal.

Nichrome is used to make heating elements in toasters, irons, and other appliances. Other metals are used as resistors in light bulbs. When electrons are forced by a high voltage through such resistors, the electric energy is changed to heat. If the temperature is high enough, the filament in a light bulb will glow.

The longer a wire, the greater its resistance. If the resistance of 300 m of wire is 1 ohm, 600 m of the same wire will have a resistance of 2 ohms. *The resistance of a wire varies directly with its length.* How do you think resistance would affect the current in an appliance if long extension cords were joined to reach a distant source?

Temperature is another factor that affects resistance. *The resistance increases as a conductor becomes hotter and decreases as it becomes cooler.* Think of what happens when you plug in an electric iron. Have you ever noticed the lights dim for an instant when the iron is plugged in? They dim because when the wires inside the iron are still cool, the wires have a low resistance and draw a large current. Less current is available for the lights. Then, when the wires become hot, their resistance increases, and the current decreases. The room lights then regain their brightness.

In recent years, scientists have discovered that when certain substances are cooled to near absolute zero ($-273°C$), they have no resistance to current at all. In fact, once started, a current will continue to travel in a circuit even after the current source is removed.

Why should cold metals offer less resistance to currents than hot metals? One theory states that electrons moving through a wire are retarded by the vibration of the atoms of the wire. The atomic movement increases when the metal is warmed and decreases when it is cooled. In other words, when a conductor is cooled, its atoms are less likely to get in the way of moving electrons. In some substances, the atoms seem to be slowed to a point where they do not hinder electron flow at all. Such substances are called *superconductors*.

The amount of resistance in an electric circuit can be controlled by a device known as a *rheostat*, or variable resistor. Connected to a light circuit, a rheostat can be used as a "dimmer." When the resistance is raised, the amount of current is reduced and the lights dim. When the resistance is lowered, the flow is increased, and the lights get brighter again.

There is a relationship between volts, amperes, and ohms. A potential difference of one volt causes a current of one ampere to move through a resistance of one ohm. See Fig. 26-5. You have seen that two major factors (voltage and resistance) affect the amount of electrons that flow through a wire. Therefore, if there is a change in either volts or ohms, there must also be a change in amperes.

If 1 volt "pushes" 1 ampere through 1 ohm, 5 volts are needed to produce five times as much current (5 amperes). If the resistance were then doubled to 2 ohms, twice as much voltage would be needed. That is, it would now take 10 volts to push the 5-ampere current through a 2-ohm resistance.

The relationship between voltage, current, and resistance can be stated in a formula known as *Ohm's law:*

$$\text{volts} = \text{amperes} \times \text{ohms}$$

Using symbols

$$V = A \times \Omega$$

In this example, 10 volts = 5 amperes × 2 ohms.

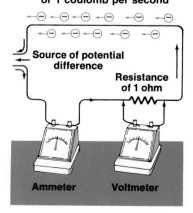

26-5 How many amperes are pushed through this conductor by one volt of potential difference?

6.25 billion billion electrons
or 1 coulomb per second

Source of potential difference

Resistance of 1 ohm

Ammeter   Voltmeter

chapter 26   current and circuits   483

(a) How large a current will 120 volts send through a resistance of 2 ohms?
(b) What is the resistance of an electric toaster if 120 volts sends a current of 5 amperes through it?

solution:

(a) Ohm's law formula is rearranged to read

$$amperes = volts/ohms$$
$$= 120/2$$
$$= 60$$

(b) Rearranging the formula,

$$ohms = volts/amperes$$
$$= 120/5$$
$$= 24$$

You pay for the energy you use. The electric energy that you use in your home is measured in units called *kilowatt-hours* (kwh). To see how the kilowatt-hours are arrived at, look at some other units used to measure electricity.

Each electric device in your home uses energy at a rate measured in a unit of power called the *watt*. One watt is produced when 1 volt pushes a current of 1 ampere through a conductor. The word equation for the relationship is stated as follows:

$$watts = volts \times amperes$$

or using symbols

$$W = V \times A$$

Since the watt is a fairly small unit, it is more convenient to use *kilowatt*, equal to 1000 watts. This unit still does not tell how much energy is used. You still need to know how long the energy was used. The number of kilowatts, along with the time (in hours), gives the energy in kilowatt-hours (kwh). As the name suggests, the kilowatt-hour is one kilowatt used for one hour.

A meter records the number of kilowatt-hours used. Your bill is computed by multiplying the number of kilowatt-hours by the cost per kilowatt-hour. See Fig. 26-6. If, for example, you use 1200 kwh in one month at a cost of 5¢ per kwh, the total bill would be 1200 × 0.05 = $60.00.

26-6 What units are shown on an electric meter that measures the amount of electric energy used in your home?

sample problem

Find the total wattage of the following appliances: (1) a toaster that draws a current of 5 amperes on a 115-volt line; (2) an electric range that uses a 5-ampere current on a 230-volt line; and (3) ten 100-watt light bulbs.

solution:

Step 1. For the toaster,

$$\begin{aligned} \text{watts} &= \text{volts} \times \text{amperes} \\ &= 115 \times 5 \\ &= 575 \end{aligned}$$

Step 2. For the range,

$$\begin{aligned} \text{watts} &= \text{volts} \times \text{amperes} \\ &= 230 \times 5 \\ &= 1150 \end{aligned}$$

Step 3. Ten bulbs at 100 watts each,

$$\text{watts} = 10 \times 100$$
$$= 1000$$

Step 4. The total energy is

575 + 1150 + 1000 = 2725 watts, or 2.725 kilowatts

26-7 Is the voltage the same in all parts of a series circuit? the current?

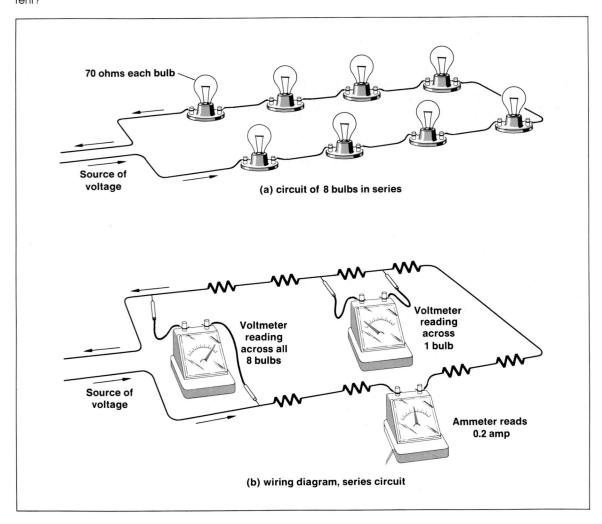

(a) circuit of 8 bulbs in series

(b) wiring diagram, series circuit

paths for electrons

Electrons can flow in a series circuit. Figure 26-7 shows a circuit of eight electric light bulbs in a single circuit. The same current must pass through all eight lamps in turn, since there is only one path for the electrons to follow. If the current that goes through one lamp is 0.2 ampere, then 0.2 ampere is the current that goes through each lamp. This type of circuit is called a *series circuit*.

In a series circuit, the total resistance is found by adding the resistance of all the separate parts. For example, if the resistance of each bulb is 70 ohms, the total resistance of the string of lights will be 8 × 70 = 560 ohms.

What do you suppose would happen if one bulb in this series circuit burned out? The electrons could not flow through any of the bulbs because the circuit has been broken. The current in a series circuit is the same everywhere along the circuit, and the total voltage is the sum of the voltages across each resistance (bulb).

sample problem

(a) How many volts are needed to pass a current of 0.2 ampere through eight bulbs in a series if each bulb has a resistance of 70 ohms? (b) In the same circuit, how many volts are needed to pass the current through *each* bulb?

solution:

(a) Step 1. Total resistance is 8 × 70 = 560 ohms
    Step 2. Using Ohm's law,

$$volts = amperes \times ohms$$
$$= 0.2 \times 560$$
$$= 112$$

(b) Step 1. Applied to one bulb,

$$volts = amperes \times ohms$$
$$= 0.2 \times 70$$
$$= 14$$

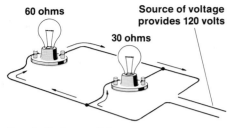

60 ohms

Source of voltage provides 120 volts

30 ohms

Two bulbs in parallel
(wiring diagram shown below)

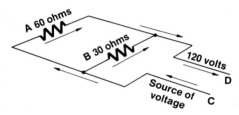

A 60 ohms

B 30 ohms

120 volts

D

Source of voltage C

26-8 Through which bulb will the greater amount of current flow?

Electrons can flow in a parallel circuit. A *parallel circuit,* as shown in Fig. 26-8, is the type used most commonly in homes. The current divides at C and forms two branches, A and B. (There could be any number of branches.) These branches join together again at D. The following statements are true of all parallel circuits:

1. The total current in the circuit equals the sum of the currents in each of the separate branches.

2. If the circuit is broken in any branch (by removing a bulb, for instance) electrons can still pass through the other branches because each branch has its own complete path for electrons.

3. As more branches are added to a parallel circuit, the total resistance decreases. The more pathways that are opened for the current, the less total resistance the current meets.

4. In a parallel circuit, the voltage across all of the branches is the same. Compare the circuit in Fig. 26-8 to a river flowing downhill from C to D. The difference in level between C and D is the same regardless of which path is taken. In electric circuits, this difference in "level" is the voltage.

activity

Series and parallel bulbs. Arrange three 6-volt bulbs into a series circuit. Arrange three other 6-volt bulbs into a parallel circuit. Power each circuit with a 6-volt battery.

Questions:
1. Which group of three bulbs gives the brighter light?
2. What happens when one of the bulbs is removed from its socket in the series circuit? in the parallel circuit?
3. Arrange two 6-volt bulbs into a series circuit. Compare its total brightness to that of the three-bulb series circuit.

A short circuit is dangerous. Have you ever used an electric appliance only to be startled by a shower of sparks, and a room plunged into sudden darkness? What caused this to happen?

Suppose a person is using an electric toaster in which the resistance of the nichrome wire is 23 ohms and the voltage at the plug is 115 volts. Then, according to Ohm's law (amperes = volts/ohms), the normal current flowing is 115/23, or 5 amperes.

Now suppose that the insulation on the cord to the toaster became so worn that the bare copper wires touched each other. The wires "shorted," or a *short circuit* was formed. The electrons would then flow from one copper wire to the other, instead of going through the nichrome in the toaster. A short circuit, a sudden and large drop in resistance in the electric circuit, would result.

If the resistance of the short circuit is only 1 ohm, instead of 23 ohms, the current would become 23 times as great (115 amperes). This large current produces intense heat in the copper wires and would quickly burn the covering of the cord. If the short circuit lasts more than an instant, it could set the house on fire.

Fuses and circuit breakers protect homes from short circuits. You cannot take a chance of having a house burn down any time a defective appliance might be used. Is there any protection against short circuits? That protection is built in by using *fuses* or *circuit breakers* in every circuit. Any current that goes through the circuit must first go through the fuse or circuit breaker.

A fuse contains a short wire, made of an alloy of lead that melts easily. If the circuit carries a load that is too large, the fuse melts, breaking the circuit and stopping the current.

A fuse stamped "20" will carry up to 20 amperes safely all day without getting hot. If the current is much greater than 20 amperes, however, the fuse wire melts and breaks the circuit. A 115-ampere current would "blow" the 20-ampere fuse instantly.

Many homes and industries use circuit breakers instead of fuses. Circuit breakers act like switches to open a circuit if the current becomes too great. They are safer and easier to use than fuses, and cheaper in the long run since they can be reset and used over again.

Circuit breakers are also used in the light circuits of many cars. If a short circuit occurs while someone is driving at night, the lights blink on and off. In older cars,

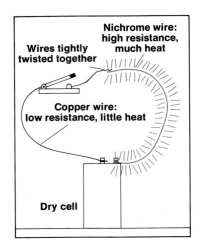

26-9 What happens to a high-resistance wire when a current flows in it?

26-10 What rule can you state about the heat of the nichrome wire from these two diagrams?

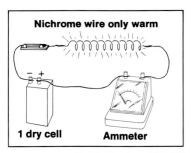

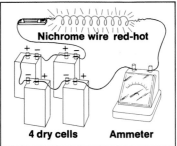

when a fuse burned out, the lights went out and did not relight, leaving the driver in danger.

If the cause of a "blown" fuse is a wiring defect, the defect must be repaired before the fuse is replaced or the circuit breaker is reset. Otherwise, the fuse will melt again as soon as the defective appliance is used. An electrician can tell what size fuses or circuit breakers should be used for most home circuits.

heat and light from electricity

Electricity produces heat. Earlier in the chapter, you saw that an electric current produces heat when it passes through a conductor. If that conductor is a copper wire in an extension cord, carrying normal household currents, very little heat is produced. However, if the wire is nichrome used inside a toaster, then the nichrome wire gets red hot, even if the same current is used.

Why is it that the same amount of current heats some wires very much and others very little? You have already seen that resistance is caused by collisions between atoms of the metal and the electrons of the current. Wire size, length, and composition affect resistance. The resistance causes heat. The more resistance the electrons meet in being pushed along, the more heat is produced.

The difference in resistance between two wires can be shown by connecting lengths of copper and nichrome wire to each other and to a dry cell as shown in Fig. 26-9. The same amount of current flows in each wire, but different amounts of heat are produced. The heat produced in each wire depends upon its resistance. *The greater the resistance, the greater the heat for a given current.*

The amount of heat produced also depends on the amount of the current. When more electrons flow through a conductor, the conductor gets hotter. To show this, a current from a single dry cell is passed through a thin nichrome wire. An ammeter is used to measure the current, as shown in Fig. 26-10. Now, instead of one cell, several cells in series are used. Two things will happen: (1) the ammeter shows that there is more current, and (2) the wire gets much hotter.

Exact tests in the lab show that the heat increases as the square of the current. That is, when the current is doubled, four times as much heat is obtained for a given resistance.

Three factors determine the amount of heat produced: (1) the resistance of the conductor, (2) the size of the current, and (3) time. The longer a heater is turned on, of course, the more heat it gives off.

Tests show that the number of calories of heat produced can be found by using the following formula:

$$\text{calories} = 0.24 \times \text{amperes}^2 \times \text{ohms} \times \text{seconds}$$

sample problem

How many calories of heat are produced when there is a current of 5 amperes through a toaster of 20 ohms resistance for 2 min?

solution:

Step 1. Change 2 min to 120 sec

Step 2. Using the formula,
$$\text{calories} = 0.24 \times \text{amperes}^2 \times \text{ohms} \times \text{seconds}$$
$$= 0.24 \times 5^2 \times 20 \times 120$$
$$= 14,400$$

activity

Hot and cold wires. Twist together the ends of two short lengths of copper and nichrome wires (about 24 gauge). Connect the free ends of the wires to a dry cell for a few seconds. CAUTION: (1) *Use a pair of pliers to attach the nichrome wire to the cell because the wire may become too hot for your fingers.* (2) *Do not leave the wires attached too long, or you may damage the dry cell.*

Describe what happens to the copper and nichrome wires.

Electricity produces light. The first type of electric light was the *electric arc*. An arc is obtained when a small gap is part of an electric circuit. If there is enough voltage, the current will jump the gap, producing a blinding light.

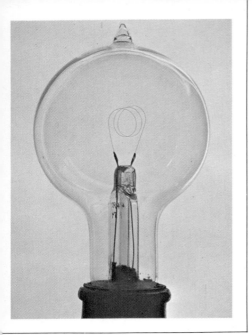

26-11 You can see the thin filament in this photo of an early light bulb. Why are most modern bulbs frosted so that the filament cannot be seen?

26-12 Mercury vapor lamps are replacing other types of highway lighting in many parts of the world. How can you tell that this photo is a time exposure?

The electric arc is used in some types of searchlights and movie projectors. However, for several reasons, the arc is not practical for most lighting purposes.

During the 1880's, Thomas Edison and others developed the *incandescent light bulb*. The basic idea behind this type of electric light is to heat a wire or filament to the point where it glows brightly. Can you see why the filament must be made of a substance that has a very high melting point? Tungsten is best for this purpose. See Fig. 26-11.

To keep the filament from burning, the air is pumped out of the bulb. Instead of air, a gas such as nitrogen or argon is used because it does not support combustion.

Even though it is widely used, the incandescent bulb is not an efficient source of light. Less than 2% of the energy used is released in the form of light waves. The rest of the energy is radiated as heat. Large light bulbs are more efficient than small bulbs. For example, one 150-watt bulb gives as much light as ten 25-watt bulbs (totaling 250 watts). The efficiency of light sources was discussed in Chapter 23.

Incandescent bulbs would be more efficient if a larger current was forced through them. The bulbs would glow more brightly, like photoflood lamps, but would not last as long as common bulbs. The added current would make the wire so hot that it would burn out in a short time.

At night, cities are brightly lit with signs of many colors. People refer to all of these lights as "neon" lights. However, the only lights with neon in them are the orange-red lights. The other colors are produced by different gases.

How do these lights work? Gases at normal pressures act as insulators, but at low pressures they become good conductors. A high voltage strips electrons from atoms of the gas, making the gas a conductor.

The color of the light depends on the kind of gas used. Sodium, for example, produces a bright yellow light, neon a red light, and mercury vapor a blue-green light.

The *mercury-vapor lamp* has largely replaced all other types of lamps for the lighting of streets and highways. This lamp gives more light with less current and less glare than many other types of lamps. See Fig. 26-12.

Fluorescent lamps work somewhat like mercury-vapor lamps. They use high voltage and a low-pressure mercury-vapor arc to produce ultraviolet light.

Remember, ultraviolet light has frequencies beyond the range of your eyes. The inside wall of the fluorescent light bulb is coated with powdered minerals called *phosphors*. These minerals convert the ultraviolet light to visible light. This process is known as *fluorescence*. The color of the light produced in this way depends on the kind of mineral powder used to coat the inside of the bulb.

Fluorescent lamps are even more efficient than neon lamps and are better for normal lighting purposes. It is often important to use lamps that give true color values to objects they light up. Such a light must be a great deal like sunlight if the objects are to look natural. Some fluorescent lamps produce light of this quality.

Even the best lighting is still not efficient. There is much room for improvement, and research is being done in this field. Some day light may be produced by panes of *luminescent* glass. This glass absorbs light energy during the day and releases it at night when a small electric current is passed through it.

Edison
The Master Inventor

Thomas A. Edison (1847–1931) was one of the truly great inventors. By the time he was 16 years old, Edison had built his first invention. This invention grew out of a need to tap out a routine signal on a railroad telegraph every hour. Edison thought this was a waste of time, so he hooked up a clock to the telegraph to send the signal for him.

At the age of 23, Edison received $40,000 for patent rights from another invention. With this money, he set up a research lab where he continued to invent for the rest of his life. The light bulb, phonograph, motion picture projector, typewriter, electric generator, telephone, and diode are only a few of over 1100 inventions Edison patented.

Edison soon became known as a genius. Yet he smiled at this label, defining genius as "1% inspiration and 99% perspiration."

Edison's definition of genius was at least partly correct when applied to his own life. Edison worked with such persistence that his co-workers were amazed. Once on the track of an idea, Edison seldom quit. When 10,000 tests with a storage battery failed to produce results, he was not discouraged. "After all," Edison said, "I've disposed of 10,000 ways that won't work." His work finally resulted in the widely used Edison storage battery.

/ummary

1. An electric current is a stream of moving electrons.
2. The path followed by an electric current is called a circuit.
3. An ampere is a flow of one coulomb through a wire per second.
4. A potential difference pushes electrons through conductors.
5. Electrons can travel in a conductor but not in an insulator.
6. The resistance of a conductor depends upon its thickness, composition, length, and temperature.
7. Ohm's law states that volts are equal to amperes times ohms.
8. A watt is a unit of power found by multiplying volts times amperes.
9. The electric energy that you use in your home is measured in kilowatt-hours.
10. A series circuit is a single pathway for all electrons in the circuit.
11. A parallel circuit has two or more branches for electrons.
12. A short circuit is a sudden sharp drop in resistance.
13. Fuses and circuit breakers protect homes from the dangers of short circuits.
14. The amount of heat produced by a current in a wire depends on the resistance of the wire, the amount of current, and the length of time the current is on.
15. Electric currents produce light by heating a wire or by going through a low-pressure gas.

review

Match each item in the left column with the best response in the right column. *Do not write in this book.*

1. conductors
2. superconductors
3. insulators
4. rheostat

a. all current goes through each resistance
b. subtances that have a low resistance
c. device that prevents circuit overloading
d. nonconductors

5. ampere
6. ohm
7. volt
8. circuit breaker
9. parallel circuit
10. series circuit

e. property of some subtances near absolute zero
f. total resistance drops as more resistances are added
g. unit of current
h. unit of electric "pressure"
i. unit of electric resistance
j. variable resistor
k. voltmeter

questions

Group A
1. What is the difference between an open and a closed circuit?
2. What is an electric conductor? an insulator? Give examples of each.
3. What is the unit of electrical resistance? of electric current?
4. What two factors determine the size of the current in an electric circuit?
5. What is the difference between volt and ampere?
6. What is the meaning of the number "30" stamped on a fuse?
7. How is a fuse like a switch?
8. Suppose a 10-ampere and a 20-ampere fuse are both wired in series in an overloaded line. Which will be likely to blow first? Explain.
9. What is a circuit breaker?
10. How does an electric current produce light in an incandescent bulb?
11. How does an electric current produce light in a neon tube?
12. How does an electric current produce light in a fluorescent lamp?

Group B
13. In what way are voltage and water pressure alike?
14. Explain why copper wire is used in extension cords and nichrome wire in electric toasters.
15. What three factors determine the total amount of heat that is produced when a current flows in an electric iron?
16. How much does the heat produced in a wire increase when the current through it is tripled?
17. Compare the cost of using 100 watts for 1 hr with 50 watts for 2 hr.
18. If all Christmas tree lights go out when one bulb burns out, are they connected in series or parallel? How do you know?
19. Why does the total resistance of a parallel circuit decrease when more branches are added to the circuit?
20. Is it always better to use a 30-ampere fuse instead of a 20-ampere fuse to protect a home-wiring circuit? Explain.
21. List four factors that affect the resistance of a conductor. Explain how the resistance is affected by each factor.
22. What is a superconductor?
23. Discuss the efficiency of incandescent bulbs.

problems

1. How many volts are needed to push 7 amperes through an electric heater that has a resistance of 16 ohms?
2. How many amperes will a 6-volt battery send through a 3-ohm circuit?
3. What is the resistance of a toaster if 110 volts produce a 5-ampere current through it?
4. A light bulb draws 0.5 ampere on a 10-volt line. (a) What is the resistance of the bulb in ohms? (b) If you plugged the bulb into a 110-volt line, what would be its current? (c) How much extra resistance would you have to put in series with the bulb so that it would draw only 0.5 ampere on the 110-volt line?
5. A circuit has four similar lamps in parallel, each drawing 1 ampere on a 110-volt line. (a) What is the total current used by all four lamps? (b) What is the resistance of each lamp? (c) What is the total resistance of all four lamps in parallel?
6. About how much current is sent through 900 m of No. 10 copper wire by a 3-volt battery?
7. An electric iron draws 5.5 amperes on a 115-volt line. What is its wattage?
8. A theater uses 50 light bulbs of 200 watts each for 4 hr. Calculate the cost at $0.06 per kilowatt-hour.
9. How much does it cost to light your school room for 5 hr at $0.06 per kilowatt-hour? Assume ten fluorescent fixtures of 80 watts each.
10. An immersion heater is dipped into 1 L of water for 5 min. If the resistance of the heater is 40 ohms and it draws 3 amperes, how much heat is produced? Calculate the rise in temperature of the water, remembering that 1 cal warms 1 g of water 1 C°.

further reading

Asimov, Isaac, *How Did We Find Out About Electricity?* New York: Walker, 1973.

Basic Wiring. Morristown, N.J.: Silver Burdett, 1976.

Burch, Monte, *Basic House Wiring.* New York: Harper & Row, 1976.

Demers, Ralph, *The Circuit.* New York: Viking, 1976.

Dunsheath, Percy, *Giants of Electricity.* New York: Thomas Y. Crowell, 1967.

Graf, Rudolf F., *Safe and Simple Electrical Experiments.* New York: Dover, 1973.

Sootin, Harry, *Experiments with Electric Currents.* New York: Norton, 1969.

Wade, Harlan, *Electricity.* Raintree Pubs. Ltd, 1977.

sources of
electric currents

For several thousand years, scientists searched for the "invisible fluid" that they thought was an electric current. Scientists knew all about static charges during all those years, but static charges could not be controlled. These charges could be built up, but they were released in a spark or flash. Until these charges could be released in a controlled way, however, there seemed no way in which the energy could be put to work.

Then, in 1798, Alessandro Volta made an important discovery. With Volta's discovery, electric energy could be controlled and used.

Volta used chemical action to produce an electric current. Chemical means are still among the most common sources of electric current today even though a number of other ways have since been found to produce current. Cells and batteries for flashlights, radios, or cars are common examples of these chemical sources. In this chapter, you will learn about the five major sources of electricity.

In addition to chemical action, electricity can also be produced by heat, light, crystals, and magnets

objectives:

☐ To understand the action of chemical cells.

☐ To compare series with parallel wiring of cells.

☐ To describe the action and use of the storage battery.

☐ To identify the five sources of electric current.

cells and batteries

Cells are made with different metals. Volta made the first simple electric cell by dipping two different metals, called *electrodes*, into a liquid that was able to conduct a current. Such a liquid, as you have seen earlier, is called an *electrolyte*.

To understand how a cell works, observe what happens when zinc and copper are dipped into a solution of dilute sulfuric acid. See Fig. 27-1. Atoms of zinc dissolve from the zinc plate. The zinc atoms go into the acid solution as positive ions. This action is shown in the following equation:

$$Zn \rightarrow Zn^{++} + 2 \text{ free electrons}$$

The free electrons collect on the zinc plate, giving it a negative charge. Therefore, the zinc becomes the *negative pole,* or *negative electrode,* of the cell.

Meanwhile, hydrogen ions from the acid are taking electrons from the copper electrode. This action turns the hydrogen ions into molecules of hydrogen gas as follows:

$$2\,H^+ + 2 \text{ free electrons} \rightarrow H_2 \uparrow$$

The loss of electrons from the copper leaves it with an excess of positive charges. Thus, the copper becomes the *positive electrode* of the cell.

When these two electrodes are connected by an outside conductor, such as a wire, electrons move from the zinc (where they are plentiful) to the copper (where they are scarce). Meanwhile, the chemical action within the cell supplies the energy to keep the electrons flowing. The current goes from the negative electrode through the wire to the positive electrode. Since the current goes in only one direction, it is called a *direct current* (dc).

An electric cell is a device by which chemical energy is changed into electrical energy. A cell always consists of two electrodes and an electrolyte, which acts chemically on at least one of the electrodes. Flashlight cells are often referred to as *dry cells.* They are not really dry, but are covered in a way to keep the liquid from leaking out. *A* **battery** *is a collection of two or more cells that are hooked together.*

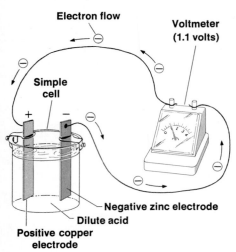

27-1 What controls the size of the voltage produced by the set of metal plates?

Electron flow

Voltmeter (1.1 volts)

Simple cell

Negative zinc electrode

Dilute acid

Positive copper electrode

Voltage depends on the elements used. When different electrodes are dipped into electrolytes, they produce different voltages. For example, zinc and carbon, the electrodes used in common dry cells, produce 1.5 volts. The voltage of a cell depends only on the elements used, not on its size. A single element cannot have a voltage unless it is compared to another element.

Different elements have different voltages. The voltages for some elements are listed in the *electrochemical series* shown in Table 27-1.

Table 27-1:

Electrochemical Series*	
Element	Value
Lithium	−3.04
Potassium	−2.92
Calcium	−2.87
Sodium	−2.71
Magnesium	−2.37
Aluminum	−1.66
Manganese	−1.18
Zinc	−0.76
Iron	−0.44
Nickel	−0.25
Tin	−0.14
Lead	−0.13
Hydrogen	0.00
Copper	+0.34
Mercury	+0.79
Silver	+0.80
Gold	+1.50

*In this series, hydrogen is given a rating of zero, a standard value against which all other elements are compared.

When any two of these elements are put into an electrolyte, the voltage shown on the voltmeter is the potential difference between them. For instance, copper with a value of +0.34 and zinc with a value of −0.76

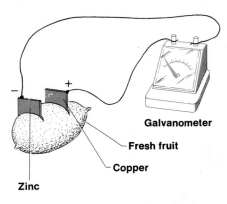

27-2 What factors cause the lemon to act like a dry cell?

produce a potential difference of 1.10 volts [0.34 − (−0.76) = 1.10]. To get the greatest voltage, one electrode made of an element near the top of the list is combined with another near the bottom of the list. However, cost, capacity, useful life, and other problems greatly limit the choices of elements.

The series of elements listed in Table 27-1 is often called an *activity series*. The most active elements are on top, decreasing in activity towards the bottom. Each element in the list will replace any element below it in a compound. For example, we have seen that zinc replaces the hydrogen in sulfuric acid in a single replacement reaction.

$$Zn + H_2SO_4 \rightarrow ZnSO_4 + H_2 \uparrow$$

activity

Lemon cell. Make a simple cell by placing two different metals, such as copper and zinc, into a lemon. (The lemon juice serves as an electrolyte.) The tiny current that results can be measured by a galvanometer, as shown in Fig. 27-2. If you have a very sensitive voltmeter, compare your readings with those in Table 27-1. Use other metals and find the values produced by each pair.

Cells may be placed in series or in parallel. You will recall that in the last chapter you saw that bulbs could be wired in series and in parallel. Cells can be wired together in a like manner. In a series hookup, the positive pole of one cell is wired to the negative pole of the next cell.

When the cells are connected in series, the voltage is increased. For instance, three dry cells, each of 1.5 volts, attached in series, form a 4.5-volt battery.

If the same three dry cells are connected in parallel, however, they form a battery that produces only 1.5 volts. In the parallel arrangement, the positive pole of one cell is connected to the positive pole of the next, and the negative pole to the negative pole, as shown in Fig. 27-3. In a sense, this is like making a cell with a larger positive pole and a larger negative pole. The advantage of this arrangement is that it lengthens the life of the battery.

Hands-On Internet Exploration

1. From the menu, choose **#8 - Windows.**

2. Click on **Internet.**

3. Click on **Netscape 1.22.**

Option 1
4. Start your exploration by clicking on the **OPEN** button. Then type:
   http://www.yahoo.com
and press **Enter.**

Click on a subject area that interests you. Try to find some information that may be useful or interesting to you.

Option 2
4. Explore the SCCC homepage. Check out what interests you. Visit the library and some of the links to the Internet Resources.   You can also find YAHOO there.

*   *   *   *   *   *   *

Some other sites that you might like to try: (Click on the **OPEN** button and type the address).

Another subject directory (like Yahoo)   **http://galaxy.einet.net/galaxy.html**

http://www.uwm.edu/Mirror/inet.services.html

Educational hotlists: **http://sln.fi.edu/tfi/hotlists/hotlists.html**

Educational Exhibits & Museums: **http://sln.fi.edu/tfi/jump.html**

Bev Tice-Deering's homepage: **http://www.sccd.ctc.edu/~ticedeer/**

Seattle Times (newspaper):   **http://www.seatimes.com**

The Emerald Web (Seattle)   **http://www.cyberspace.com/bobk/**

The White House   **http://www.whitehouse.gov/**

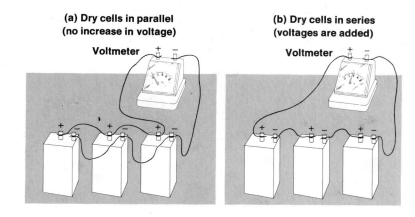

(a) Dry cells in parallel
(no increase in voltage)

Voltmeter

(b) Dry cells in series
(voltages are added)

Voltmeter

27-3 What advantage, if any, is gained by connecting the cells in parallel?

Since the voltage depends only on the kinds of poles used and not on their size, the voltage is not affected in a parallel hookup.

Modern cells and batteries have many uses. The lead storage battery has been a very useful source of electric current. Its special value is that it can be used more than once.

A storage battery can be made by dipping two clean lead plates into a jar of dilute sulfuric acid. Almost at once, a thin film of lead sulfate will form on each plate. If a current from two or three dry cells (in series) is passed through the solution for a few minutes, bubbles of gas will form around each plate. The brown coating that forms on the positive plate shows that electrolysis of water is taking place. Oxygen, appearing at the positive plate, reacts with the lead to form brown lead dioxide (PbO_2). At the negative plate, hydrogen is set free. When these reactions take place, the cell is being *charged*.

The cell can be *discharged* by connecting it to a small flashlight bulb, or a doorbell, as shown in Fig. 27-4. The cell can be recharged again by connecting the positive pole of a battery of two dry cells to the positive plate of the storage cell. A voltmeter shows that the potential difference of this charged cell is about 2 volts. A car battery is made up of six cells of 2 volts each connected in series.

The action described above takes place over and over again in a car's storage battery. The charging current

27-4 Is this lead storage battery being charged or discharged? How do you know?

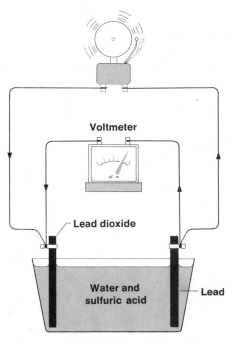

Voltmeter

Lead dioxide

Water and sulfuric acid

Lead

Lead + lead dioxide + sulfuric acid ⟶ lead sulfate + water + energy

comes from a *generator* or *alternator*. The *ammeter*, the red light on the instrument panel, shows if the battery is charging or discharging at any given time.

Each cell of a car battery has a set of plates made of spongy lead and a set of lead dioxide plates. The sulfuric acid acts on these plates while the cell is discharging, forming lead sulfate and water. The chemical equation for the reaction is

$$Pb + PbO_2 + 2 H_2SO_4 \rightarrow 2 PbSO_4 + 2 H_2O + energy$$

| lead | + | lead | + | sulfuric | $\rightarrow$ | lead | + | water |
| | | dioxide | | acid | | sulfate | | |

To charge the battery, the electrons are sent through it in the opposite direction. This reverses the chemical reaction:

$$2 PbSO_4 + 2 H_2O + energy \rightarrow Pb + PbO_2 + 2 H_2SO_4$$

| lead | + | water | | $\rightarrow$ | lead | + | lead | + | sulfuric |
| sulfate | | | | | | | dioxide | | acid |

Notice that energy must be put into the cell to charge it. This energy is then stored in the cell in the form of chemical energy, which is later released. The above chemical reactions are *reversible*. A reversible reaction is one in which the products are changed back to the original substances. Cells of this kind may last for years.

How do you know when a storage battery is charged? Notice from the above equation that water is formed during the discharge of the cell. Water is not as dense as sulfuric acid. Therefore, the specific gravity of the solution drops as the cell discharges. If the specific gravity of the cell is about 1.25 to 1.30, the cell is fully charged. If the value is less than 1.15, the cell is discharged. Distilled water must be added from time to time to replace the water that is lost by evaporation and by electrolysis.

Keep in mind that a high or low specific gravity reading tells only if the cell is charged, not the condition of the cell. That is, a new cell in good condition could be discharged, or an old cell in poor condition could be charged.

In a storage battery, chemicals are used to produce an electric current. This is the opposite of using an electric current to make chemicals as was discussed in earlier chapters. You have already studied the use of an electric current to break down water into hydrogen and oxygen, and in the production of aluminum. In fact, when you charge a storage battery, you use an outside source of electric current to produce chemical energy in the battery.

activity

Electrolytes and current. How do varying concentrations of an electrolyte affect the amount of a current? Set up the equipment as shown in Fig. 27-5. Pour 250 mL of distilled water into the jar. Then dissolve 0.25 g of table salt (NaCl) in the water and read the ammeter. Add more salt in 0.25-g portions and record the ammeter reading each time. Plot the current reading against the salt levels on a graph. How do these two factors compare?

There are many kinds of chemical cells. Most flashlight cells are made of carbon and zinc. The carbon-zinc cell produces 1.5 volts and is fairly simple to make. This cell is useful for those devices that operate on low voltages.

In recent years, newer types of cells using different substances have been made. One type, called the *alkaline cell,* has a positive pole of manganese and a negative "jacket" of zinc, with a highly alkaline electrolyte. The major advantage of this cell is that it works well when a high charge is needed for a long time. Under these conditions, the alkaline cell will give up to ten times the service of a standard carbon-zinc cell of equal weight.

Another type is the *nickel-cadmium* cell, having a positive pole of nickel oxide and a negative pole of cadmium. This cell produces about 1.2 volts. Not only is it an efficient source of energy, but it can be recharged. In this respect, the nickel-cadmium cell is like the lead-acid storage battery in cars. The cell can be recharged many times without breakdown.

As shown in Table 27-2, batteries of even greater load and efficiency are made with lithium and nickel fluoride. Storage batteries using these substances have 10 to 15 times more energy per gram than do the lead-acid batteries used in cars.

The energy output of these cells is measured in watt-hours per kilogram. Table 27-2 shows that the real energy produced by each of these cells is still far below its

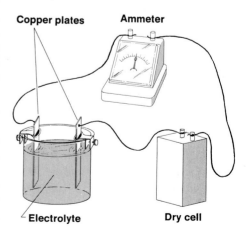

27-5 How does the concentration of the salt solution affect the amount of current produced in a cell?

Copper plates   Ammeter

Electrolyte   Dry cell

potential limit. Why is there a difference between the potential and the actual energy output?

Table 27-2:

		Actual Energy (watt-hours per kg)	Theoretical Energy (watt-hours per kg)	
Type	Typical Use			Rechargeable?
Lead-acid	car ignition	20	180	yes
Zinc-carbon	flashlight	30	330	no
Nickel-cadmium	TV, tools	40	230	yes
Mercury batteries	hearing aids	100	250	no
Silver-oxide	hearing aids, watches	100	290	no
Lithium-nickel fluoride	electric cars	330	1650	yes

Output of Chemical Cells

One cell not shown in Table 27-2 is the *fuel cell.* The fuel cell is different from any other cell in that it consumes fuel in much the same way as gasoline is used in a car. The fuel is fed to the cell while it is in use.

The fuel cell is also very efficient. It converts up to 80% of its potential energy to an electric current. Compare this value with 20% efficiency for a car engine.

Fuel cells are now used only in space missions. With their efficient use of energy, fuel cells may someday be used to power cars and other devices. At the present time, such cells are far too costly for home use.

Cells have many uses. Today just about anything that runs with a plug-in cord can also run on a battery. Because of batteries, you can use electric devices in places where there is no electric outlet. Also, batteries are safer because there is less danger of shock when batteries are used for power.

Shavers, power tools, movie cameras, power typewriters, tape recorders, toothbrushes, toys, TV sets, fire

unit 5 electromagnetic nature of matter

27-6 A prototype of an electric car and one available today. What are some of the benefits and limitations of an electric car?

27-7 X-ray photo of a battery-powered pacemaker implanted in the patient's chest. What is the function of this device?

alarms, carving knives, and even some small cars can now be battery-powered. See Fig. 27-6. Modern batteries are smaller, last longer, and are more reliable than earlier models.

One of the most amazing uses of cells is to provide a small electric current to keep a defective heart beating. The heart beat is controlled by a *pacemaker*. When the heart's own pacemaker breaks down, the battery-operated pacemaker can be implanted in the chest to do the job. The power for this system comes from a few tiny cells, each about as large as a fingernail. The pacemaker can keep the heart beating for years before the cells need to be replaced. See Fig. 27-7.

The chemical action that takes place inside dry cells can sometimes also occur outside cells. Underground or underwater steel structures, for instance, may corrode because there is an electric current from the structure to the soil or water.

One effective way of stopping this form of corrosion is to reverse the current. It is possible to direct the flow of electrons to the steel from its surroundings. This reverse current is produced by placing blocks of zinc near the steel. Since zinc is higher in the electrochemical series, it acts like a negative pole of a cell. The current is then sent to the steel, which serves as a positive pole.

In the course of saving the steel, the zinc corrodes. When it is used up, it is simply replaced. The cost is low

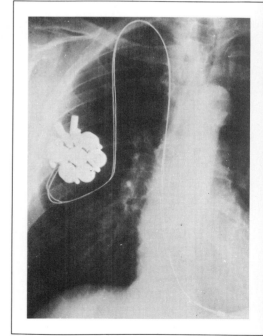

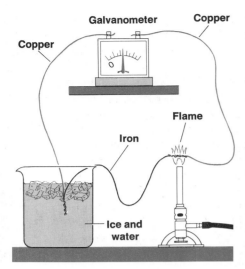

27-8 How is an electric current produced in a thermocouple?

when compared to the cost of replacing the structure itself.

Zinc electrodes are used for structures such as gas lines, piers, and ship hulls. This method can be used to protect other metals. Metals other than zinc can also serve as electrodes.

other sources of electricity

The thermoelectric effect produces an electric current. Heat can be changed to an electric current by a method called the *thermoelectric effect.* When two different metals are joined to form a circuit, as in Fig. 27-8, electrons will flow if heat is applied at one junction of the metals. This effect is even greater if the other junction is cooled. Such a device is called a *thermocouple.* When several such devices are connected in series, the arrangement is called a *thermopile.*

In this setup, the voltage depends on the temperature difference at the hot and cold ends of the wires. The greater the difference, the greater the voltage. Thus, the thermocouple can also be used as a thermometer. The range of such a device depends on the types of metals used. Iron and copper are useful up to 275°C, while platinum and rhodium are useful up to 1600°C.

A thermocouple is a very useful measuring device for several reasons. (1) It can be made highly sensitive, measuring differences as small as 0.001°C. (2) The meter can be set far from the object being measured. For instance, a thermocouple can be placed in the wall of a jet engine and the results shown in the pilot's instrument panel. (3) The thermocouple can cover a range far greater than that of the common glass thermometer.

Thermocouples are also useful in many homes. A thermocouple can be used as a safety switch in a furnace. Most gas furnaces have a *pilot light* that ignites the main gas jets whenever the furnace is turned on.

Have you ever wondered what might happen if the pilot light went out? Does the house fill with the explosive gas? No. A thermocouple, placed in the flame of the pilot light, protects you. It acts as a switch to shut down the main gas line if the flame goes out. The flame of the pilot light makes a small electric current in the thermocouple.

This current keeps a valve in the main gas line open. When the flame goes out, the current stops, the valve closes, and gas stops flowing into the house.

The photoelectric effect produces an electric current. You may recall that when a beam of light strikes certain metals, electrons are knocked out of the metal. These electrons can produce an electric current. This action is known as the *photoelectric effect.* Electric "eyes," which can be used to open doors, use photocells. A beam of light is sent across the path where people must walk to reach the door. As long as the beam is not stopped, light will strike the cell, producing electric current, and the door will stay shut. However, when someone breaks the light beam by walking across its path, the circuit is no longer complete. Thus, the current stops and activates the door opener.

The piezoelectric effect produces an electric current. When certain crystals are subjected to a force, tiny currents may be produced. This is called the *piezoelectric* (pee-*ay*-zo) *effect.* Rochelle salt crystals and quartz show this effect very well and are used in some microphones and phonographs. In addition, crystals are used to produce the carrier wave of radio stations. These crystals can produce very stable frequencies so that the station does

27-9 Why does the light flash when the crystal is tapped with a gavel?

not drift from one place to another on the radio dial. See Fig. 27-9 on p. 507.

Electromagnetic induction can produce an electric current. In 1831, Michael Faraday of England and Joseph Henry of the United States both found that a magnet could be used to produce an electric current.

Push a magnet into a coil of copper wire as shown in Fig. 27-10. What does the meter show when you do this? You will find that the meter needle goes one way when the magnet is pushed in and the other way when the magnet is pulled out. The needle moves any time there is a current produced. Since the needle alternates back and forth when the magnet moves in and out, the current is called *alternating current* (ac). If the magnet does not move, the meter shows no current.

This activity shows that *a magnet moving near a conductor produces a current in that conductor.* This method is used by power plants to make the electric current that is used in homes, schools, and factories.

Current is produced only when there is motion between the magnet and the coil of wire. However, it does not matter which moves, the magnet or the coil. Faraday called this process *electromagnetic induction.* The current produced in this way is *induced current.*

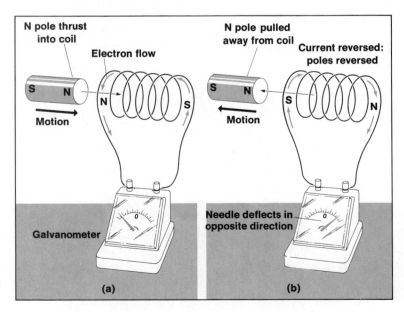

27-10 What happens when a magnet is thrust into or pulled out of a coil of wire?

unit 5   electromagnetic nature of matter

Alessandro Volta
The Search for Electric Current

Luigi Galvani was startled. He had dissected frogs many times before, but this was the first time he had ever seen anything like that! When he held the knife and probe in a certain way, the leg twitched. It moved! How could a dissected frog's leg do that?

Galvani repeated his probes, and soon he could make the leg twitch almost any time he chose to. After long months of further testing, Galvani reported his findings. He said that the movement was caused by something inside the tissues called "animal electricity."

Galvani was a respected scientist, and his findings were widely accepted by his colleagues. However, Alessandro Volta, a young Italian professor of physics, was one notable exception. Volta claimed that the source of the electricity was not the frog, but rather the different metals used in the knife and probe. Volta called it "contact electricity."

Galvani returned to his lab and after much effort produced the effect by using only similar metals. This result seemed to prove Galvani's idea. However, Volta refused to give up his "contact electricity" idea. He insisted that the "similar" metals were still unlike.

Galvani did not give up either. He got rid of the metals altogether. Galvani now produced the twitch by touching the nerve of the frog leg to the end of a muscle. This test seemed to prove once and for all that the "animal electricity" theory was correct.

In the face of this proof, what could Volta do? Volta went one step further and got rid of the frog's legs. He produced an electric current by using only unlike metals, and an acid instead of a frog.

Volta's finding was important for two reasons: (1) He had produced, for the first time ever, a strong, constant current, not just something barely able to twitch a frog's leg; (2) his finding was very simple to duplicate.

Volta's discovery of an electric current shook the world. It also destroyed Galvani, who felt such a loss of prestige that he gave up. Shortly after Volta's discovery, Galvani died, a broken man.

However, the story did not end there. Had Galvani been able to look into the future, he would have seen that his theory was not wrong after all. Today, scientists know that electric currents are produced by the cells of the body. For example, modern medical science uses electric currents from the heart to check its function. In addition, a small electric current regulates the beating of the heart. Brain waves are also a measure of electric current.

summary

1. Electric currents can be produced from chemical cells.
2. The voltage of a cell depends on the elements used as electrodes.
3. Current that goes in one direction only is called direct current (dc).
4. Cells can be connected in series or in parallel.
5. A storage battery can be recharged by reversing the current through its cells.
6. Cells can be made of many different substances and have many different uses.
7. Heat can be changed to an electric current by a method called the thermoelectric effect.
8. Light can be changed into an electric current by a method called the photoelectric effect.
9. Forces acting on certain crystals can produce an electric current by a method called the piezoelectric effect.
10. Magnets can induce an electric current in a wire by a method called electromagnetic induction.
11. The electric current produced by magnets and coils is alternating current (ac).

review

Match each item in the left column with the best response in the right column. *Do not write in this book.*

1. battery
2. fuel cell
3. electrolyte
4. electric current
5. photoelectric
6. thermoelectric
7. piezoelectric
8. voltage

a. current produced when light strikes certain substances
b. complete path of electric current
c. device of coil and magnet to produce electric current
d. solution that conducts an electric current
e. current produced by squeezing certain crystals
f. most efficient chemical source of current
g. flow of electrons along a conductor
h. current produced by heating ends of different metals
i. two or more cells joined together
j. potential difference

questions

Group A
1. List three essential parts of a simple cell.
2. What is the difference between a cell and a battery?
3. Who made the first useful chemical cells?
4. What is the function of the lemon in a lemon cell?
5. How is voltage related to the size of the electrodes used in a cell?
6. Is a dry cell really dry inside? Explain.
7. Why does the fluid in a storage battery have a higher specific gravity when it is charged than when it is discharged?
8. In what two ways does a car's storage battery lose water?
9. What is one advantage of an alkaline cell over a carbon-zinc cell?
10. In what way is a nickel-cadmium cell superior to an alkaline cell?
11. List five sources of voltage.
12. How are thermoelectric currents produced?
13. What action produces the piezoelectric effect?
14. Describe the photoelectric effect.

Group B
15. Which are the negative and positive poles of a zinc-copper electric cell? Describe how they become charged.
16. If lead and zinc are used as the electrodes in a cell, which one will be negative? Explain.
17. What is the advantage of connecting several dry cells in series? What is the advantage of connecting them in parallel?
18. What specific chemical reaction occurs when a storage battery is charged?
19. How is a fuel cell like a storage battery? How is it different?
20. Compare the actual energy output of a lithium-nickel fluoride battery with a lead-acid battery.
21. How do zinc blocks retard the corrosion of underground or underwater steel structures?
22. Describe how an electric current is produced by induction.

problems

1. Four 2-volt cells are connected in series to form a battery. (a) What is the voltage of the battery? (b) If the resistance of the electric circuit totals 100 ohms, how much current is produced?
2. Four 2-volt cells are connected in parallel to form a battery. (a) What is the voltage of the battery? (b) If the resistance of the electric circuit totals 100 ohms, how much current is produced?

further reading

Asimov, Isaac, *How Did We Find Out About Electricity?* New York: Walker, 1973.

Basic Wiring. Morristown, N. J.: Silver Burdett, 1976.

Demers, Ralph, *The Circuit.* New York: Viking Press, 1976.

Graf, Rudolf F., *Safe and Simple Electrical Experiments.* New York: Dover, 1973.

Leavy, Herbert, and Joy, M. R., *Small Appliances: How to Buy, Use, Repair and Maintain.* New York: Drake, 1977.

Wade, Harlan, *Electricity.* Raintree Pubs. Ltd., 1977.

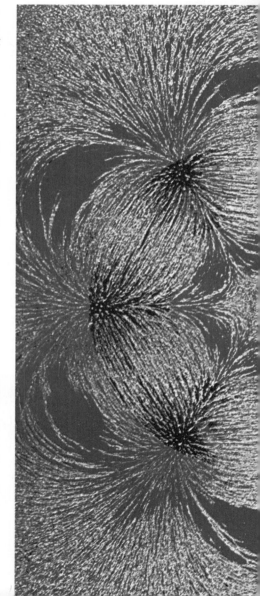

magnetism and electromagnetism

In the last chapter, you learned that there is a relationship between magnetic forces and electric currents. In this chapter, you will take a closer look at that relationship and its uses.

objectives:

- [ ] To describe the functions of magnets and magnetic fields.
- [ ] To discover from data the action of like and unlike magnetic poles.
- [ ] To discover the link between electric currents and magnetic fields.
- [ ] To relate electromagnetic induction with the production of alternating and direct currents.
- [ ] To apply induction to such devices as relays, motors, buzzers, and bells.
- [ ] To analyze how coils and transformers are used.

magnetism

Like poles repel and unlike poles attract. Over 2000 years ago, people were aware of a kind of stone that attracted bits of metal. They called it *lodestone*. Today this mineral is called *magnetite*. See Fig. 28-1.

It was found that when a steel needle or bar was stroked with this natural magnet, it too became mag-

28-1 What is one of the surprising properties of this mineral?

28-2 On the basis of these diagrams, can you state the rule of magnetic poles?

28-3 What do the iron filings show about the force of a magnet?

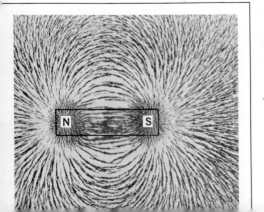

netized. When this needle was placed on a pivot so that it could turn freely, it would always swing around until it pointed in a north and south direction. Although early scientists could not explain why the lodestone worked the way it did, it was put to good use.

The magnetic *compass* was soon used by early sea captains when traveling out of sight of land. The first compasses were simply magnetized needles on pivots. The end of the needle that points north is called a north (N) pole. The other end is a south (S) pole.

How do magnetic poles affect each other? You can easily find out by doing the tests shown in Fig. 28-2. When two N poles are close together, they *repel* each other. On the other hand, when the N pole of one magnet comes near the S pole of another, the two poles *attract* each other. This action can be expressed as a law: *Magnetic poles repel each other if they are of the same kind, and attract each other if they are different.* In other words, like magnetic poles repel, unlike magnetic poles attract. How does this law compare with the action of static charges studied in Chapter 27?

As far as the force between two poles is concerned, the well-known inverse-square law holds true. If the distance between the poles is doubled, the force becomes about one fourth as much. At three times the distance, the force is only one ninth, and so on.

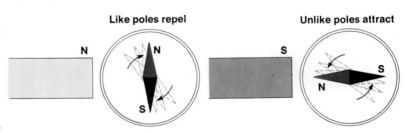

Like poles repel   Unlike poles attract

All magnets have force fields around them. When iron filings are sprinkled on a pane of glass placed over a bar magnet, a pattern of lines like that shown in Fig. 28-3 results. This pattern shows that the magnetic effect extends out into space around the magnet. The space in which the magnetic force is felt is described as a *magnetic field*.

Because the iron filings form "lines" around the magnet, the magnetism is described as being in *lines of force.*

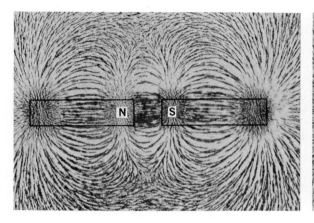

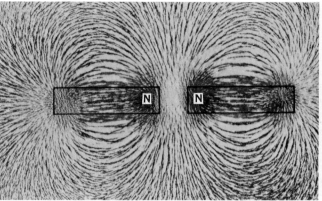

These lines show the direction and strength of the magnetic field. The more dense the lines, the stronger the magnetic field. Where is the field strongest?

In Fig. 28-4 (left), two magnets are placed with their opposite poles near each other. The pattern shows that the lines of force go from one pole to another, unlike, pole. Because the lines act like stretched rubber bands, you see that unlike poles attract.

In Fig. 28-4 (right), the magnets are placed with their like poles together. The lines of force as shown by the iron filings indicate that like poles repel.

Magnetic lines of force can go right through many substances. However, some kinds of matter can be used to shield out magnetism. A magnetic compass is of little use inside a submarine, for instance, because the inside is shielded from the earth's magnetic field by the steel walls of the vessel.

28-4 What relationship between magnetic poles is shown by these photos of iron filings?

28-5 What is meant by the statement in the diagram, "Glass is transparent to magnetism"?

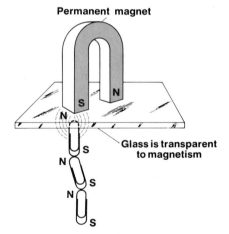

Permanent magnet

Glass is transparent to magnetism

activity

"Stop" and "go" materials for magnets. Figure 28-5 shows what happens when a glass plate is placed between a magnet and a box of paper clips. Find out what happens when other items such as cardboard, wood, plastic, copper, iron, steel, and lead are used instead of glass. Make a table showing what you discovered. Classify the items into those that are "transparent" to the magnetic force and those that are "opaque" to it.

The earth has a magnetic field. The earth acts like a huge magnet. Even before the year 1600, when Sir William Gilbert made many tests with magnets, it was known that the earth was magnetized. Gilbert studied and described the way in which the earth acts as a magnet in one of his first books. The earth behaves much as it would if a huge bar magnet were buried deep below the surface. This imaginary buried magnet is slightly tilted from the polar axis of the earth. That is, the magnetic poles are not in line with the geographic poles. See Fig. 28-6.

A compass needle points to the magnetic pole, not to the true or geographic pole. In San Diego, for instance, a compass points about 15 degrees east of true north, while in Boston it points about 15 degrees west of true north. This angle between the magnetic pole and the geographic pole is called *declination*. Navigation charts must be constantly updated because the magnetic poles are slowly shifting.

28-6 Can you see why the south magnetic pole is located near the north geographic pole? Which pole of a compass needle points north?

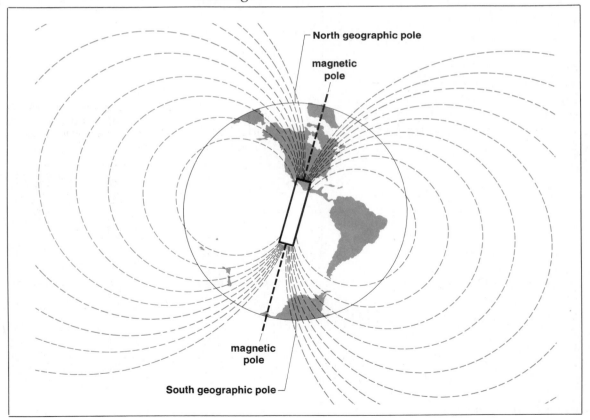

North geographic pole

magnetic pole

magnetic pole

South geographic pole

unit 5 electromagnetic nature of matter

SEATTLE CENTRAL COMM. COLL
1701 BROADWAY
SEATTLE   WA 98122

REGISTRATION APPOINTMENT

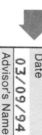

ARGUETA VICTOR M
1518 NW 53RD ST #1
SEATTLE   WA 98107

Quarter	
9344 - SPRING-94	
Name	Student I.D.
ARGUETA VICTOR M	533298157
Date Time	Place Room
03/09/94 11:00A-12:00P RM1105	
Advisor's Name	Phone Room

Program Code	Current Program Status
0020	LIBERAL ARTS
Res Status	Fee Status
2 NON-RES	02 NON-RESIDENT
	Past Due Amount

MUST BE PRESENTED AT TIME OF REGISTRATION

33

There is a link between magnetic forces and electric currents. In 1819, a Danish teacher named Oersted saw something that surprised him. He happened to notice that a compass needle moved whenever a current was turned on in a nearby wire. After many tests, Oersted found that *an electric current always produces a magnetic field.* Why do you suppose this is true?

When you recall that an electric current is made of moving electrons, you may begin to understand what causes matter to act like a magnet. All matter is made of atoms, with electrons revolving in orbit as well as spinning on their own axes. Therefore, each electron within an atom can be thought of as a tiny electric current.

As is true of all electric currents, each of these electrons has a magnetic field. In most matter, however, about the same number of electrons spin in one direction as spin in the opposite direction. Thus, the magnetic fields cancel each other out. In iron, and a few other substances, however, more electrons are spinning in one direction than in the opposite direction. As a result, not all of the magnetism is cancelled out.

In a substance that can be made into a magnet, the atoms group themselves into regions called *domains.* Within the domains, the magnetic fields of the atoms are lined up so that they all point in one direction. The domains themselves, however, are not lined up. They are in a random arrangement so that their forces cancel each other. However, when a magnet is brought near, the domains all line up so that their north poles point one way and their south poles point the other. See Fig. 28-7.

The action of domains helps to explain why the N pole cannot be separated from the S pole of a bar magnet by cutting the magnet in half. If you tried, you would find that each half was still a complete magnet, with an N and S pole. Regardless of how many times you cut the magnet, and no matter how tiny the pieces become, each piece is still a complete magnet. Within each little piece, all the atomic N poles are pointing one way, while the S poles are pointing the other way. In addition to iron and nickel, cobalt and a number of alloys can also be made into magnets.

If a bar magnet is hammered sharply, the pattern of domains will be broken up and the bar demagnetized. Heating a magnet has the same effect. Why would you expect this to happen?

28-7 The magnetic domains are random in (a) and aligned in (b). Which diagram shows the pattern in a magnet?

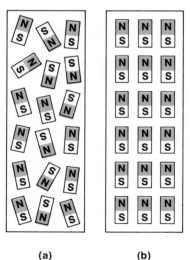

(a)   (b)

chapter 28 magnetism and electromagnetism   517

Each magnetic substance can be heated to a point where it becomes as nonmagnetic as paper or wood. For iron, this point is 750°C; for nickel, 340°C; and for cobalt, 1100°C.

Once magnetized, some alloys tend to stay that way. It is hard to magnetize them in the first place, but they are also hard to demagnetize. These alloys make very good *permanent magnets*. One of the best of these alloys is *Alnico*, a steel alloyed with aluminum, nickel, and cobalt. Some Alnico magnets can lift over 1000 times their own weight.

Other substances, such as soft iron, are very easy to magnetize, but they lose their magnetism just as easily. These metals make good *temporary* magnets and are used as cores for electromagnets. When a magnetic substance is described as "soft," it means that it loses its magnetism easily.

How can you explain the earth's magnetic field? You know that all magnetic fields are caused by electric currents. Therefore, you can be fairly sure that an electric current is causing the earth's magnetic field. One theory states that slow movements of liquids in the earth's interior produce electric currents. These currents, in turn, produce the earth's magnetic field

electromagnetism

Electricity can be used to make magnetism. In the last chapter, you learned that magnetism can be used to make an electric current. In fact, it is the most common of all forms of electric current and is used in homes and schools.

The reverse can also happen: an electric current can be used to make a magnet. A wire carrying a current has a magnetic field around it. If the wire is turned into a coil, the magnetic field inside the coil becomes much stronger than that formed by a straight wire.

The magnetic force becomes many times stronger if a soft iron core is placed into the coil. The coil and iron core together are called an *electromagnet*. If the core is long enough, it can be bent so that the two opposite poles are brought close together. A very strong field is produced in the gap between the poles. See Fig. 28-8.

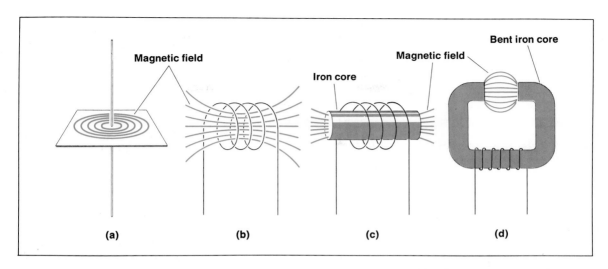

(a)   (b)   (c)   (d)

Magnetic field

Iron core

Magnetic field

Bent iron core

A simple electromagnet is shown in Fig. 28-9. Since the core is made of soft iron, the magnetic force is lost quickly when the electric current is turned off. Thus, whatever is being held by the magnet can be dropped simply by shutting off the current.

Three things can be done to make an electromagnet stronger: (1) a larger current can be used in the coil; (2) the number of turns of wire forming the coil can be increased; and (3) a more permeable core, such as certain iron alloys, can be used. *Permeable* means how well a substance can concentrate the magnetic lines of force.

The strength of a magnetic field is measured in a unit called the *tesla* (T). The magnetic pull of the earth is about 0.00005 T at the equator and twice as much (0.0001 T) at the poles. Why do you suppose the field is greater at the poles? A toy horseshoe magnet produces a field of about 0.3–0.4 T, and huge electromagnets may produce 2–9 T.

In Chapter 26, you learned that when certain substances are cooled to near absolute zero (−273°C) they become superconductors. Having no resistance, superconductors can carry huge currents. As a result, they can be used to make *supermagnets* hundreds of times stronger than electromagnets at normal room temperatures.

One of the first supermagnets, weighing only 4 N, was small enough to fit easily into a person's hand. Yet this magnet produced a field of 4.3 T. A common magnet of the same strength would have weighed 170,000 N. Such

28-8 Can you explain the four steps showing how an electromagnet works?

28-9 What are the main parts of an electromagnet?

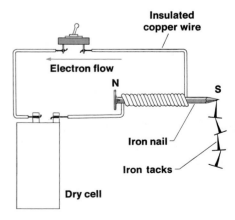

Insulated copper wire

Electron flow

N

S

Iron nail

Iron tacks

Dry cell

chapter 28   magnetism and electromagnetism   519

magnets may meet vital needs in the future. For instance, a train without wheels could float above the track, lifted by a magnetic field.

activity

Electromagnets. Set up three simple electromagnets, A, B, and C. A should have 15 turns of wire and one cell; B, 30 turns of wire and one cell; C, 15 turns of wire and two cells attached in series.

Find the strength of each hookup by counting the total number of pins or paper clips that are picked up.

Questions:
1. Which magnet is the strongest?
2. Which has the greater effect on the strength of the magnet: increasing the number of turns or increasing the voltage? Why?

28-10 (left) What is the left-hand rule of an electromagnet?

28-11 (right) Which letters in the diagram represent the opening and closing of the second circuit in this relay?

The left hand rule is used to find the north pole. There are times when it is vital to know which end of a magnet is the north pole. Scientists have found that the pole can be determined quickly by using a simple guide called the *left-handed rule*. This rule is shown in Fig. 28-10. If you wrap the fingers of your left hand around the electromagnet so they point in the direction in which the current

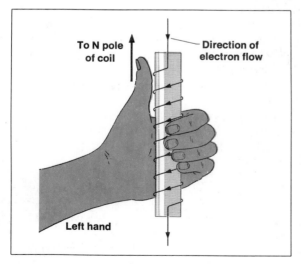

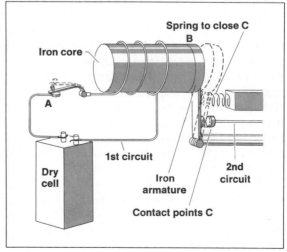

flows (that is, from the minus to the plus terminals of a battery), your outstretched thumb will point toward the N pole of the coil. If you reverse the direction of the current, the poles of the coil will also reverse.

A common device in which an electromagnet is used is the *relay*. See Fig. 28-11. When the switch at A is closed, the electromagnet attracts a small block of iron, B, called an armature. This opens the switch C in a different circuit. When switch A is opened, the magnetic force is turned off and the spring again closes contact C. This setup is a very simple form of electric relay. The relay can operate even with a very small current in the main line. A device like this relay turns on the starting motor in a car when you turn the ignition key.

Electric doorbells and buzzers show how an electromagnet can be used to break an electric current. When a current is in the circuit, the electromagnet attracts the armature. This action opens (or breaks) the circuit at the fixed contact point, and stops the magnetic attraction. Then, the spring pulls the armature back to close the circuit. Can you see why the armature vibrates back and forth, opening and closing the circuit as you press the button? This rapid vibration causes the well-known buzzing or ringing sound. See Fig. 28-12.

Electric motors and generators are alike. The magnetic field of an electric current makes possible the electric motor. The electric motor is really a magnetic motor; its rotor (movable magnet) is turned by magnetic forces.

A movable electromagnet, called the armature or rotor, spins because it is constantly attracted and repelled by fixed magnets, called the *field magnets*. As shown in Fig. 28-13, the poles of the armature are first attracted to the opposite poles of the field magnet. Then, when the armature reaches the opposite poles, the current reverses its flow. This change in direction also changes the poles of the armature. Thus, the armature is again repelled by the poles of the field magnet. At each half-turn of the armature the current is reversed, resulting in a constant spinning motion.

The current is reversed by contacts called *brushes*, which slide over a *split-ring commutator*. One half of the commutator is attached to one end of the armature wire. The other half is attached to the other end of the armature wire. The commutator turns between the brushes that are

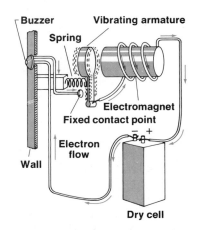

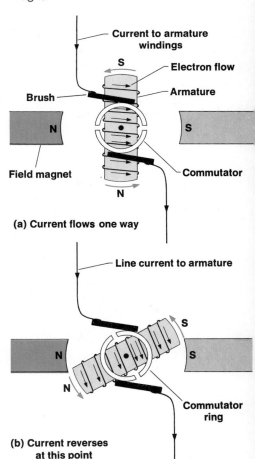

28-12 Why does the armature vibrate when the current is turned on?

28-13 What happens to the S pole of the armature in (a) when it reaches the N pole of the field magnet?

(a) Current flows one way

(b) Current reverses at this point

attached to the outside source of electric current. As a result, with every half-turn of the commutator, the armature wire receives current in the opposite direction.

Although there are many types of electric motors designed for specific uses, they all have one important thing in common. *All motors use a magnetic force to change electric energy into useful mechanical energy.*

You can also see the relationship between magnetism and electricity by comparing an electric motor with an *electric generator.* A generator changes mechanical energy into electric energy, which is the opposite of how an electric motor functions.

An electric generator, or *dynamo*, produces an electric current when coils of insulated copper wire in the armature cut through magnetic lines of force. When this happens, an electric current is set up in the wire. This process is called *electromagnetic induction;* the currents are called *induced currents.* Refer back to Fig. 27-10 (p. 508).

An outside source of energy is needed to turn the armature and to force it through the magnetic field of the magnets. In an electric power station, the energy comes from falling water, from burning fuel, or from nuclear reactors.

Thus, the input of a generator is mechanical energy and the output is electrical energy. In an electric motor, the input and the output are reversed. In other words, in the generator, magnetic forces plus motion produce a current. In the motor, an electric current plus magnetic forces produce motion. Generators and motors are built very much alike and can sometimes be used both ways.

A generator does not produce an electric current without the use of some energy. In other words, electric energy cannot be made without work being done. This is another case of the law of conservation of energy.

Generators are commonly built to produce alternating current (ac). Look at the loop of wire turning in a magnetic field, shown in Fig. 28-14. Note that in part **A** the armature wire ABDC is moving parallel to the lines of force. In this position, the lines of force are not being "cut" and no voltage is induced. The graph in **C** shows that, at this point, the voltage is zero.

When the armature loop makes a quarter turn to position B, it is cutting directly across the magnetic lines of force. Now positive voltage is induced, as shown on the graph in **C**.

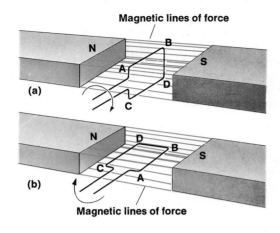

Magnetic lines of force

(a)

(b)

Magnetic lines of force

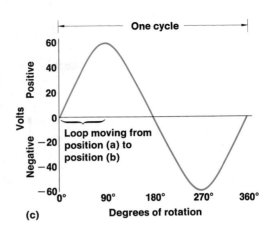

One cycle

Loop moving from position (a) to position (b)

(c)

Degrees of rotation

Another quarter turn brings the armature wire again parallel to the lines of force, and once more the voltage falls to zero. Now, as the loop rotates further, it cuts the magnetic field again, but in the opposite direction. This change of direction induces a negative voltage, as shown on the curve.

Because the induced voltage reverses, the current in the wire reverses as well. A current that flows first in one direction and then in the other is called *alternating current*, or ac. Much of the electrical energy used in this country is in the form of ac.

Diagram C shows one full cycle of ac, induced when the armature wire passes both a N and a S pole. The frequency of ac in the United States is 60 cycles per second. An armature having only one pair of poles would have to spin at a speed of 60 turns per second to produce this frequency.

Alternating current can be changed to direct current by a rectifier. Although ac is fine for most uses, *direct current* (dc) is also needed. For instance, dc is used to charge batteries and for electroplating. Some motors run only on dc current. *A device that changes ac to dc is known as a* **rectifier.**

The commutator of a generator acts as a rectifier. The commutator keeps a direct current in the outside line as it spins. See Fig. 28-15. Car generators are usually of the ac type or alternators.

28-14 Compare the quarter turn of the rotating loop in (a) and (b) with the cycle shown on the graph in (c). Why does the voltage change back and forth?

28-15 What effect does this splitting-ring commutator have on the alternating current of the coil?

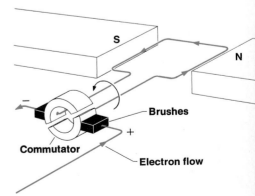

Brushes

Commutator

Electron flow

chapter 28 magnetism and electromagnetism 523

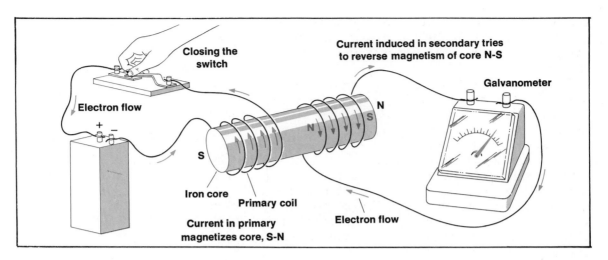

Closing the switch

Current induced in secondary tries to reverse magnetism of core N-S

Electron flow

Galvanometer

+ −

N

S

N

S

S

Iron core

Primary coil

Current in primary magnetizes core, S-N

Electron flow

28-16 Why is it important to turn the switch on and off in this device? What causes the current in the secondary coil?

Currents are induced in transformers and induction coils. An electric current is produced when a bar magnet is thrust into or pulled out of a coil of wire. The same result can be obtained by using an electromagnet and switching the current on or off.

Suppose two coils of wire are wrapped around a common iron core. Only one, called the *primary coil,* is attached to a dry cell. See Fig. 28-16. When the switch in the primary circuit is closed, the iron core becomes magnetized. This has the same effect on the *secondary coil* as if a magnet were thrust very rapidly into it. Therefore, a pulse of current flows briefly in the secondary circuit, as shown on the meter. This is called an *induced current.* The device is called an *induction coil.*

When the switch in the primary circuit is opened, the meter needle again moves, but in the opposite direction. The effect is the same as if a magnet were being pulled out of the secondary coil. A current is induced in the secondary coil only when a change occurs in the magnetic field through the secondary coil. In addition to opening and closing the primary circuit with a switch, the same effect can be achieved by feeding the primary circuit with an ac source.

The induction coil can produce high voltage. The more turns of wire there are in the secondary coil, the higher the voltage of the induced current. In Fig. 28-16, the primary and secondary coils are shown wrapped around a common iron core. If the secondary coil has twice as many turns as the primary, the voltage of the secondary is twice that of the primary. However, the current in the secondary is only half that in the primary.

The induced voltage depends upon the number of lines of force cut per second by all the turns of wire in the secondary. Thus, a very high voltage is induced in the secondary coil when it has many turns.

A special kind of induction coil, called a *spark coil* or *ignition coil*, is used in a car. The primary has only a few turns of insulated wire wrapped around an iron core while the secondary coil has thousands of turns of wire. The voltage produced in the secondary is high enough to produce a spark at the points of the spark plugs. This spark ignites the fuel-air mixture in the engine's cylinders.

A *transformer* is much like an induction coil. It contains two coils of insulated wire, wound around a soft iron core. Alternating current is fed into the primary (P) coil, which produces a changing magnetic field. As this field cuts through the secondary (S) coil, ac is also induced in the secondary.

If, as shown in Fig. 28-17, there are twice as many turns in the S coil as in the P coil, the voltage induced will be twice the input voltage. If such a transformer were plugged into a 115-volt line, it would have an output of 230 volts. This type of transformer is called a *step-up transformer*.

The above description makes it seem as if transformers create energy. However, you know this cannot be true. Although the voltage is doubled in the secondary coil, its current is cut in half (power in watts = volts × amperes). In fact, the current is *less* than half because some energy is lost as heat.

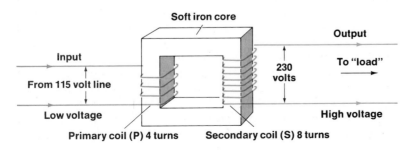

Soft iron core

Input

From 115 volt line

Low voltage

Primary coil (P) 4 turns

230 volts

Secondary coil (S) 8 turns

Output

To "load"

High voltage

28-17 Why is the voltage doubled in the secondary coil? What happens to the amperes?

chapter 28 magnetism and electromagnetism   525

28-18 Why is it necessary to step up the voltage in power lines?

A *step-down transformer* has fewer turns in the secondary coil than in the primary coil and thus reduces the voltage. If the secondary coil has one tenth as many turns as the primary coil, the output voltage will be only one tenth as much. However, its current becomes nearly ten times as large as that in the primary.

The cost of sending electric energy is lowest if the voltage is high and the current is low. This is because *losses in the wires are proportional to the square of the current.* High voltage results in low current, which, in turn, causes lower heat loss in the transmission lines. See Fig. 28-18.

Step-down transformers can be used whenever low voltages are needed, as in electric bells and chimes. On the other hand, a step-down transformer may also be used whenever a large current is wanted. For example, the large current needed for electric welding is obtained from step-down transformers. Welding transformers use a very large wire in the secondary coils. This wire can handle the high currents produced without too great a heat loss.

sample problem

A step-up transformer is used on a 120-volt line to produce 3000 volts across the secondary coil. (a) If the primary has 80 turns, how many turns must the secondary have? (b) If the current in the primary circuit is 20 amperes, what is the power produced? (c) What is the current in the secondary circuit?

solution:

Step 1. Find the voltage ratio between the two coils.

$$\text{voltage ratio} = \frac{\text{secondary coil}}{\text{primary coil}}$$

$$= \frac{3000 \text{ volts}}{120 \text{ volts}} = 25$$

Step 2. Since the voltage across the secondary is increased 25 times over that of the primary, the number of turns must also be increased by the same factor, or

(a) 80 × 25 = 2000 turns

(b) power = volts × amperes
    = 120 × 20
    = 2400 watts

(c) current in the secondary = $\dfrac{\text{watts}}{\text{volts}} = \dfrac{2400}{3000} = 0.8$ ampere

Michael Faraday — Davy's Greatest Discovery

When Michael Faraday of England invented the electric generator in 1831, he appeared before a government committee to ask for funds to build a large model of his invention. Faraday showed his small model to the committee and explained how it worked.

"Of what practical use is your device?" they asked.

Faraday pointed out that since his invention was new, he could not foretell the future uses of such a device. He predicted, however, that "in a hundred years you will be taxing something like this." The committee turned down Faraday's request.

Despite this loss, Faraday kept up his research. Countless uses were found for the electric generator, and just as he predicted, electric power companies have paid billions of dollars in taxes.

Faraday received his start in science when he attended a lecture given by Sir Humphrey Davy. Faraday was so impressed with that lecture that he applied for work as an assistant to Davy. He was accepted and within a few years began making major discoveries of his own. Faraday became so successful that his fame exceeded that of Davy. In fact, it has often been said that "Faraday was Davy's greatest discovery."

summary

1. Like poles of a magnet repel and unlike poles attract.
2. All magnets have force fields around them.
3. The earth has a magnetic field.
4. There is a relationship between magnetism and electric currents.

5. The left-hand rule states that if you wrap the fingers of your left hand around an electromagnet so that they follow the direction of the current, your outstretched thumb will point toward the N pole.
6. When a coil of insulated wire cuts through magnetic lines of force, an electric current is induced in the wire.
7. An electric motor changes electrical energy to mechanical energy.
8. A generator changes mechanical energy to electrical energy.
9. A rectifier changes ac to dc.
10. Induction coils and transformers are electromagnetic induction devices.

review .

Match each item in the left column with the best response in the right column. *Do not write in this book.*

1. lodestone
2. domain
3. magnetic pole
4. tesla
5. left-hand rule
6. electromagnet
7. supermagnet
8. electric motor
9. electric relay
10. transformer

a. part of a magnet where magnetic force is strongest
b. magnet produced by direct current
c. device for changing electrical energy to mechanical energy
d. primary coil
e. device for changing voltage
f. magnetic switch
g. way of finding the N pole of an electromagnet
h. natural magnet
i. produced near absolute zero
j. secondary coil
k. magnetic region of aligned atoms
l. unit of magnetic field strength

questions

Group A
1. State two laws that apply to magnetic poles.
2. Name three metals that can be strongly magnetized.
3. Why must compass readings be corrected when they are used for navigation?
4. Why is it that not all elements can be strongly magnetized?
5. In what two ways can a permanent magnet be weakened?
6. To make a permanent magnet, would you use a substance that is easy or difficult to magnetize? Give a reason for your choice.
7. What three factors affect the strength of an electromagnet?
8. What is a unit of magnetic strength called?
9. How strong is the magnetic field of the earth at the equator? at the poles?
10. What is a supermagnet?
11. What is an electric relay and how is it used?

12. Is a spark coil like a step-up or a step-down transformer? Justify your answer.
13. What happens to the strength of the magnetic field around a current-carrying wire when that wire is coiled into a loop?

Group B
14. Why do we say certain materials are transparent to magnetic lines of force?
15. Describe the earth's magnetic field and suggest one of its causes.
16. What is a magnetic field and how would you study it?
17. Why is a magnetic compass of little value inside a submarine?
18. Why is the magnetic field greater at the poles of the earth than at the equator?
19. What are magnetic domains and how do they affect the magnetism of matter?
20. If atoms of an element have an excess number of electrons spinning in one direction, what can you predict about the magnetic property of the element?
21. When does electromagnetic induction occur?
22. What factors determine the amount of voltage produced by electromagnetic induction?
23. Explain why an electric motor might also be called a magnetic motor.
24. How can an alternating current be converted to a direct current?
25. Discuss the construction and use of step-up and step-down transformers in the transmission of power.
26. Explain the construction and action of a spark coil.

problems

1. The primary coil of a transformer has 200 turns of wire at 1100 volts and 10 amperes. (a) What is the power produced in the primary coil? (b) What is the voltage across the secondary of 20 turns? (c) How many amperes are in the secondary circuit? (Assume 100% efficiency.)
2. Suppose you get a 220-volt output from a 2640-volt line carrying 1 ampere of current. If there are 120 turns on the primary coil, how many turns of wire are there on the secondary? What is the current in the secondary?

further reading

Asimov, Isaac, *How Did We Find Out About Electricity?* New York: Walker, 1973.
Lefkowitz, R. J., *Forces in the Earth: A Book About Gravity and Magnetism.* New York: Parents', 1976.
Simple Experiments in Magnetism and Electricity. Seven 32-pg. booklets, Charles Edison Foundation, 101 So. Harrison Street, East Orange, N.J. 07018.
Wade, Harlan, *Electricity.* Milwaukee: Raintree Pubs. Ltd., 1977.

29

electronics

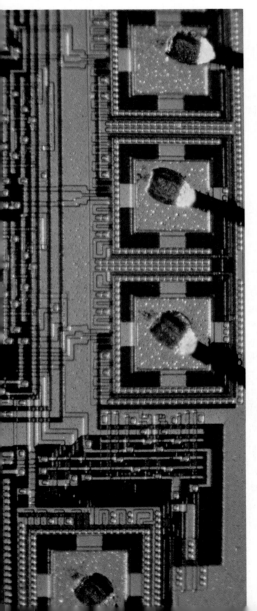

So far, your study of electricity has been concerned with currents and with devices that produce and use these currents. Electronics, a more specialized branch of physical science, is concerned with the following major topics: (1) radio and radar waves; (2) the motion of electrons through a vacuum, a gas, or a semiconductor; and (3) the photoelectric effect. In addition, electrostatics, discussed in Chapter 25, is sometimes also included in the field of electronics.

objectives:

☐ To describe how radio waves are broadcast.

☐ To relate the discovery of the vacuum tube and transistor to the present field of electronics.

☐ To explain the function of the cathode-ray tube and its uses.

☐ To explain the photoelectric effect.

☐ To describe the basic functions of a computer.

radio waves

Radio stations produce "carrier" waves. Whenever you turn on your radio, you can tune in any given station just by turning a dial. This happens even though your radio antenna is receiving radio waves from many broadcasting stations at the same time. The waves come from commercial broadcasts, "ham" operators, and many other "short wave" sources. Yet, these signals are not mixed together on your radio. How is it that you can "tune in" to any given signal from the many that the radio receives?

The answer to that question can be found by looking first at the radio broadcasting station, and then at your radio receiver. A radio station transmits radio waves of a single *radio frequency* (or *r-f*). This single frequency matches a single point on the radio dial.

In order to send music or voice, the sound waves produced in the studio are changed to an electrical pattern. This pattern, in turn, goes through a modulator that "shapes" the radio waves sent by the transmitter. **Modulation** *is the process of shaping or coding the radio waves so that they carry the pattern of the sound waves.* In a sense, the radio waves "carry" a coded message (an *audio frequency* or *a-f*). That is why the radio waves are called *carrier* waves.

One method of broadcasting is called *amplitude modulation,* or AM. In this method, the amplitude (height) of the carrier wave is changed to match the pattern of the sound wave. See Fig. 29-1.

29-1 How is it possible for a carrier radio wave to produce sound in your radio? How do waves **(a)** and **(b)** combine to form **(c)**?

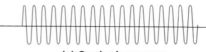

(a) Carrier frequency

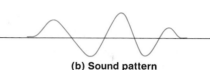

(b) Sound pattern

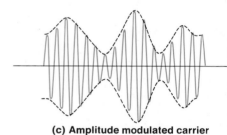

(c) Amplitude modulated carrier

(d) Frequency modulated carrier

activity

Static transmitters. Turn on the AM band of a radio. Set the dial between stations and turn up the volume. Then do the following activities: (1) Turn on and off any electric switch in the room. (2) Rub a plastic rod with wool and cause a spark near the radio (or shuffle across a rug and touch your hand to a grounded object). (3) Short-circuit a dry cell by attaching a wire to one post and scraping the other end of the wire across the other post.

Questions:
1. What effect do the above actions have on the radio?
2. What do you hear if these tests are repeated with the radio tuned to the FM band? to the short-wave band?
3. How does distance affect the results of the tests?

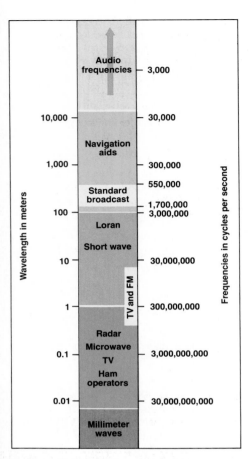

Wavelength in meters

Frequencies in cycles per second

Audio frequencies — 3,000

10,000 — 30,000

Navigation aids

1,000 — 300,000

550,000

Standard broadcast — 1,700,000 / 3,000,000

Loran

Short wave

10 — 30,000,000

TV and FM

1 — 300,000,000

Radar

Microwave

0.1 — 3,000,000,000

TV

Ham operators

0.01 — 30,000,000,000

Millimeter waves

29-2 Does the standard AM broadcasting band have a higher or lower frequency range than that of TV or FM?

Another broadcast method, called *frequency modulation*, or FM, is also used. In transmitting by FM, the amplitude of the carrier wave is not changed, but its frequency is changed to carry the sound pattern. See Fig. 29-1.

To keep stations from interfering with each other as the static transmitter does in the activity above, each station must broadcast only its carrier waves at its own assigned frequency. The countries of the world have agreed to set aside certain frequency bands for standard broadcasts, amateurs, police, airline, military, and other uses. See Fig. 29-2.

So far, you have seen that each station broadcasts at its own assigned radio frequency. However, this still does not explain why all these different signals do not interfere with each other in the radio receiver. The answer lies in the fact that every electronic circuit has its own natural frequency. When the natural frequency of the radio's circuit matches that of a broadcasting station, you hear the program that is broadcast by that station, but not by the others. The natural frequency of a radio can be changed, or "tuned," by adjusting a device called a *variable capacitor*.

AM waves reflect off the ionosphere. The region of air from about 80 km to about 600 km above the surface of the earth can reflect standard AM radio waves. This region is called the *ionosphere*.

Some of the air in this region is ionized by cosmic rays and ultraviolet light from the sun. The lower part of the ionosphere acts like a mirror, reflecting the longer wavelengths of radio broadcasts back to earth. As a result, the reflected signals can be received at very long distances from the station. See Fig. 29-3. However, FM radio and TV use a carrier wave of very high frequency that goes right through the ionosphere. These signals are not reflected back to earth. For this reason, FM and TV signals can be received only if the transmitter and receiver are in a straight line from one another and are not blocked by the earth.

Radar sends and receives its own signals. The concept of *radar* originated in 1922. In that year, the U.S. Navy observed that high-frequency radio waves were cut off when ships passed through the line of transmission. In 1934, the Navy bounced high-frequency pulses off an airplane 24 km away. However, radar was not widely used until the early years of World War II, when it was developed by British scientists.

The word "radar" comes from the first letters of "*ra*dio *d*etecting *a*nd *r*anging." Briefly, radar functions as follows. A powerful radio transmitter sends out pulses of high-frequency waves. A receiver picks up "echoes" reflected from objects that are hit by those waves.

The distance to an object is found by measuring the time it takes for the pulse to hit the object and for the echo to return. The direction of the object is determined by the direction in which the antenna is pointed when the signal is received.

The constant stream of reflected signals produces a glowing light on the radar screen. This screen is similar to the picture tube of a TV set. See Fig. 29-4.

Today, radar has many uses. It is used to predict weather, to track space flights, and to guide ships and planes through clouds and fog. Radar is even used to study migrating birds.

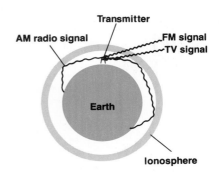

29-3 Why is it possible to hear AM radio broadcasts at longer distances than FM or television broadcasts?

tubes and transistors

The electron tube was discovered by Edison. Most of the equipment needed to send and receive radio signals use *electron tubes* and *transistors*. The electron tube is a result of a discovery made but ignored by Thomas Edison in 1883. Edison was working on ways to improve his newly invented light bulb.

Edison was trying to find a way to keep a black deposit from forming on the inside surface of the bulb. This deposit was caused by the evaporation of the metal of the hot filament. He tried all kinds of things to keep that deposit from forming. He even put a charged metal plate inside the bulb, as shown in Fig. 29-5.

Edison noticed that when this plate had a positive charge, a current from the hot filament would flow to it. Today, scientists know that this current is made up of electrons given off by the hot filament. The

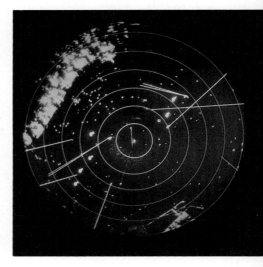

29-4 This photo shows the image formed on a radar screen. What are some ways in which radar is used?

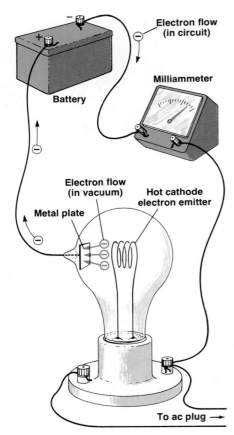

Battery

Electron flow
(in circuit)

Milliammeter

Electron flow
(in vacuum)

Metal plate

Hot cathode
electron emitter

To ac plug →

29-5 What is the Edison effect as shown in this diagram? What is the purpose of the battery? Why is an ac current needed?

electrons are attracted to the positive plate. The current is called the *Edison effect*, even though Edison never made any use of his discovery.

Some years later, an English scientist named John Fleming found a use for the Edison effect. He made a vacuum tube like Edison's. However, instead of the hot filament, Fleming used a wire with a negative charge (cathode). The cathode gave off electrons just as Edison's hot filament did. The electrons were attracted to the positive plate. Thus, the first practical electron tube was built. A tube that contains only two electrodes is called a *diode* (dye-ode).

The diode and triode are basic electron tubes. Since the diode allows current to pass in one direction only, it acts as a *rectifier*, which changes the back-and-forth alternating current (ac) into a pulsing direct current (dc) going in one direction only.

The radio waves striking the antenna are ac and send an ac voltage to the plate. Since the current can flow in one direction only (while the plate is positive), the ac current is changed to a pulsing direct current in the diode. However, none of the "coding" is lost when the ac is changed to dc. In other words, the pulsing dc has the same modulated pattern sent out by the broadcasting station. Today, the diode is usually replaced by a simple transistor.

In 1906, an American inventor named Lee De Forest put a third electrode, called a *grid*, in a vacuum tube. This new tube was called the *triode*, or three-element tube. The grid often consists of a wire mesh between the cathode and the plate.

When a negative charge is put on the grid, as shown in Fig. 29-6, the grid repels electrons coming from the cathode, keeping many of them from reaching the positive plate. If the grid is given a high enough negative charge, none of the electrons will reach the plate. On the other hand, if the grid is made positive, a larger flow of electrons will be attracted from the hot filament to the plate. In other words, the grid acts like a "valve" controlling the amount of current through the tube.

The number of electrons that can pass from the cathode to the plate in a triode depends upon the following three factors:

1. The number of electrons given off by the cathode. The hotter the cathode, the more electrons are "boiled off."
2. The positive voltage on the plate. The higher the plate voltage, the greater the attraction of electrons from the cathode.
3. The kind and amount of charge on the grid. Very small changes in the grid voltage can amplify (have very large effects on) the electron flow between the cathode and plate.

Often, in electronic circuits, weak electric currents must be amplified or strengthened. For example, the radio waves from a broadcasting station strike an antenna. They induce very weak electric currents that are far too weak to operate the speakers on a radio or television set. However, they are strong enough to change the charge on the grid of a triode and thereby control the larger cathode-plate currents.

The antenna is connected to the grid, giving the voltage on the grid the same pattern as the radio signals. As was stated earlier, even slight changes in the charge on the grid produce large changes in the electron flow reaching the plate from the cathode. Therefore, the current reaching the plate is many times greater than the strength of the signal picked up by the antenna. This strengthening of the signal is what is meant by amplifying.

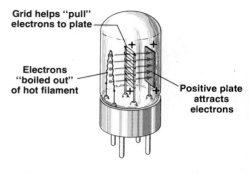

Grid helps "pull" electrons to plate

Electrons "boiled out" of hot filament

Positive plate attracts electrons

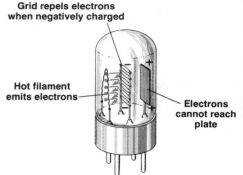

Grid repels electrons when negatively charged

Hot filament emits electrons

Electrons cannot reach plate

29-6 How does the charge on the grid control the flow of electrons from the cathode to the plate?

Transistors make circuits smaller. In 1947, a device called the *transistor* was built. It soon proved to be highly useful and was able to do the same job as vacuum tubes.

Transistors are much smaller than vacuum tubes. Therefore, products that use transistors can be made smaller. Since they do not use a hot filament, transistors do not need a source of heat. They save energy and need no warm-up time.

The theory and construction of transistors are quite complex. For the most part, transistors are made of thin wafers of silicon or germanium, both of which are semiconductors. In one type of wafer, certain impurities are added so that free electrons are given off. This type of wafer is called the *negative* or *N-type* semiconductor. If other types of impurities are added to the wafer, it becomes a collector of electrons and

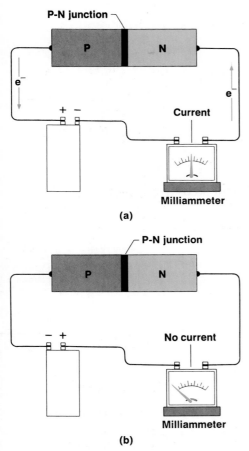

P-N junction

(a)

P-N junction

(b)

29-7 Why is there a flow of electrons in **(a)** but not in **(b)**?

29-8 What happens between C and D in the cathode-ray tube when air is pumped out?

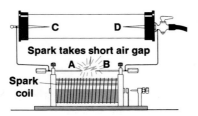

is called a *positive* or *P-type* semiconductor.

A transistor is made by placing a P-type and N-type semiconductor together to form what is called a *P-N junction*. A P-N junction can be used in place of a common diode as a rectifier. Like a diode, a P-N junction permits a charge to flow in only one direction. See Fig. 29-7.

By making N-P-N or P-N-P junctions, transistors can be used as triodes. Transistor radios, for instance, use simple circuits containing P-N-P triodes.

The emitter of the transistor, which gives off the free electrons, can be compared to the cathode of a vacuum tube. The collector is like the plate. If a third wafer is used, it acts like the grid. Like the vacuum-tube triode, this type of transistor can both rectify and amplify the current.

In recent years, electronic circuits have become amazingly small and compact. More and more circuits are packed into less and less space. Today, tens of thousands of transistors fit onto a "chip" the size of a match head. If such a chip were built using vacuum tubes, it would need a space as large as a classroom.

This tiny chip forms the heart of a *microprocessor*, the central processing unit, or logic, of a computer. The microprocessor is used to control a wide range of products such as microwave ovens, video games, car engines, electronic watches, and home computers.

cathode ray tubes

Cathode rays flow in low-pressure gases. Dry air at normal pressure is not a good conductor of electricity. To send a current through the short gap between A and B in Fig. 29-8, for instance, requires a potential difference of about 3000 to 4000 volts per centimeter. At very low pressures, however, air becomes a good conductor. When most of the air is pumped out of the tube, the electrons travel across the long gap from C to D more easily than from A to B.

The discovery that low-pressure gases are good conductors was made by Sir William Crookes, a British scientist, in about 1875. Crookes sealed a wire electrode in each end of a glass tube. A pump was attached to the tube so that air could be removed. As

Crookes slowly pumped the air from the tube, he found that the current increased.

At first, a thin, crackling arc jumped from one electrode to the other. Then, as more air was removed, the arc spread until the whole tube glowed with a pale pink light. Then a strange thing happened. When the vacuum was increased still more, the glow became fainter and fainter, and finally disappeared. However, Crookes saw that the end of the tube opposite the cathode (the negative electrode) began to glow. This glow meant that some kind of unseen rays were coming from the cathode. Crookes called these rays *cathode rays*. Cathode rays are actually streams of electrons.

The cathode is somewhat like an "electron gun." Electrons "boil off" the hot cathode and are pulled through ring-shaped positive electrodes called *anodes*. As a thin beam of electrons shoots through the length of the tube, it is controlled by two sets of parallel deflecting plates. The plates of each set are oppositely charged. Depending on the kind and amount of charge on the plates, the electron beam is pushed up and down or from side to side. See Fig. 29-9.

After passing between the deflecting plates, the cathode ray hits the end of the tube and causes a special mineral coating to glow, or *fluoresce*. The horizontal deflecting plates control the side-to-side motion of the beam across the face of the screen. The vertical deflecting plates control the up-and-down motion of the beam. Electromagnets can be used instead of plates to control the electron beam.

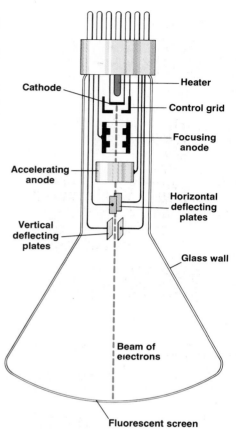

29-9 How is the beam of electrons made to move from side to side or up and down?

The television tube is a form of cathode-ray tube. The electron beam is as thin as a pencil lead. It sweeps back and forth across the end of a TV tube like a brush. In 1 sec, the electron beam makes 15,708 strokes on the screen so quickly that the first stroke has not faded away by the time the last line is drawn.

The strength of the electron beam determines the brightness of the light it produces. The strength is controlled by the changing strength of the radio waves received by the television antenna. These waves are modulated to carry the light pattern of the scene being televised. As a result, you see an image of this scene on your TV receiver.

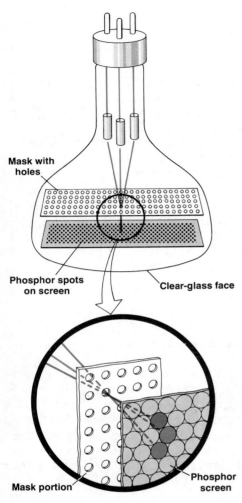

Mask with holes

Phosphor spots on screen

Clear-glass face

Mask portion

Phosphor screen

29-10 What is the purpose of the three color beams in this television picture tube?

Programs in color are sent and received by blending the three primary colors: red, green, and blue. The light entering the camera is separated by filters into the three primary colors. Three separate signals, one for each color, are broadcast.

On the receiving end, the color TV picture tube has three electron guns, producing one electron beam for each of the primary colors. The inner face of the tube is coated with three different minerals, arranged in a series of tiny dots. Each incoming electron beam hits the proper dots. See Fig. 29-10. The dots are blended together to reproduce the same color that is picked up by the camera. If you look closely at a TV screen with a magnifying glass, you can see the dots.

In some television models, the dot pattern is no longer used. Instead of tens of thousands of dots, several hundred parallel lines are used. This system of lines is less complex, less expensive, and also brighter than dots.

Color TV programs are *compatible* with black and white sets. That is, a color program can be received in black and white on a noncolor TV set.

TV signals are limited, like FM, to a range no farther than 100–150 km from the sending station. The high-frequency waves of TV do not reflect from the ionosphere. Therefore, for long-distance transmission, the program is carried by cables, or sent by a series of relay towers or by earth satellite stations.

Cathode rays led to X rays. While working with cathode rays in 1895, Wilhelm Roentgen (*rent*-g'n) noticed that a nearby fluorescent mineral was glowing from the effects of his cathode-ray tubes. He found that this ray went right through many kinds of matter. Not knowing its nature, Roentgen called it simply X *ray*. Scientists now know that X rays are electromagnetic radiations similar to light, only of much shorter wavelength. (Refer to Fig. 24-1, page 445.)

X rays are formed when an electron beam strikes a metal target. See Fig. 29-11. If a higher voltage is used, X rays of higher frequency are produced. More than a million volts will produce X rays that can penetrate several centimeters of steel. Such radiation can be used to check metal castings and welds for hidden flaws.

X rays are widely used in medicine. Some are used to detect broken bones, cavities in teeth, diseases of the lungs, and for many other uses.

Intense doses of X rays can destroy body tissue. Many early workers suffered serious burns and even death before this danger was recognized. Studies have shown that in addition to causing burns and killing body cells, exposure to X rays may produce unwanted **mutations.** *Mutations are changes in the factors that control heredity.* Extreme care should be taken to protect the body from an overdose of X rays.

As soon as scientists found that X rays could destroy body cells, they put this knowledge to use. They found that certain types of cancer could be destroyed or checked by the proper use of X rays. X rays are commonly used today to treat cancer.

devices using photoelectric tubes

The photoelectric effect is used in solar batteries. The *phototube,* or *photoelectric cell,* sometimes called an electric eye, is a clever use of an electron tube. The tube is based on the photoelectric action of light discussed in an earlier chapter.

Recall that when metals, such as cesium, are struck by light, some electrons are knocked loose, leaving the metal with a positive charge. The energy of a single photon is enough to detach an electron from its atom. This reaction is called the photoelectric effect.

Since light is the source of the energy of the photocell, sunlight can be used to operate such a cell. *Solar batteries* operate on the energy of sunlight and have worked for years in spacecraft without any other source of energy.

The photoelectric effect is used in sound films and TV cameras. The photoelectric effect is used to produce sound in movies. If you examine a section of sound movie film, you will notice an uneven dark streak along one edge. This is the record of the sound, called the *sound track.*

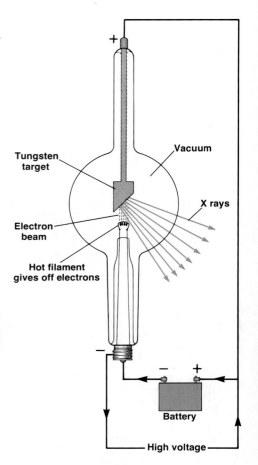

29-11 What is emitted when an electron beam strikes a dense target?

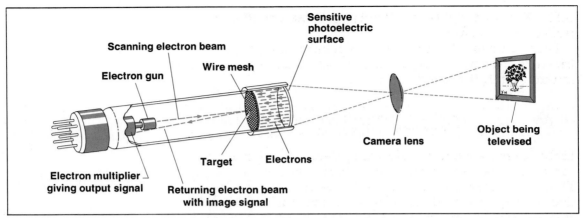

29-12 What causes the flow of electrons in a television camera?

A tiny beam of light shining through this sound track strikes a phototube. The amount of light that gets through varies with the changing density of the sound track. This changing light causes the phototube to produce a changing current. This current is amplified and then changed to sound in loudspeakers located near the movie screen.

The television camera also depends upon the photoelectric effect. In the TV camera, light forms an image on a surface that is coated with thousands of particles of a photoelectric substance. Each particle of this substance acts like a tiny photoelectric cell. The substance receives a varying charge, depending upon the brightness of the light that strikes it. An electron beam within the tube scans this photoelectric surface at great speed. The beam produces an electric current that varies in the same way as the pattern of light from the scene being televised. This varying current is amplified and broadcasted. See Fig. 29-12.

There are many other uses of the photoelectric effect. A phototube circuit can be arranged in such a way that a counting device operates each time a beam of light is broken. These interruptions might be caused by traffic on the highway or packages on a moving belt.

As stated in the chapter on light, doors can be opened and closed by photocells when an approaching person breaks a light beam. Street and building lights can be turned on as darkness approaches by using photocells. The foul lights on bowling alleys are also controlled by photocells.

Even burglar alarms can be operated by photoelectric cells. When a burglar breaks an unseen beam of infrared light, the alarm is set off.

the computer

Early humans used simple computers. It is likely that early humans used their fingers and toes for counting. They may have used sticks and stones as well. Primitive people also used markings on cave walls as simple aids to adding and subtracting, and to keep a record of belongings. The *abacus* was the first of all computers. The abacus is known to have been used in China as a counting device 3000 years ago. Addition and subtraction are performed by simply moving beads with certain values toward or away from the center board. See Fig. 29-13.

There are analog and digital computers. Electronic computers in common use today can be divided into two basic groups: *analog* and *digital.* Computers in each group can solve a wide range of complex problems in math and logic. An analog computer solves problems by measuring things. It may measure the expansion or contraction of a solid, liquid, or gas; the increase or decrease of voltage or current in an electric circuit; or a change in the loudness of a sound.

A thermometer is a simple analog computer. As it gets warmer, the mercury column rises. As it gets colder, the mercury drops. A slide rule is another common example of an analog computer. You can multiply or divide with a slide rule by making it longer or shorter.

A digital computer provides solutions to problems by counting things. These machines process data and do other math-related tasks. Well-known examples of digital computers are adding machines, cash registers, and common desk calculators. The following list gives some of the functions of modern computers:

1. Digital computers are amazingly accurate. In microseconds, a computer can divide 29.2473051 by 8.90677162. You would probably settle for 29.2 ÷ 8.9.

29-13 How can a Chinese abacus be compared to a computer? What type of computer is it?

Table 29-1:

Decimal Digit	Binary Code
0	000000
1	000001
2	000010
3	000011
4	000100
5	000101
6	000110
7	000111
8	001000
9	001001

Table 29-2:

Alphabetic Characters			
A	010001	N	011110
B	010010	O	011111
C	010011	P	100000
D	010100	Q	100001
E	010101	R	100010
F	010110	S	100011
G	010111	T	100100
H	011000	U	100101
I	011001	V	100110
J	011010	W	100111
K	011011	X	101000
L	011100	Y	101001
M	011101	Z	101010

2. Computers are automatic. Once the human operator provides the machine with a set of step-by-step instructions (a program) and supplies the needed data, the working of the machine is totally automatic.

3. A computer can complete a problem in an extremely short time. It is capable of doing a task in a few minutes that would take years to complete with pencil and paper.

4. Unlike the human memory, the computer never forgets. A computer can "remember" facts and instructions, and can recall them in a split second.

Computers use the binary, or base-two, number system. The operation of a digital computer depends only on the position of a switch. You flip a switch and a light goes on. Flip it again and the same light goes off. There is no half-way position. A switch can only be on or off.

An "on-off" device can be used to convey data in coded form. Most digital electronic computers are based on a *base-two*, or *binary*, number system. This system uses only two numerals, "1" and "0," to carry all data and do all arithmetic processes.

The binary system is quite different from our common decimal, or base-ten, system. The decimal system has ten numerals as digits: 0, 1, 2, 3, 4, 5, 6, 7, 8, 9. The base is the total number of digits used in the numbering system. The *highest* number that can be represented by a single digit is one less than the base. In the decimal, or base-ten, system the highest numeral is 9; in the binary, or base-two, system the highest numeral is 1.

The binary system may seem awkward to those used to the decimal system. However, the binary system is ideal for the computer because of the simple way in which 1 and 0 can be shown. You can compare the decimal and binary systems as shown in Table 29-1. The English alphabet can also be expressed in a binary code. See Table 29-2.

Computers have three basic functions. All digital computers have three functions in common: *input, processing,* and *output.* Figure 29-14 shows how these functions are related.

1. *Input.* The input section receives data to be processed by the computer. Coded data is fed into this sec-

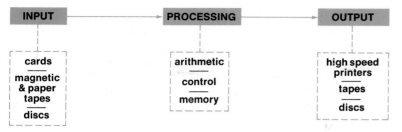

INPUT	PROCESSING	OUTPUT
cards	arithmetic	high speed printers
magnetic & paper tapes	control	tapes
discs	memory	discs

29-14 What are the basic functions of a digital computer? How are they related?

tion by means of punched cards, magnetic tape, perforated paper tape, or some other device. The data to be processed include human-produced coded data and step-by-step instructions about what to do with this data.

2. *Processing.* This function includes control, memory, and arithmetic. Control makes sure that the data and instructions are sent to the right places.

Once in the computer, the data and instructions are sent to the memory unit to await recall by the control unit whenever needed.

The arithmetic section does the needed calculations as directed by the control function. This section is capable of working problems in logic.

3. *Output.* The computer processes data in the form of electric signals only a computer can use. The output is the unit that translates the electric signals into records that people can read. Data or solutions to a problem may appear (1) in a binary code on punched cards or on magnetic tape, (2) decoded into numerals and letters of the alphabet, or (3) printed into words.

Needed in Many Areas: The Science Technician

If you like math, chemistry, and machinery, you may be well suited for work as a science technician. Science technicians often do research on new products, test them, and control how they are produced. Technicians work out ways to make better products with greater efficiency.

Technicians also give advice to salespeople on how to install and maintain complex machinery and products, and also write technical manuals that explain how the

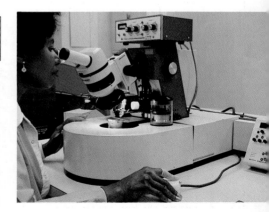

machinery and products work. Some of the jobs available to science technicians are listed in the following table.

Areas	Jobs and Duties
Aerospace	Build aircraft, rockets, and spaceships.
Agriculture	Design, test, and maintain farm equipment.
Air conditioning	Install and service heating and refrigeration units.
Biomedical	Design, build, and test medical equipment.
Ceramics	Work with kilns and related equipment.
Chemical	Produce and test chemicals for countless products.
Civil	Build highways, bridges, and dams.
Electrical	Work with motors, appliances, and other equipment.
Environmental	Work for clean air and water, and increase worker safety.
Industrial	Plan efficient use of personnel, material, and machines.
Mechanical	Work with autos, diesels, tools, and machines.
Metallurgy	Produce and test metals and alloys.
Instrumentation	Work with devices used in satellites, weather forecasting, ocean studies, and environmental and medical research.

summary

1. Radio stations produce modulated carrier waves.
2. Modulation is the process of tailoring the carrier wave so that it carries the pattern of the sound waves.
3. Radio receivers can be tuned to one station by a variable capacitor.
4. Radar sends out high-frequency pulses and detects the reflected "echoes" from objects.
5. Electron tubes make use of the Edison effect.
6. The diode and the triode are basic electron tubes.
7. Transistors, thin-film, and integrated circuits are modern advances of the electron tube.
8. The cathode-ray tube produces a stream of electrons (cathode ray) in low-pressure gases.
9. A type of cathode-ray tube is used in TV sets and X-ray machines.
10. Photoelectric cells change light energy to electric energy.
11. Analog computers solve problems by measuring.

12. Digital computers solve problems by counting.
13. Computers use the binary (or base-two) number system.
14. Computers perform the functions of input, processing, and output.

review

Match each item in the left column with the best response in the right column. *Do not write in this book.*

1. amplifier
2. binary
3. cathode rays
4. modulation
5. phototube
6. radar
7. slide rule
8. transistor
9. tuning
10. X ray

a. adding a sound wave to a carrier wave
b. adjusting the frequency of a receiver
c. base-two number system
d. device that produces an electric current from a beam of light
e. device used to strengthen weak electrical signals
f. form of analog computer
g. form of digital computer
h. radiation of great penetrating power
i. radio detecting and ranging
j. semiconductor device that functions like a vacuum tube
k. streams of high-speed electrons
l. switches arranged in parallel

questions

Group A
1. What topics are most often included in the study of electronics?
2. What effect does the ionosphere have on AM radio waves?
3. How does radar work?
4. What is the Edison effect? How was it discovered?
5. Name the electrodes in a triode. What are their functions?
6. To which electrode of a triode is the antenna current fed? Explain.
7. In what ways are transistors better than vacuum tubes?
8. How is the electron beam in a TV set controlled?
9. What is meant by saying a color TV program is "compatible"?
10. Why are relay stations necessary for long-distance TV transmissions?
11. How are X rays produced?
12. What is the photoelectric effect?
13. Into what basic types can computers be classified? Give two examples of each type.
14. Why is the binary system so suitable for the computer?
15. What three basic functions are common to all computers?

Group B
16. What is the difference between AM and FM?
17. What is meant by modulation?
18. What is meant by "tuning in" a radio station?
19. Describe how radar would detect a rocket 300 km away, due north.
20. How does a diode convert ac to dc?
21. Explain how the grid is used to help amplify the current in a triode tube.
22. What three factors affect the amount of electron flow in a triode?
23. What parts of a transistor act like the cathode, plate, and grid of a vacuum tube?
24. Discuss the operation of a cathode-ray tube.
25. Describe how color is formed on the screen of a color TV set.
26. How is the photocell used in a motion-picture projector?
27. Explain how a TV camera works.
28. In what ways do the two basic types of computers solve problems?
29. Study the binary number system shown in Table 29-1 and write the binary numbers for the base-ten numbers 10, 11, 12, 13, 14, and 15.
30. Using the binary code for the alphabet in Table 29-2, decode the following word(s): 010011 011111 011101 100000 100101 100100 010101 100010 100011.

further reading

Bender, Alfred, *Science Projects with Electrons and Computers.* New York: Arco, 1978.

Berger, Melvin, *The Stereo Hi-Fi Handbook.* New York: Lothrop, 1976.

Corbett, Scott, *Home Computers: A Simple and Informative Guide.* Boston: Little, Brown, 1980.

Englebardt, Stanley L., *Miracle Chip: The Micro Electronic Revolution.* New York: Lothrop, 1979.

Goldberg, Joel, *How to Make Printed Circuit Boards.* New York: McGraw-Hill, 1980.

Greenbaum, Louise G., *Looking Forward to a Career: Electronics.* Minneapolis: Dillon Press, 1975.

King, Dennis A., *Beginners I.C. Breadboard Projects.* Jackson, Cal.: PMS King, 1977.

Marston, R. M., *One Hundred and Ten Electronic Alarm Projects.* Rochelle Park, N.J.: Hayden, 1977.

Math, Irwin, *Morse, Marconi and You.* New York: Charles Scribners, 1979.

Page, Robert M., *The Origin of Radar,* Westport, Conn.: Greenwood, 1979.

glossary

abrasive. A substance used to grind or wear away other substances.

absolute zero. Lowest temperature theoretically possible, -273.16°C.

acceleration. Time rate of change of velocity.

acid. A substance that yields hydronium (hydrogen) ions in solution.

acid rain. Rain that results when raindrops react with oxides of sulfur and nitrogen pollutants to form sulfuric and nitric acids.

acoustics. The science of sound.

actual mechanical advantage (AMA). The ratio of the resistance force of a machine (R) to the effort (E) force.

adhesion. The force of attraction between unlike molecules.

aeration. Process of purifying filtered water with oxygen of the air.

alcohol. A compound containing a hydrocarbon group and one or more OH groups.

alkali. Strong soluble base; usually refers to any base.

alkane. Straight or branched-chain hydrocarbon in which the carbon atoms are connected by only single bonds.

alkene. Straight or branched-chain hydrocarbon in which two carbon atoms are connected by a double covalent bond.

alkyne. Straight or branched-chained hydrocarbon characterized by two carbon atoms connected by a triple covalent bond.

alloy. A material composed of two or more metals in a solid solution, mixture, or compound.

alpha particles. Positively charged helium nuclei given off by radioactive elements.

alternating current. Current that flows first in one direction and then in another.

ammeter. Instrument used to measure rate of flow of electrical current.

amorphous. Without definite shape.

ampere. The unit of electrical current.

amplifier. Radio tube or circuit used to strengthen feeble electric currents.

amplitude. Maximum distance the particles in a wave are displaced from the normal position of rest.

amplitude modulation (AM). A mode of broadcasting in which the amplitude of the carrier wave is changed to indicate the pattern of the sound wave.

analgesic. Substance used in medicine to relieve pain.

analog computer. A computer that solves problems by measuring things.

anesthetic. Substance used in medicine to desensitize reaction to pain.

angle of incidence. The angle at which light strikes an object.

angle of reflection. The angle at which light is reflected from an object.

anneal. The process of cooling glass slowly to keep it from becoming too brittle.

anode. The positive (+) terminal of an electrolytic cell.

antibiotic. Chemical substance obtained from certain molds and bacteria. Used for destroying or inhibiting the growth of many infection-producing organisms.

antiparticles. Mirror images of ordinary atomic particles.

antiseptic. A substance that can be safely used on the body to stop the growth of harmful bacteria.

atmosphere. The envelope of air surrounding the earth.

atom. The smallest unit of an element that can exist either alone or in combination with other atoms.

atomic mass. The mass of an atom based on the standard mass of 12 for a carbon atom.

atomic number. The number of protons in the nucleus of an atom. Also equal to the number of electrons in a neutral atom.

atomic pile. Device in which controlled fission of radioactive material produces new radioactive substances and energy.

audio-frequency (a-f). Frequency of sound to which the ear responds.

barometer. Instrument used to measure the pressure of the atmosphere.

base. Substance that yields hydroxide (OH) ions in solution.

battery. Two or more electrical cells connected in series or in parallel.

beats. Pulsation of two sound waves caused by interference.

beta rays. High speed electrons given off by the nuclei of radioactive atoms.

binary number system. A system that uses only the two numbers 0 and 1.

biodegradable. Materials that can be broken down into simpler compounds by natural processes.

bleaching. The operation by which color is partially or wholly removed from a colored material.

block and tackle. A system of pulleys capable of lifting a heavy weight with a relatively small force.

boiling point. The temperature at which the molecules of a liquid have enough energy to overcome the attractive forces between molecules within a liquid and escape into the air.

bond. One pair of shared electrons connecting two atoms; represented by the symbol −.

branched-chained hydrocarbons. Hydrocarbon compounds in which the carbon atoms are linked together with branches attached to the main string of carbon atoms.

breeder reactor. A nuclear reactor that produces new fissionable material at a greater rate than the fuel can be used up.

buoyancy. The force that pushes up on objects when they are placed in a fluid.

calorie. The amount of heat necessary to raise the temperature of one gram of water one Celsius degree.

candle. Unit of brightness equal to a standard candle.

capacitor. A combination of conducting plates separated by an insulator; used to store an electrical charge.

capillary action. The rise (or fall) of liquids in fine, hairlike tubes.

carding. Spreading a tangled mass of fibers into a filmy, weblike sheet.

carrier waves. High frequency electromagnetic waves used for radio broadcasting.

catalyst. Chemical substance used to alter the speed of a chemical reaction.

cathode. The negative (−) terminal of an electrolytic cell.

cathode rays. Streams of high-speed electrons leaving a negative electrode.

Celsius scale. The thermometer scale that has the freezing point of water as zero and the boiling point of water as 100.

center of curvature. The center of a sphere of which a mirror or a lens is a segment.

center of gravity. The point at which all of the weight of a body may be considered to be concentrated.

centi-. Prefix meaning 1/100.

centigrade scale. See Celsius scale.

centripetal force. Inward force that pulls a revolving body out of a straight-line path.

chain reaction. Any self-sustaining nuclear reaction.

chemical change. Change in which a new substance is produced.

chemical equation. An expression that tells by formulas and symbols what is used and what is obtained in a chemical reaction.

chemical formula. An expression used to identify compounds by showing the kind and number of atoms present.

chemical properties. Properties that describe the behavior of substances reacting with other substances.

chemical symbol. A single capital letter, or a capital letter and a small letter used as an abbreviation for an element.

chord. A pleasing combination of two tones whose frequency ratios are small whole numbers.

circuit. See electric circuit.

closed pipe. Resonant air column closed at one end.

coefficient of linear expansion. The fractional increase in length of a solid per degree rise in temperature.

coefficient of volume expansion. The fractional increase in volume of a substance per degree rise in temperature.

coherent waves. Light waves in which the crests and troughs are all in step; property of laser light.

cohesion. The attraction of like molecules of a substance for each other.

color. Property of light that is determined by its wavelength.

combustion. Rapid oxidation accompanied by heat and light; burning.

combustion turbine. Engine that burns fuel and uses exhaust gases to turn a turbine.

commutator. Device on electric generators that changes ac produced in the armature to dc in the external circuit.

complementary colors. Two colors that produce white light when combined.

compound. A substance composed of two or more different elements that are chemically united.

compound machine. Machine consisting of two or more simple machines.

compression. Region of a longitudinal wave in which vibrating particles are closer than their normal distance.

compression ratio. The ratio of the volume within the cylinder of a diesel engine when the piston is at the bottom to the volume when the piston is at the top of the stroke.

concave lens. A lens that diverges parallel light rays.

conduction. Transfer of heat from molecules to adjoining molecules.

conductor. Any substance that permits electricity or heat to move through it readily.

conservation of energy, law of. Energy cannot be created or destroyed, but can only be changed from one form to another.

conservation of matter, law of. Matter cannot be created or destroyed by ordinary chemical means.

conservation of matter and energy, law of. The total amount of matter and energy in the universe does not change.

constant proportions, law of. Every compound always contains the same proportion by mass of the elements of which it is formed.

control rod. A rod of neutron-absorbing material used to regulate the reaction in a nuclear reactor.

convection. The transfer of heat energy by movement of liquids and gases (fluids).

convex lens. A lens that converges parallel light rays.

corrosive. Attacking metals and mineral material by chemical action.

cosmic rays. Radiations from outer space that have enormous energy.

covalent bonding. Bonding in which atoms share a pair of electrons.

crest. A region of upward displacement in a transverse wave.

critical mass. The smallest mass of material needed for a chain reaction to occur.

crystal. Particles of a solid having a definite geometrical form or structure.

current. See electrical current.

decibel. Unit of sound intensity or loudness.

definite proportions, law of. Every compound always has the same proportion by mass of the elements that compose it.

density. Mass per unit volume of a substance.

destructive distillation. Process of breaking up a material by heating it in a closed container without air or oxygen.

detector. Radio tube or other device that separates the audio-frequency wave from the carrier wave, a process called detection.

detergent. A substance that removes dirt.

deuterium. Heavy hydrogen; the hydrogen isotope that has a neutron in its nucleus.

diesel engine. Internal combustion engine in which the fuel is ignited in the cylinder by the heat of compression of air.

diffraction grating. Device used in a spectroscope to produce a spectrum.

diffusion. The random mixing of substances by molecular motion; scattering of light rays.

digital computer. A computer that solves problems by counting things.

diode. Electronic tube containing only two electrodes: a cathode and a plate.

direct current. Current that flows in one direction only.

distillation. Process of evaporating a liquid followed by condensing of vapors in a separate container.

domain. A microscopic magnetic region composed of a group of atoms having a magnetic field aligned in a given direction.

Doppler effect. Change in frequency of a wave owing to relative motion between the wave source and the observer.

drugs. Chemicals that cure or prevent diseases or relieve pain.

Edison effect. The flow of current from a hot cathode to a positively charged plate.

efficiency. Ratio of useful output work of a machine to its input work.

elasticity. The property of matter that causes it to resume its original shape when a distorting force is removed.

electric cell. Device for producing an electric current by chemical action.

electric circuit. Complete conducting path of an electric current.

electric field. The region in which an electric force acts on a charge placed in the region.

electric generator. Device that changes mechanical energy to electrical energy by electromagnetic induction.

electric induction. Process whereby a charged object produces an opposite charge on a nearby object.

electric motor. Device for changing electrical energy into mechanical energy.

electrical current. Flow of electrons along a conductor owing to potential difference in a circuit.

electrochemical series. The numerical listing of potential differences of various substances, as compared to hydrogen.

electrode. Terminal of an electrolytic cell; plate or "pole."

electrolysis. Chemical change in matter produced by an electrical current.

electrolyte. A substance capable of conducting an electric current when dissolved or melted.

electromagnet. An iron core that becomes a magnet when current is passed through a coil wrapped around it.

electromagnetic induction. Process of producing a current by moving a magnet near a conductor or vice versa.

electromagnetic spectrum. Range of electromagnetic radiations, such as light, heat, ultraviolet, and X-rays.

electron. Negatively charged particle in an atom.

electron cloud. The aggregate of electrons about the nucleus of an atom.

electronics. The branch of science relating to the flow of electrons.

electroplating. Depositing a metal on a surface by means of an electric current.

electroscope. Device to detect electrical charges.

electrostatic induction. Charging an object by another object, without the two objects touching each other.

electrostatics. A branch of electronics dealing with charges at rest.

element. Substance that cannot be broken up into simpler substances by ordinary chemical means.

energy. That which produces change in matter; the capacity to do work.

energy, kinetic. Energy that is due to motion.

energy, potential. Energy that is due to the position or condition; stored energy.

enzymes. Organic catalysts.

equilibrium. A condition in which all the forces acting on an object balance each other.

eutrophication. The process by which a lake becomes rich in nutrients but deficient in oxygen.

evaporation. The change in phase from a liquid to a gas.

external combustion engine. Engine that burns fuel outside the cylinder.

farsightedness. Defect of the eyeball or lens whereby light rays are focused behind the retina.

fission, nuclear. Splitting of the atomic nucleus into fragments of more or less equal mass.

flotation. Process used to separate certain ores from worthless rock by preferential wetting.

fluid. Gas or liquid phase of matter.

fluorescence. Emitting light, usually of a longer wavelength when exposed to high-energy short-wave radiations.

focal length. Distance from a mirror or lens to its principal focus.

focal point. The point to which a lens or mirror converges parallel rays of light.

focus. A point at which light rays meet or from which light rays diverge.

force. That which can change the state of rest or motion of a body; a push or pull.

force, resultant. A single force that has the same effect as two or more forces acting together.

forced vibrations. Vibrations produced in matter by direct contact with a vibrating object.

fossil. Evidences of early forms of life as recorded in rock.

four-stroke-cycle engine. The most common type of gasoline engine; intake, compression, power, exhaust.

fractional distillation. Process of separating a mixture of liquids having different boiling points.

freeze separation. A method of obtaining fresh water from salt water in which the salt water is first frozen and then melted.

freezing point. Temperature at which a liquid changes to a solid at normal pressure.

frequency. A number of vibrations, waves, or cycles per second.

frequency modulation (FM). A mode of broadcasting in which the frequency of the carrier wave is changed to indicate the pattern of the sound wave.

friction. Force opposing the motion of a body.

fuel cell. A cell in which chemical energy is changed to electrical energy by a continuous supply of fuel.

fulcrum. The point or support on which a lever pivots.

fundamental. The tone produced when a string vibrates as a whole (instead of in segments).

fungicide. A chemical that kills non-green microscopic plants known as fungi.

fuse. A safety device placed in an electric circuit that melts when the current increases beyond a safe load.

fusion, nuclear. Merging of two or more atomic nuclei to form a single heavier nucleus.

galvanometer. Instrument for detecting very small currents.

gamma rays. Electromagnetic radiations from radioactive substances, similar to high-energy X rays.

gauss. A unit of magnetic strength.

generator. See electric generator.

gram. Unit of mass in the metric system.

gravity. The attraction of the earth for a given object.

greenhouse effect. Process by which the atmosphere traps short-wavelength solar radiations and re-radiates these as longer wavelengths.

grid. An element of an electron tube that controls the flow of electrons moving from the cathode to the plate.

half-life. Time required for half the atoms of a sample of a radioactive element to disintegrate.

hard water. Water that does not make suds with soap easily because of certain minerals dissolved in it.

harmony. Combination of musical tones that are pleasing to the ears.

heat. The energy produced by the motion of molecules of a substance.

heat of fusion. Amount of heat absorbed by a substance in melting.

heat of vaporization. Amount of heat absorbed by a liquid in changing to a gas.

hydraulic. Refers to any device that operates because of pressure applied to a liquid.

hydrocarbon. Compound containing only hydrogen and carbon.

hydroelectric power. Electric power generated by falling water.

hydrogen bond. A weak chemical bond between a hydrogen atom of one molecule and the negative end of a molecule of the same substance.

hydrogenation. The chemical addition of hydrogen to a substance.

hydrometer. Instrument used to determine the specific gravity of liquids.

ideal mechanical advantage (IMA). The ratio of the distance of the effort force (E) in a machine to the distance the resistance force (R) moves.

illuminated. An object from which light is reflected.

image. Reproduction of an object formed with lenses or mirrors.

incandescent. Hot enough to radiate visible light.

inclined plane. A ramp or slanting surface used as a simple machine.

index of refraction. Ratio of speed of light in a vacuum (air) to its speed in a given transparent substance.

indicator. A substance that changes color in the presence of acids and bases.

induced current. A current caused by electromagnetic induction.

induction. See electric induction, electromagnetic induction, electrostatic induction, and magnetic induction.

inertia. Property of matter that resists change in motion.

infrared. Radiation below the visible light spectrum with wavelengths longer than those of red light.

inorganic matter. Nonliving matter.

input work. The product of the effort force (E) and the distance through which it acts.

insulator. A material that resists the flow of heat or electricity.

intensity. Loudness of a sound.

interference. The effect produced by two waves when they are "in phase" (strengthened) or "out of phase" (destroyed).

internal combustion engine. Engine that burns fuel inside its cylinders.

inverse-square law. A relationship in which the strength of a quantity decreases as the square of the distance from the source.

ion. Atom or group of atoms that has lost or gained electrons.

ion-exchange. Process of replacing an ion of one atom with another of like sign.

ionic bonding. The bonding between atoms produced by transfer of electrons from one atom to another.

ionosphere. Layer of the atmosphere directly above the stratosphere that reflects radio waves.

isomers. Compounds whose molecules have the same number and kind of atoms but a different arrangement.

isotopes. Atoms of the same element that differ from each other only in the number of neutrons in the nucleus.

jet engine. Engine that gets thrust from high-speed exhaust gases.

joule. The unit of work. Product of one newton of force acting through a distance of one meter.

Kelvin scale. Celsius scale on which the freezing point of water is 273 degrees and the boiling point of water is 373 degrees.

kiln. A special brick furnace for making lime, cement, and bricks.

kilo-. Prefix meaning 1000.

kilowatt. Unit for measuring electrical power, equal to 1000 watts.

kilowatt hour. Unit of electrical energy equivalent to using 1000 watts of power for one hour.

kindling temperature. The lowest temperature at which a substance begins to burn.

kinetic energy. Energy that matter has because of its motion.

kinetic theory of matter. All atoms and molecules are in constant motion.

laminated. Substances built up in layers.

laser. Light amplification by stimulated emission of radiation.

leavening agent. A substance that causes dough to rise.

lever. Rigid bar free to turn about a fixed point called a fulcrum.

light. A form of energy that radiates from atoms when they are violently disturbed by heat or electricity; a visible portion of the electromagnetic spectrum.

lignite. A brownish-black solid fuel; second stage in the development of coal.

lime. Calcium oxide.

limewater. A clear, water solution of calcium hydroxide.

lines of force. Imaginary lines representing the strength and direction of the magnetic, electric, or gravitational force.

liquid. Matter that takes up a definite amount of space but has no definite shape.

liter. The unit of volume in the metric system.

litmus. A dye that is used as an indicator for acids and bases.

lodestone. A rock that is a natural magnet.

longitudinal wave. A wave in which the particles of the medium vibrate to-and-fro along the direction of wave travel.

loudness. The effect of the intensity of sound energy on the ears.

lox. Liquid oxygen.

lumen. The unit of illumination.

machine. A device that changes the amount, speed, or direction of a force.

machines, law of. Neglecting friction, the output work from any machine equals its input work.

magnet. A piece of iron, steel, nickel, cobalt, or alloy that attracts other pieces of iron, steel, nickel, cobalt, or alloys.

magnetic field. Space around a magnet in which a magnetic force can be detected.

magnetic induction. Process whereby a magnetic pole produces an opposite pole in a nearby substance.

magnetism. A force by which a magnet attracts or repels pieces of iron, steel, or another magnet.

magnification. Apparent enlargement of an object by an optical instrument.

major chord. Any combination of three tones whose vibration ratios are 4, 5, and 6.

mass. A measure of the amount of matter in an object; a measure of inertia of an object.

mass defect. The difference between the mass of a nucleus and the sum of the masses of its component particles.

mass number. The whole number closest to the atomic mass of an atom.

matter. Anything that has weight and occupies space.

mechanical advantage. Number that shows how many times a machine multiplies the force applied to it.

melting point. The temperature at which a solid changes to a liquid.

mercerizing. A process of treating stretched cotton with sodium hydroxide.

metallic mixture. Crystals of one metal scattered throughout the mass of another metal.

metallurgy. The science of taking useful metals from their ores, refining them, and preparing them for use.

metals. Elements that have luster, conduct heat and electricity, and usually have a positive valence number.

metamorphic rock. Rock formed by the effect of heat and pressure on other rocks.

meter. The unit of length in the metric system.

microprocessor. The central processing unit (logic) of a computer.

milli-. Prefix meaning 1/1000.

mirage. An optical illusion caused by the bending of light rays as they pass through air layers of differing density.

miscible. Capable of being mixed.

mixture. Material containing different substances that have not been chemically united.

model. A way of explaining how or why things behave the way they do.

moderator. Agent used to slow up neutrons in a controlled atomic chain reaction.

modulation. Addition of a low-frequency wave to higher-frequency carrier waves.

molecule. The smallest particle of a substance that has the properties of that substance.

moment. Product of force and perpendicular distance from the fulcrum; torque.

momentum. Product of velocity and mass.

monomer. A single molecule or single unit structure of a polymer.

mordant. A substance used to make certain dyes cling to cloth fibers.

mortar. Hard mass made from calcium hydroxide, sand, and water.

musical tone. Sound produced by regular vibrations in matter.

mutations. Changes in the factors that control heredity.

natural frequency. The one frequency at which an object will vibrate freely when disturbed.

nearsightedness. Defect of the eyeball or lens causing light rays to focus in front of the retina.

negative charge. Electric charge resulting from a gain of electrons.

neutralization. The reaction of the hydronium (hydrogen) ions of an acid and the hydroxide ions of a base to form water.

neutralizing. Combining an equal amount of positive and negative electricity.

neutron. A neutral particle inside the atom.

newton. The unit of force in the metric system.

nitrogen fixation. The process by which bacteria that grow on the roots of plants change the nitrogen of the air into nitrogen compounds.

noise. Sound waves of irregular vibration in matter.

nonelectrolytes. Solutions that do not conduct an electric current.

nonmetals. Elements that are very poor conductors of heat and electricity, and usually have a negative valence number.

nuclear chain reaction. A series of rapid nuclear fissions begun by the splitting of an atomic nucleus.

nuclear reactor. Device in which controlled fission of radioactive material produces new radioactive substances and energy.

nucleus. The positively charged center of an atom containing protons and neutrons.

octave. The interval between a given musical tone and one that is double or half the frequency.

ohm. Unit of electrical resistance.

Ohm's law. Volts equal amperes times ohms.

opaque. Not able to transmit light.

open pipe. Resonant air column open at each end.

organic matter. Material from an animal or plant source.

oscillator. Radio tube or circuit that produces steady, alternating currents, usually at high frequency.

oscilloscope. Type of cathode-ray tube used to make visible the pattern of electrical vibrations, sound waves, etc.

output work. The product of the resistance force and the distance through which it acts.

overtone. Tone of higher pitch than the fundamental, made by the shorter segments of the vibrating object.

oxidation. The process whereby oxygen unites with other substances.

oxidizing agent. Substance that gives up oxygen to other substances.

ozone. An isotope of oxygen, O_3.

parallel circuit. An electrical circuit containing two or more branches through which the current can flow.

parallel forces. Forces that act in either the same or in opposite directions.

pendulum. A suspended mass that can swing freely.

percussion instrument. An instrument that is played by striking it with some kind of stick or mallet.

period. The time required for one complete cycle or vibration; a horizontal row of elements in the periodic table.

periodic table. An arrangement of elements according to their atomic numbers and properties.

photoelectric effect. The release of electrons from certain metals by the action of light.

photometer. Instrument for measuring light intensity.

photon. A single packet (quantum) of light energy.

photosynthesis. The food-making process in green plants.

phototube. Electronic tube that generates an electrical current when exposed to light.

physical change. A change that does not produce a new substance.

pickling. Removing the surface impurities from a metal with an acid bath.

piezoelectric effect. Producing tiny currents by applying pressures or distorting certain crystals.

pigment. A chemical that has color because it reflects light of only certain wavelengths.

pitch. The effect of the frequency of sound waves on the ears; the distance between the threads of a screw; the angle of the blades of an airplane propeller.

plane mirror. Flat mirror, without curved surfaces.

plasma. A gaseous mixture of highly ionized particles.

plastic. A substance containing giant molecules that can be formed or shaped by molding.

pneumatic. Refers to any device that operates because of pressure applied to a gas.

polarized light. Light that vibrates in only one plane.

pollutant. An impurity in water, land, or air caused by the activities of people.

polymer. A giant molecule formed by smaller molecules joining together.

positive charge. Electric charge resulting from a loss of electrons.

positron. The antiparticle of the electron.

potable water. Water that is fit to drink.

glossary

potential energy. Energy stored in matter because of its position or condition.

power. A measure of the amount of work that can be done in a unit of time.

precipitate. An insoluble solid that separates from a solution.

pressure. Force acting on a unit area of surface.

primary coil. Coil of insulated wire attached to the source of electrical current.

primary colors. Red, green, and blue-violet light; mixes to produce white light.

principal focus. Point at which parallel rays of light converge after passing through a lens or hitting a concave mirror.

prism. A three-sided piece of glass that separates light into its colors.

proton. Nuclear particle with a positive charge.

quality. The effect of a tone on the ear based on the overtones present in the sound wave.

radar. Device used to locate objects by means of high-frequency radio waves.

radiation. Transfer of energy in waves through space.

radical. A group of atoms that acts as if it were a single atom.

radio frequency (r-f). Electromagnetic radiations produced by very rapid reverses of current in a conductor.

radioactivity. The giving off of alpha and beta particles and gamma rays as certain atoms decompose spontaneously.

radioisotopes. Isotopes that are radioactive.

rarefaction. Region of a longitudinal wave where the vibrating particles are farther apart than normal.

real image. Image that can be projected on a screen; it is always inverted.

reciprocating engine. Engine that converts heat energy to mechanical energy by the back-and-forth motion of a piston in a cylinder.

rectifier. Device for changing alternating current to direct current.

reduction. Any chemical action that removes oxygen from a compound.

reflection. A wave or ray striking and bouncing off a surface.

refraction. The bending of light waves as they pass at an angle from one material to another.

resistance, electrical. Opposition of a substance to an electric current passing through it.

resonance. The ability of anything to be set in vibration by absorbing energy of its own natural frequency.

respiration. The exchange of oxygen and carbon dioxide in living things that results in the making of energy.

resultant force. A force produced by the effects of two or more forces.

reverse osmosis. Method of obtaining fresh water in which salt water is forced under pressure through a thin membrane.

ring compound. Compound in which the ends of the chain of carbon atoms are linked together.

roasting. Heating in the presence of air.

rocket engine. Device propelled by exhaust gases that can operate beyond the atmosphere.

salt. A compound formed when the positive ions from a base and the negative ions from an acid react.

saturated compound. Organic compound that has only single covalent bonds between atoms.

saturated solution. Solution in which the concentration of solute is the maximum possible under the existing conditions.

scientific law. A general statement that describes the behavior of nature.

screw. An inclined plane wrapped around a cylinder; a simple machine.

secondary coil. Insulated coil of wire in which current is induced.

semiconductor. A material such as germanium or silicon whose electrical resistance is between conductors and insulators.

series circuit. Electric circuit that contains only a single path through which the current can flow.

short circuit. A sudden and massive drop in resistance in an electrical circuit.

slag. The waste product from the smelting of ores.

slow oxidation. An oxidation process such as decay or rusting that produces some heat but no light.

smog. A mixture of fog with industrial gases, automobile exhaust gases, and smoke.

soapless detergent. A cleaner that lowers the surface tension of water.

soft water. Water that easily makes suds with soap because of its low mineral content.

solid. Matter that takes up a definite amount of space and has a definite shape.

solid solution. One metal dissolved in another.

solute. Any material that is dissolved in a solution.

solution. Liquid formed by dissolving a solute in a solvent.

solvent. The liquid in which a solute dissolves.

sound. A form of energy produced by the vibration of matter.

specific gravity. Density of a substance divided by the density of water.

specific heat. The number of calories required to raise the temperature of one gram of substance one Celsius degree.

spectroscope. Instrument that uses a prism or grating to separate light of different wavelengths.

spectrum. See electromagnetic spectrum.

speed. Rate of motion of an object, regardless of direction.

spinneret. Small metal plate with tiny holes used for forming strands of synthetic fibers.

spontaneous combustion. Combustion resulting from the accumulation of heat from slow oxidation.

standard candle. A standard measure of light brightness. One standard candle radiates about 12.5 lumens.

standard pressure. Pressure equivalent to 760 mm of mercury.

standard temperature. Temperature of 0°C.

static electricity. Accumulation of positive or negative charges on an object.

steel. Iron containing carefully regulated amounts of carbon and other elements.

sterling silver. An alloy that is 92.5% silver and 7.5% copper.

storage battery. An electrical battery that can be repeatedly recharged.

STP. Standard conditions of temperature and pressure, 0°C and 760 mm.

straight-chain hydrocarbons. Hydrocarbons in which the carbon atoms are linked together in long, straight chains.

stroboscope. Device that permits moving objects to be viewed as though standing still.

sublimation. The passing of matter from a solid to a gas without passing through a liquid phase.

superconductor. A conductor that has no resistance to electrical current.

supersonic. Faster than the speed of sound.

surface tension. The drawing together, or skin effect, of the surface of a liquid.

suspension. A mixture formed by mixing a liquid with particles that are much larger than ions or molecules.

symbol. A shorthand way of writing the name of an element.

temperature. A measure of the average motion (average kinetic energy) of molecules.

tempering. The heating of steel, followed by rapid or slow cooling to produce the degree of hardness desired.

thermocouple. A device composed of dissimilar metals, joined to form a circuit in which current flows when heat is applied at a junction of the metals.

thermoelectric effect. A process by which heat is changed into electricity.

thermonuclear reaction. See fusion, nuclear.

thermoplastic. Can be softened by heat.

thermosetting. Cannot be softened by heat.

thermostat. An automatic temperature regulator.

torque. Product of force and the perpendicular distance from a fulcrum; a moment.

tracers. Radioactive or "tagged" atoms used in scientific research.

transformer. Device for changing the voltage of an alternating current.

transistor. An electronic device that can be used in place of a vacuum tube to control electric current.

translucent. Able to transmit light, though not enough to see through clearly.

transparent. Able to transmit light so that objects can be seen clearly through it.

transverse wave. Wave in which the particles vibrate at right angles to the direction of wave travel.

triode. Radio tube containing grid, cathode, and plate.

trough. Valley of a wave.

turbine. An engine turned by the force of a gas or liquid acting against its vanes or blades.

ultrasonic. Sound with a frequency above 20,000 vps.

ultraviolet. Radiations above the visible light spectrum with wavelengths shorter than those of violet light.

unsaturated compound. Organic compound with a double or triple bond between two carbon atoms.

vacuum. A space that contains no gas, liquid, or solid.

valence number. A number representing the combining power of an element with other elements.

vector quantity. A quantity that has both magnitude and direction.

velocity. Speed in a definite direction.

virtual image. An image that cannot be projected on a screen.

volatile. Easily vaporized.

volt. Unit of electrical pressure.

voltmeter. Instrument to measure voltage.

volume. Measure of the space occupied by an object.

watt. Unit of electrical power.

wavelength. Distance between any two corresponding points of two consecutive waves.

wedge. Double inclined plane used to split or separate objects.

weight. A measure of the force of attraction of the earth for an object.

wheel and axle. Simple machine consisting of a large wheel rigidly attached to a smaller one.

word equation. A brief statement that identifies the reactants and the products formed in a chemical reaction.

work. The result of a force moving an object a certain distance. It is determined by multiplying the force by the distance through which the force acts.

X rays. Invisible radiation of great penetrating power, produced in vacuum tubes.

xerography. An electrostatic process of copying printed material without liquid inks or chemicals.

picture credits

Tire & Rubber Co. **248** (top) Firestone Industrial Rubber Products (bottom) Goodyear Tire & Rubber Co. **249** Goodyear Tire & Rubber Co. **252** Dow Chemical Co. **253** (top) UPI (bottom) Goodyear Tire & Rubber Co. **254** R. Landesman. **259** American Textile Manufacturers Inst. Inc. **260** (top) Australian Information Service (bottom) National Cotton Council. **261** (top) American Textile Manufacturing Inst. Inc. **154** (bottom left) Wool Bureau (bottom right) Institute of Textile Technology. **262** (top) USDA (bottom) American Textile Manufacturing Inst. **263** (top and center) ATMI (bottom) The Sanforized Co. **264** Paola Koch/Photo Researchers. **266** Leo Choplin/ Black Star. **267** Int. Paper Company. **267** Johns-Manville Corp. **268** Monsanto. **269** (top) DuPont (bottom) HRW photo by Russell Dian. **273** HRW photo by Russell Dian. **277** Zahl/Nat. Audubon Society/Photo Researchers. **278** Howard Sochurek. **279** Alexander Lowry/Photo Researchers. **282** Al Kaplan. **288** NASA. **295** Dave Nadig/Photo Researchers. **300** Earl Roberge/Photo Researchers. **304** Joern Gerdits/Photo Researchers. **305** Bechtel Power Corp. **322** Hermann Eisenbeiss/Photo Researchers. **331** Exxon. **342** Exxon. **346** Tennessee Valley Authority. **350** Freda Leinwand/ Monkmeyer. **353** NASA. **361** Peter Kaplan/Photo Researchers. **368** NASA. **372** NASA. **373** Grant Heilman. **383** Woodfin Camp & Associates. **384** (top) NASA (bottom) © Joel Gordon. **387** General Electric Research & Development Center. **388** Fundamental Photos. **389** Fred Anderson/Photo Researchers. **401** RCA. **403** Rohn Engh/Photo Researchers. **404** DPI. **414** DuPont. **417** (top) Martin Adler Levic/Black Star (bottom) Linda Lindroth. **424** N. Y. Eye & Ear Infirmary, photo by John Goeller. **444** NASA. **457** Robert Isaacs/Photo Researchers. **461** Jack Fields/Photo Researchers. **462** Russ Kinne/Photo Researchers. **469** Philadelphia Museum of Art. **477** George Holton/Photo Researchers. **485** Con Edison. **497** Carl Frank/Photo Researchers. **505** (top left) Electric Fuel Propulsion Corp. (top right) General Motors (bottom) Medtronic Inc. **507** Central Scientific Co. **513** Manfred Kage/Peter Arnold. **514** Int. Minerals & Chemicals. **526** © Harald Sund/The Image Bank. **530** © Phillip Harrington/The Image Bank. **533** Exxon. **541** HRW photo by Ken Karp. **543** General Electric Solid State Applications Operation, Syracuse, N.Y.

Drawings on pages 13, 48, 83, 103, 121, 185, 241, 255, 318, 338, 368, 410, 440, 474, 493, 509, and 527 by Jerry F. M. Phelan.

chapter opener photos

1. Science and measurement: glassware used in a chemistry laboratory
2. Properties, changes, and composition of matter: rust illustrates a chemical change in matter
3. Structure of atoms and molecules: a micrograph illustrates the structure of a crystal
4. Kinetic theory of matter: ice forming on the surface of a stream illustrates the different phases of matter
5. Acids, bases, and salts: salt evaporation ponds near San Francisco
6. Nuclear reactions: the sun, whose energy is the result of nuclear fusion reactions
7. Common gases of the atmosphere: hot air balloons rise when the air inside them is heated
8. The chemistry of water: ocean waves
9. Environmental pollution: a stream polluted by the run off from a textile mill
10. Chemistry and the home: chemistry plays an important part in cooking, flavoring, and preserving foods
11. Metallurgy: copper flotation tanks
12. Fossil fuels: coal, the most common solid fossil fuel
13. Organic compounds: the exterior of an organic chemicals plant
14. Rubber and plastics: a plastic dome shelter
15. Natural and synthetic fibers: a close-up of natural fibers
16. Motion and its causes: a high-speed photo of a bullet in flight
17. Using force and motion: many forces act on a moving skier
18. Forces in solids, liquids, and gases: a water strider is supported by the water's surface tension
19. Work, energy, and power: a large derrick
20. Heat energy: a computer thermogram of Florida taken from space
21. Engines: the first launch of the space shuttle *Columbia*
22. Wave motion and sound: energy transfer from a vibrating tuning fork
23. The nature of light: optical fibers
24. Color: a color composite photo of the Black Hills region of South Dakota taken by a Landsat satellite
25. Electrostatics: an electric discharge between two poles
26. Current and circuits: a display of Christmas lights at Rockefeller Center in New York City
27. Sources of electric currents: high-tension wires carry electricity from power plants
28. Magnetism and electromagnetism: iron filings show the magnetic field around a magnet
29. Electronics: a microchip circuit

Appendix: Using the Book

Your textbook, Modern Physical Science, has been written and designed to make the study of physical science easier for you. A tool is only useful if you know how to use it. This book is a tool to help you in your study of physical science. To use the book correctly, you should (1) be familiar with the parts of the book and (2) know the purpose of each part of the book.

This appendix displays the various parts of a chapter. It also tells you the purpose of each part and how each part can help you in your study of Modern Physical Science.

1 A list of objectives at the beginning of each chapter tells you the important topics covered in that chapter.

2 Major topic heading

3 A boldface sentence at the beginning of each section tells you the main idea of that section.

4 Sample problems provide a mathematical approach to specific concepts.

5 Tables present information in a concise way for easy reference.

6 Activities give you the chance to further explore a concept by making observations.

1

objectives:

☐ To distinguish between physical and chemical properties, and between physical and chemical changes.

☐ To distinguish between atoms and molecules, and between elements, mixtures, and compounds.

2

ways to identify matter

3

Matter has physical and chemical properties. In day-to-day living, you come in touch with many materials that you learn to know. Some materials, such as a piece of iron, are heavier than others, such as an equal-size piece of cork. In order to tell one material from another, scientists look for certain traits that help identify a particular material. These traits are called *properties*.

4

sample problem

A piece of brass has a mass of 32.0 g and a volume of 3.8 cm³. Find the density of brass.

solution:

Step 1. Write the formula for density from its definition,

density = mass/volume

Step 2. Substitute the values given in the problem in the formula,

density = 32.0 g/3.8 cm³
= 8.4 g/cm³

5

Table 2-1:

Distribution of Elements in the Earth's Crust			
oxygen	49.5%	sodium	2.6%
silicon	25.8%	potassium	2.4%
aluminum	7.5%	magnesium	1.9%
iron	4.7%	hydrogen	0.9%
calcium	3.4%	titanium	0.6%
all other elements 0.7%			

6

activity

Making and separating a mixture. Grind together about 20 g of table salt and about 15 g of powdered charcoal with a mortar and pestle until you obtain a gray powder. Now see if you can find a way to separate this mixture into the two original substances.

```
================ careers in ================
                 CHEMISTRY
================================================
```

Career opportunities in the area of chemistry can be divided into three general job groups: chemical technician, chemist, and chemical engineer. Sometimes it is hard to decide in which group a given job belongs.

<u>7</u>

7 Special features give you information about career opportunities based on a knowledge of physical science. Other features deal with famous scientists or current scientific problems.

summary

<u>8</u>

1. Density equals the mass of a substance divided by its volume.
2. A physical change does not produce a new substance.
3. A chemical change produces a new substance.

8 A summary at the end of each chapter lists the concepts covered in that chapter.

review

<u>9</u>

Match each item in the left column with the best response in the right column. *Do not write in this book.*

1. density
2. chemical change
3. formulas
4. physical change
5. symbols

   a. oxygen and hydrogen
   b. MgO and H_2SO_4
   c. sharpening a pencil
   d. Fe and Pb
   e. mass per unit volume
   f. burning coal
   g. conservation of energy

9 The end-of-chapter review covers key words used in the chapter.

questions

<u>10</u>

Group A

1. What is meant by properties of matter? Give five examples of physical properties.
2. What is density? How is it usually expressed?

Group B

12. What is wrong with this statement: "Lead is heavier than aluminum"? How can it be corrected?
13. Describe how you would find the density of a small irregular solid.

10 Questions serve as a self-test for each chapter. Group A asks for recall of specific information. Group B requires application and interpretation.

problems

<u>11</u>

1. The density of a sample was found to be 11.2 g/cm^3. Could you find the volume of the sample from this information? What do you need to know to find its mass?
2. What is the mass in grams of water that fills a tank 100 cm long, 50 cm wide, and 30 cm high? in kilograms?

11 Problems test your skill in dealing with a problem quantitatively.

further reading

<u>12</u>

Chorley, Richard J. (Ed.) *Water, Earth, and Man.* New York: Methuen, Inc., 1979
Keough, Carol, *Water Fit to Drink.* Emmaus, Penn. Rodale Press, 1980.
Steel, E. W., et al., *Water Supply and Sewage.* New York: McGraw-Hill, 1979.

12 A list of books and articles at the end of each chapter provides suggestions for further reading.

appendix

index

(Note: Page numbers in **boldface** type include illustrations
and those in lightface include definitions)

babbit (alloy), 204
background radiation, 157
Baekeland, Leo, and Bakelite, 250
Bakelite, 250, 254
baking powders, 180
baking soda (see sodium
bicarbonate)
balloon, floating of, 333–334
barbiturates, 182
barium, atomic number of (table),
36; valence number (table), 43
barium sulfate, formula for and uses
of (table), 83
barometer, 335–**336**
base(s), 68, 74, **75**–79; and acids,
reactions between, 79–81; caustic
effects of, 75; as electrolytes,
63; formulas for, 75; hydroxide ions
in, 75; properties of, 74–**75**
basic-oxygen furnace, **193**
batteries, 497; lead storage, 501–
503; solar, 554; uses of, 504–506.
See also electric cells.
Becquerel, Henri, and discovery of
radioactivity, 87–88, 103
benzene, 235, 271
Bernoulli principle, 311
beta particles, **89**, 92, 93
binary system (table), 542
biodegradable materials, 147, 176
biological sciences, 4
blast furnace, 113, **191**, 197
bleaches, 269–270
block and tackle, 307
blow molding, **251**
boiling, 55
boiling point(s), 17; of liquid nitrogen,
112; of liquid oxygen, 112; of
petroleum products (table), 219;
of water, 125
bond(s), chemical, 44–47, 230;
double covalent, 233; triple
covalent, 234
bonding, covalent, 46–**47**, 125, 228,
230; in hydrocarbons, 229–235;
ionic, 44–45, 81
Boyle's law, 336–337
brass, 203, 204; rate of expansion
(table), 360; specific heat of
(table), 357
breeder reactor, **99**
bricks, 168–169; glass, 171
bright-line spectrum, **454**
broadcasting, radio, 532–533, 549
bromine, atomic number of (table),
36; from seawater, 126; symbol for
(table), 36; valence number
(table), 43
bronze, 203
building materials, 164–173;
artificial, 167–173; natural, 165–167

bullets, neutron, **96**–97
bullion, silver, **200**
buoyancy, 331–334, 338–339
butane, 220, 229, 231; structural
formula for, 230
by-pass jet, **382**

cable television, 551–552
cadmium, uses of (table), 201
calcium, atomic number of (table),
36; distribution in earth's crust
(table), 20; mass number (table),
36; from seawater, 126; symbol for,
23; valence number (table), 43;
voltage (table), 499
calcium hydroxide, 75, 118, 130;
formula for (table), 24; making of,
77–78; uses of, 78
calcium sulfate, 130–131; formula for
and uses of (table), 83
calorie(s), 356–357
camera(s), 436–437; eye as, **437**–
438; television, 437, 555
"cap" rock, 216, **217**
capacitors, **471**–474
capillary action, **326**
carbon, atomic number of (table),
35; in coal, 212; crystallized, 222–
223; mass number (table), 36; in
steel, 193; symbol for, 23
carbon dioxide, 77, 78, 108, 125; in
air, 116; dissolved under pressure,
61; formula for (table), 22;
importance to life, 116; percent in
air by volume (table), 109; and
photosynthesis, 116; preparation of,
116–**117**; test to detect presence
of, 116–117
carbon dioxide fire extinguisher, 175
carbon dioxide molecule, **116**
carbon-14, half-life and uses of
(table), 94
carbon-14 dating, 90–91
carbon monoxide, and air pollution,
148–149, 150, 154; breathing of,
221; as reducing agent, 191
carbon residue, 72–**73**
carbonic acid, 69, 117; formula for
(table), 69
carding machine, **260**, 263
careers, in chemistry, 27–28; in
health services, 138–140; in
mechanics and repair, 384; in
production of goods, 300; science
teachers, 64; science technicians,
543–544; service jobs, 300; in
textile industry, 273–274; in
transportation, 350

Carlson, Chester, and xerography,
473
carrier waves, **531**
Carver, George Washington, short
biography of, **121**
catalysts, 112, 118, 120, 145, 172
catalytic converters, 149
cathode, 196, 534
cathode-ray oscillator, 539
cathode-ray tubes, **536**–538, 546,
550–552
cathode rays, nature of, 539
cells (see electric cells)
celluloid, 250, 252
cellulose, 262, 266–268
cellulose acetate, 253
cellulose plastics, 252–253
Celsius scale, 112, 355
cement, 167–**168**
cement kiln, **168**
center of gravity, 290
centigram, 11
centimeter, 7
chain reaction, nuclear, 98
charcoal, heat value of (table), 215;
preparation of, **214**, **215**
chemical bond(s), 44–48, 230
chemical change in matter, 18–19,
26–27
chemical energy, 12, 47, 343, 346,
374
chemical equations, 24–26;
balanced, 70–71; ionic, 80; writing
of, 71
chemical formulas, for common
radicals (table), 47; compounds
identified by 23–24; method of
writing, 47–48; structural, 230,
231, 234–240. See also specific
compounds.
chemical properties, of matter, 18
chemical reactions, 45, 47, 69–81,
87; acid-base, 79–81; addition,
235–237; decomposition, 78;
double replacement, 74, 131;
neutralization, 80–81; reversible,
77, 502; single replacement, 70,
119, 200, 500; substitution, 236–
238; synthesis, 71
chemical symbols, 23. See also
specific elements.
chemistry, 4; careers in, 27–28;
inorganic, 227; organic, 227, 228
chlorine, 82, 137, 138, 145, 150;
atomic number of (table), 35; as
bleach, 269–270; mass number of
(table), 36; symbol for, 23;
valence number (table), 43
chlorine atom, 42–**43**, 44, **45**, 70, 94
chlorine isotopes, 94
chlorophyll, 116

index   **561**

(table), 36; symbol for (table), 36; uses of (table), 201; voltage (table), 499

nickel-cadmium cell, 503

nitric acid, care in handling of, 74; formula for (table), 69; preparation of, 74; uses of, 74

nitrogen, 21, 108; atomic number of (table), 35; in fertilizers, 178; liquid, boiling point of, 112; mass number (table), 36; percent in air by volume (table), 109; properties of, 116; symbol for (table), 35

nitrogen fixation, 116

nitrogen nucleus, **92**

nitroglycerin, 116, 240

noble (rare) gases, 120–121

noise pollution, 151

nonelectrolytes, 63

nonmetals, 41

northern lights, 54

novas (exploding stars), 535

nuclear energy, 3, 96–99, 224, 346–347

nuclear equations, 91–92, 96

nuclear fission, 97–**99**

nuclear fusion, 101–102

nuclear mass defect, 101

nuclear particle accelerators, **95**

nuclear power plant, 100; safety of, 158–**159**, 160

nuclear reactions, 91; chain, 96–98; fission, 95–96; 97–100; fusion, 100–103; nature of, 87; thermonuclear, 101

nuclear reactors, **99–100**

nucleus, atomic, **34**

nylon, 214, 253, 268, 269

obsidian, 170

octane, 232

octane number, 233

octaves, law of, 48

ohm, 482

Ohm's law, 483, 489

oil, drilling for, 216–**217**; short supplies of, 223; as water pollutant, 144–**145**. See also petroleum.

oil paint, ingredients in, 172

oil-eating bacteria, 5

orbit(s), satellite, **316**

ore(s), 189; copper, 197; iron, 190; reduction of, 197–198; roasting of, 197

ore-flotation process, 197, **198**

organic acids, 228

organic chemistry, 227, 228

organic compounds, inorganic compounds and, 228. See also hydrocarbons.

organic matter, 128

Orlon, 268

oscillator, 532; cathode-ray, 539

osmosis, 133; reverse, 133, 138

overtones, musical, 405–406

oxidation, 113–115

oxyacetylene torch, **113**

oxygen, 108–115; actual mass (table), 37; atomic number (table), 35; compressed, 113; discovery of, **109**; distribution in earth's crust (table), 20; liquid, boiling point of, 112; mass number (table), 36; percent in air by volume (table), 109; preparation of, **109, 110, 111,** 113; specific gravity, 335; symbol for (table), 22; test for identification of, **113**; in universe, 108–109; uses of, 112–113; valence number (table), 43; in water, 19

oxygen atom, 25, **46**, 91, 93–94

oxygen isotopes, 93–94

oxygen molecules, 23, 25, **46**

oxygen nucleus, **92**

ozone, 149, 151

pacemaker, **505**

paints, 72, 172–173

papermaking, **266**–267

Pascal's law, 330, 331

Pasteur, Louis, brief biography, **185**; and germ theory of disease, 185; and pasteurization, 185; and yeast, 179

peat, 211, 212

pendulum, energy transfer in, **345**

penicillin, 183

percussion instruments, 404

periodic table, 38–39, 40–41, 48, 93

pesticides, 4–5, 57, 136, 144, 156–157

petroleum, 210; boiling points of products of (table), 219; composition of, 216; fractional distillation of, 218; hydrocarbons in, 230; in nylon, 269; origin of, 215–216; products of, 216; refining of, 217–**218**. See also oil.

pH of solutions, 78, 79

phosphorus, atomic number of (table), 35; in fertilizers, 178; mass number (table), 36; symbol for (table), 35

photochemical smog, 148–149

photoelectric cell (phototube), 539

photoelectric effect, 415, 507, 539–**540**, 541

photons, 13, **415**, 422–423

photosynthesis, 116

phototube, 539

physics, 4

piezoelectric effect, 507–508, 547, 548

pigments, in face powders, 184; mixing of, 453; paint, 172; primary, **453**

pipe still, **218**

Planck, Max, and wave theory of light, 415

planets, flights to, 317

plasma, 54, 102

plaster, 169

plaster of paris, 169–170

plastics, 72, 74, 249–250; acrylic, 253; cellulose, 252–253; discovery of, 250; epoxy resins, 255; fluorocarbon, 253–254; laminated, 255; melamine, 254; molding of, 250–252; nylon, 253; phenolics, 250, 254; polyesters, 254–255; polyethylene, 253; polyurethane, 255; thermoplastics, 250–254; thermosetting, 250, 254–255

platform balance, **11**

platinum, atomic number (table), 36; symbol for (table), 36; uses of (table), 201

plutonium, 159

plutonium-239, 99

polar molecule, 125

pollutant(s), 143

pollution (see environmental pollution)

polonium, discovery of, 103

polyester, 268, 269

polyester plastics, 254–255

polymer(s), 246, 250, 268

polyunsaturated fats, 235

population growth, 160–**161**

Portland cement, 167–168

positron(s), 101, **102**, 103

potable water, 127

potassium, atomic number of (table), 36; distribution in earth's crust (table), 20; in fertilizers, 178; mass number (table), 36; from seawater, 126; symbol for (table), 36; valence number (table), 43; voltage (table), 499

potential energy, 47, 343–344

pottery, 169

power, 348; formula for, 348; hydroelectric, 346; units of, 349; and work, 347–349

precipitate, 26, 117

pressure, 327–338; air, measurement of, 335–336; average, 330; of gases, 335; of liquids, 328–333; and phase change in matter, 55–56; of solids, 327–328; water, 328–329. See also force(s).

index

index